CW01266964

THE NUCLEAR AGE

THE NUCLEAR AGE

AN EPIC RACE FOR ARMS,
POWER, AND SURVIVAL

Serhii Plokhy

ALLEN LANE
an imprint of
PENGUIN BOOKS

ALLEN LANE

UK | USA | Canada | Ireland | Australia
India | New Zealand | South Africa

Allen Lane is part of the Penguin Random House group of companies whose addresses can be found at global.penguinrandomhouse.com.

Penguin Random House UK
One Embassy Gardens, 8 Viaduct Gardens, London SW11 7BW

penguin.co.uk

Penguin Random House UK

First published in the United States of America by W. W. Norton & Company, Inc. 2025
First published in Great Britain by Allen Lane 2025

001

Copyright © Serhii Plokhy, 2025

The moral right of the author has been asserted

Penguin Random House values and supports copyright. Copyright fuels creativity, encourages diverse voices, promotes freedom of expression and supports a vibrant culture. Thank you for purchasing an authorized edition of this book and for respecting intellectual property laws by not reproducing, scanning or distributing any part of it by any means without permission. You are supporting authors and enabling Penguin Random House to continue to publish books for everyone. No part of this book may be used or reproduced in any manner for the purpose of training artificial intelligence technologies or systems. In accordance with Article 4(3) of the DSM Directive 2019/790, Penguin Random House expressly reserves this work from the text and data mining exception.

Printed and bound in Great Britain by Clays Ltd, Elcograf S.p.A.

The authorized representative in the EEA is Penguin Random House Ireland, Morrison Chambers, 32 Nassau Street, Dublin D02 YH68

A CIP catalogue record for this book is available from the British Library

ISBN: 978–0–241–58286–2

Penguin Random House is committed to a sustainable future for our business, our readers and our planet. This book is made from Forest Stewardship Council® certified paper.

MIX
Paper | Supporting responsible forestry
FSC® C018179

"We shall by a process of sublime irony have reached a stage in this story where safety will be the sturdy child of terror, and survival the twin brother of annihilation."

—WINSTON CHURCHILL, MARCH 1, 1955

CONTENTS

Preface		1
1.	Prophecy	7
2.	Fright	16
3.	Nazis and Their Friends	28
4.	Transatlantic Alliance	40
5.	Manhattan Project	52
6.	Unequal Partners	65
7.	American Bomb	79
8.	Stolen Secret	93
9.	United Nations	105
10.	Union Jack	117
11.	Stalin's Bomb	131
12.	British Hurricane	143
13.	Managing Fear	155
14.	Super Bomb	168
15.	Missile Gap	181
16.	Bombe Atomique	195

17.	China Syndrome	206
18.	Cuban Gamble	218
19.	Banning the Bomb	229
20.	Star of David	243
21.	MAD Men	256
22.	Smiling Buddha	270
23.	Star Wars	283
24.	The Fall of the Nuclear Colossus	296
25.	Giving Up the Bomb	310
26.	The Return of Fear	326
27.	Preemptive War	340

Epilogue   357
Acknowledgments   367
Notes   369
Index   409

THE NUCLEAR AGE

PREFACE

The nuclear age came into existence with the explosion of the first atomic bomb in the New Mexico desert on July 16, 1945. The man widely seen as the main creator of the bomb, the American physicist J. Robert Oppenheimer, welcomed it with the words: "Now I am become death, the destroyer of worlds." He was quoting the *Bhagavad Gita*, a 701-verse Hindu scripture, but he might have been speaking on behalf of the nuclear age itself. Conceived by scientists and novelists at the turn of the twentieth century as an era that would transform the world for the better, it arrived with an unprecedentedly fearsome explosion.[1]

Many people—scientists, engineers, and, not least, politicians—tried to realize the peaceful potential of the nuclear age. They attempted to use the power of the atom to cure diseases, dig channels, release underground gases, and transform deserts into orchards by making saltwater potable. Most of those projects failed. Some succeeded, as demonstrated by the production of electricity and the powering of icebreakers, ships, and submarines, but even the successful ones failed to deliver on the promise of the early days of the atomic era. Even the most daring nuclear undertakings fell far short of those expectations—today, only 10 percent of world electricity is produced by nuclear reactors. Names such as Three Mile Island, Chernobyl, and Fukushima remind the world about the cost of that limited success and hamper its future development.

Thus, the nuclear age became first and foremost the age of the bomb or, rather, two bombs—atomic, deployed on human beings in August 1945 in the Japanese cities of Hiroshima and Nagasaki, and hydrogen, which made a dramatic entrance by polluting a good part of the Pacific Ocean in the Castle Bravo test of 1954. The arrival of nuclear weapons changed the nature of international politics. On the one hand, the nuclear arms, and in particular the hydrogen bomb, put the survival of all humankind in question. In the 1960s, half of the world could well have been destroyed by all-out nuclear war between the two nuclear-armed superpowers, the United States and the Soviet Union. On the other hand, the vision of a nuclear holocaust helped to keep those superpowers in check, leading to the longest global peace of the modern era.

If the nuclear age has been defined by the bomb, its trajectory and main stages have been determined by the nuclear arms race, which had two components: the race to build the bomb, which is known as the horizontal proliferation of nuclear weapons, and the vertical proliferation or the race to accumulate nuclear stockpiles and the means of delivering warheads: bombers, land-based missiles, and their sea-based counterparts. The United States won the race to create the bomb, reaching the finish line first in 1945 and leaving other early nuclear aspirants—Nazi Germany, imperial Japan, the communist Soviet Union, and, last but not least, Great Britain, a competitor turned partner—far behind.

The American acquisition of the bomb triggered the next stage of the race. Other countries were eager to jump on the nuclear arms bandwagon. The Soviet Union got the bomb in 1949, Great Britain in 1952, France in 1960, China in 1964, and Israel about 1969. India acquired its nuclear capability in 1974 and exploded its first atomic bomb in

May 1998, a few weeks before Pakistan tested its own nuclear weapon. Finally, North Korea joined the nuclear club in most public fashion in 2006. The United States, Britain, China, and France, in that order, also acquired hydrogen bombs between 1954 and 1968. Today, nine countries possess nuclear weapons. That number is much smaller than many predicted in the early 1960s—John Kennedy was concerned that there might be as many as twenty new nuclear states before the end of the decade—but the current numbers are deceiving. Although there are only nine full-fledged nuclear-armed states, close to forty more have access to the requisite technology and raw materials or have the capability to produce the bomb, in some cases on very short notice.[2]

Many observers speak today about the return of the nuclear age, or the second nuclear or atomic age. What they mean is the return of the nuclear arms race. The nuclear age in fact never left us when it comes to nuclear weapons or nuclear energy. There was, however, a brief lull in the arms race starting in the late 1980s and early 1990s, when the two superpowers, the United States and the Soviet Union/Russian Federation, agreed to reduce their nuclear stockpiles, and a number of countries that possessed the bomb decided to give it up or were forced to do so. That lull ended in 1998 with the series of nuclear tests carried out by India and Pakistan. With the start of the new millennium, the United States and Russia began dismantling the late Cold War–era regime of nuclear arms control and reduction, abandoning or denouncing one agreement after another. By 2006 North Korea joined the nuclear club, challenging the US and its allies in the region and raising questions about the possibility of Japan and South Korea going nuclear. Iran meanwhile acquired its first nuclear reactor and resumed its nuclear weapons program. In 2022 Russia threatened to use nuclear weapons against Ukraine and in 2024 used for the first time in history an intermediate-range ballistic missile to attack an urban center.

"Dark clouds loom on the nuclear horizon, with threats from all directions," wrote Matthew Bunn of Harvard's Project on Managing the Atom in June 2024. He listed the following threats: "Russia's nuclear bombast in its war on Ukraine, China's construction of hundreds of nuclear missile silos, North Korea's missile testing, India and Pakistan's ongoing nuclear competition, and Iran's push toward nuclear weapons capability." Bunn continued: "In response, U.S. policy-makers are discussing whether a further American nuclear arms buildup is needed. At the same time, evolving technologies, from hypersonic missiles to artificial intelligence, are straining military balances and may be making them more unstable. The risk of nuclear war has not been so high since the Cuban Missile Crisis."[3]

In the midst of the rapidly worsening international climate, the United States, Russia, China, and a number of other members of the nuclear club expedited their efforts to "modernize" their nuclear arsenals. A more precise term for the process is nuclear rearmament. With no Cold War–era agreements remaining to regulate the nuclear arms race, and those still in place constantly questioned and undermined, we are indeed in a more dangerous situation today than we have been since the Cuban missile crisis of 1962, which was in many ways the result of an unregulated nuclear arms race. As nuclear arsenals are replenished and more countries than ever before acquire or come within range of nuclear weapons technology, the uncontrolled nuclear race increases the chances of nuclear war, accidental or not.

There have been efforts to curb the proliferation of nuclear weapons by electrocuting real or suspected spies, as was the case in the United States in 1953; by withholding nuclear weapons technology while offering electricity generated by nuclear power, as was done under the auspices of the Nuclear Non-Proliferation Treaty of 1968; by waging preemptive wars, as in Iraq in 2003; and by applying sanctions against violators, such as Iran. All these methods have failed to produce the desired results, as proliferation continued, although at a much slower pace. Even small and poor countries such as North Korea can acquire the bomb if they are determined to do so. Nuclear weapons technology is arguably easier to obtain today than it was during the Cold War, and there is growing concern that it will end up in the hands of non-state terrorists, against whom nuclear deterrence does not work. A new threat came with the reckless behavior of a state actor, the Russian Federation, whose army brought warfare to Ukraine's nuclear sites, and whose leadership threatened to use nuclear weapons against a non-nuclear state.

Is there anything we can learn from the first nuclear arms race that can help us to stop or at least control the new one? That is the question I am trying to answer in this book. Narrating the history of the nuclear arms race, I show how the "club" that now constitutes the nuclear arms community came into existence. But the matters that interest me most are in the nature of "why." What drove individual countries to commit themselves to the tremendous expense of building atomic and hydrogen bombs and then engage in the process of building up nuclear arsenals? Why did countries that possessed nuclear weapons decide to establish some control over them and even reduce their arsenals, while others decided to give up the weapons they already had?

The question of why countries go nuclear has long attracted the attention of scholars, and we now have a whole library on the issue. That literature, written mostly by political scientists, focuses on such factors as security, prestige, and domestic politics, as well as issues of political economy and the psychology of leaders. As I show here, the reasons for acquiring or giving up nuclear weapons often differed from one country to another. But no matter what the combination of factors in a particular case, the fear of nuclear attack by another state or of superior conventional forces possessed by an adversary was the most common reason. In other words, security considerations outranked other matters that concerned decision-making elites, such as the desire to gain military or strategic advantage over adversaries, maintain or achieve great-power status, or improve standing in the international community.[4]

I believe the nations' concerns about their security, represented by the emotion of fear, is paramount for understanding their decision to go nuclear. I considered fear a key factor influencing nuclear decisions in my book on the Cuban missile crisis, *Nuclear Folly*, and argue here that it was a major element in the history of the Nuclear Age as a whole. Fear can produce different reactions and perform a variety of functions in international relations, making one nation aware of danger and prodding another to increase its security or, on the contrary, prompting irrational decisions and engagement in questionable enterprises, such as the nuclear arms race.

Fear has had contrary effects in the history of the nuclear arms race, both fueling it and making it more manageable and predictable. It was fear, shared by nuclear adversaries and created by what Churchill called the "balance of terror," that made possible the signing of arms control and arms reduction agreements in the last decades of the Cold War. With the Cold War agreements gone and a new, unregulated arms race under way, fear is back, poised this time to fuel an arms race that at best will turn out to be a huge waste of money and resources, and at worst might lead to a nuclear confrontation. In this new context, we have to relearn how to manage our fear, recall how dangerous nuclear brinkmanship can be, and recognize how important it is not to become a victim of nuclear blackmail and encourage the adventurism of nuclear rivals.[5]

Chapter 1

PROPHECY

The timing could not have been more perfect or more unfortunate, depending on one's point of view. H. G. Wells, one of Britain's most renowned and bestselling novelists, completed his new work prophesying a world war one year before the actual outbreak of World War I. The novel, titled *The World Set Free*, was published in the first year of the war, 1914. Wells's new book introduced readers to the notion of a world war fought not only with airplanes—the theme of his 1907 book, *War in the Air*, but also with a more advanced and devastating weapon, the atomic bomb.[1]

As he foretold atomic war, Wells announced the arrival of the age of the atom. "In this bottle, ladies and gentlemen, in the atoms in this bottle there slumbers at least as much energy as we could get by burning a hundred and sixty tons of coal," Professor Rufus, a character in the novel, tells his audience. "If at a word, in one instant I could suddenly release that energy here and now it would blow us and everything about us to fragments; if I could turn it into the machinery that lights this city, it could keep Edinburgh brightly lit for a week."[2]

Where did Wells get these ideas? The famous novelist did not hide his source of inspiration—a book titled *Interpretation of Radium*, published in 1909 by a thirty-two-year-old chemist named Frederick Soddy. "This story, which owes long passages to the eleventh chapter of that book, acknowledges and describes itself," writes Wells in the dedication of *The World Set Free*.[3]

Soddy was not only an excellent scholar but also an effective and prolific popularizer of his own research and that of his colleagues. None of them was as important and prominent as the physicist Ernest Rutherford, age thirty-eight when Soddy's book appeared in print. The two men began working together in 1901, studying the emissions of the radioactive mineral thorium. Their experiments led Rutherford to formulate the theory of radioactive decay, which suggested that the atoms of radioactive substances disintegrate into atoms of a different type. He proposed that the atom was by no means indivisible. Its radioactive decay could produce not only new elements but also energy, and that energy, wrote Rutherford and Soddy in 1903, "must be enormous compared to that rendered free in ordinary chemical change."[4]

Rutherford was sometimes uneasy about his discovery. He was overheard saying, "Some fool in a laboratory might blow up the universe unawares." Others heard him suggesting, "Could a proper detonator be found, it's just conceivable that a wave of atomic disintegration might be started through a matter which would indeed make this old world vanish in smoke." Some who heard Rutherford saying those words believed that he was not serious. But Soddy, lecturing in 1904 to the annual gathering of the Royal Engineers, treated the possibility of a new atomic weapon with the utmost seriousness. He asserted that "the man who put his hand on the lever by which a parsimonious Nature regulates so jealously the output of this store of energy would possess a weapon by which he could if he chose destroy the earth."[5]

H. G. Wells took these and other statements of Soddy and Rutherford to heart. In *The World Set Free* he imagines what the weapon mentioned by Soddy might look like and, most important, what its effect might be. Wells's atomic bomb is "a black sphere two feet in diameter," light enough to be loaded onto an aircraft and be thrown from it by hand. "When he could look down again it was like looking down upon the crater of a small volcano," writes Wells, describing the bomb thrower's view of the resulting explosion. "In the open garden before the Imperial castle a shuddering star of evil splendour spurted and poured up smoke and flame toward them like an accusation," continues Wells. "They were too high to distinguish people clearly, or mark the bomb's effect upon the building until suddenly the facade tottered and crumbled before the flare as sugar dissolves in water."[6]

Despite this dramatic rendering of the explosion, Wells does not imagine its destructive power surpassing that of existing devices. The truly horrifying effect of the atomic bomb is unleashed by the chain of explosions that irradiate enemy territory for days, weeks, and even years in succession, making the land uninhabitable. "Once launched, the bomb was absolutely unapproachable and uncontrollable until its forces were nearly exhausted, and from the crater that burst open above it, puffs of heavy incandescent vapour and fragments of viciously punitive rock and mud . . . were flung high and far," writes Wells.[7]

The atomic war imagined by Wells pits the Allies, including Britain, against the Central European powers, Germany among them. He clearly has in mind the Entente and the Central Powers, the two opposing alliances whose conflict was then only months away. In the novel, however, these are not the only warring parties that possess the bomb. In the world described by Wells, the proliferation of nuclear weapons is unlimited. It appears that every state has them and is eager to put them to use. "Power after Power about the armed globe sought to anticipate attack by aggression," writes Wells. "They went to war in a delirium of panic, in order to use their bombs first. China and Japan had assailed Russia and destroyed Moscow, the United States had attacked Japan, India was in anarchistic revolt with Delhi a pit of fire spouting death and flame; the redoubtable King of the Balkans was mobilising."[8]

Like the later adherents of the realist school in international relations, Wells sees the world as anarchic, with each state doing its best to undermine or destroy its enemy and survive on its own. With the threat of destruction clear and present, every state launches a preemptive atomic attack on its potential enemy. The result is the destruction of the whole world. But Wells, a socialist and pacifist by conviction, sees the danger of atomic conflict as a warning that may save the world from devastation by future wars initiated by egoistic rulers. In his novel, the planet draws back from this existential threat, abolishes warlike nation-states, and creates a world government that takes control of nuclear weapons, ensuring peace.

World War I, which broke out a few months after the publication of *The World Set Free*, fulfilled some of Wells's predictions. It pitted Britain and its allies against Germany and its partners, was fought in the fields of France, involved aircraft, and produced unprecedented carnage. To prevent future wars, major powers created the League of Nations—not the world government of Wells's imagination, but a step away from four years of devastating anarchy, all the more shocking because it followed Europe's largely peaceful last decades of the nineteenth century.

The powers fighting the Great War, unlike the states involved in Wells's atomic conflict, did not have atomic bombs in their possession, but the war brought the use of poison gas, the world's first weapon of mass destruction. Unlike other wartime technological innovations, such as airplanes and tanks, which were the work of engineers rather than scientists, poison gas was a product of pure science. In Berlin, Fritz Haber dedicated his talents and the facilities of his Kaiser Wilhelm Institute for Physical Chemistry to the war effort and produced chlorine gas for the use of the German army. It was first unleashed on a large scale at the Battle of Ypres in the spring of 1915, with Haber personally overseeing its release. By way of explanation, Haber wrote that "during peace time a scientist belongs to the World, but during war time he belongs to his country." His action clearly violated the 1907 Hague Convention on Land Warfare, but that did not make him a pariah in the world of science. Haber was awarded a Nobel Prize for chemistry in 1918, the last year of the war.[9]

While the Germans were the most active and effective in the use of poison gas, they were not alone in resorting to it. As early as August 1914, the French used tear gas grenades against the Germans. In France, Victor Grignard, the 1912 Nobel Prize winner in chemistry, helped to design phosgene, a much deadlier gas than chlorine and much more difficult to detect because it is colorless, unlike the greenish clouds of chlorine. It was released against German positions in 1915. Two years later, the Germans responded with an even more potent chemical weapon, "mustard gas." None of the gases used during World War I met their makers' expectations in terms of being lethal, but they proved extremely effective in incapacitating, disorienting, and terrorizing enemy troops.[10]

Scientific research on radioactivity was also put to use during the war, but not in the way imagined by Wells. It was employed to heal rather than to kill and was brought to the front lines by a woman. In France, Marie Curie, the discoverer of radium, trained herself and her seventeen-year-old daughter, Irène, a future Nobel Prize winner for her chemical research, in anatomy and driving automobiles. They put X-ray equipment on wheels and brought it to the front in order to facilitate surgery on wounded soldiers. Curie collected money to equip and operate twenty mobile X-ray units and organize ten times as many stationary ones. It would have been out of character for Marie Curie not to put her beloved radium to work as well. She did so by inventing hollow needles filled with a radium-produced gas, later called radon, that was used for sterilization.[11]

The prewar scientific international was no more. Few indicators could better attest to the international character and openness of pre–World War I science than the list of the recipients of the Nobel Prize, the world's most prestigious award in science. In physics, between 1901, when the prize was first awarded, and 1914, when awards were suspended because of the war, the recipients included five Germans, four Frenchmen, four Dutchmen, and two Britons, with American, Italian, and Swedish scientists obtaining one prize each. In chemistry, the group of laureates was no less international: Germany and

France had five laureates each, Britain two, and Sweden, the Netherlands, and the United States one each. Overall, Germany was in the lead and France second, followed by the Netherlands, Britain, Sweden, the United States, and Italy. With Germany, France, and Britain, the countries with the most Nobel laureates, going to war in 1914, the world of international science was not just divided by national boundaries but riven by trenches.[13]

While fighting began in August 1914, conflict in the world of science was not far behind. It burst out in early October of that year, when ninety-three German artists and scientists issued an appeal "To the Civilized World." The signatories reacted to the worldwide outcry against the German violation of Belgian neutrality, which marked the opening salvo of war on the western front. They also denied the atrocities committed by German troops in occupied Belgium. "As representatives of German Science and Art, we hereby protest to the civilized world against the lies and calumnies with which our enemies are endeavoring to stain the honor of Germany in her hard struggle for existence—in a struggle that has been forced on her," reads the appeal.

Among the signatories was a score of Nobel Prize winners in physics and chemistry, including Wilhelm Roentgen, the instigator of the X-ray revolution and first recipient of the Nobel Prize in physics (1901), Phillip Lenard, who received his prize in 1905 for cathode-ray research, and Wilhelm Wien, awarded in 1911 for his work on heat radiation. Fritz Haber, who would receive the award four years later, also signed the appeal. The document produced outrage in Britain and France, where scientists demanded the expulsion of the signatories from their domestic and international academies and institutions. The Germans threatened to reciprocate. International scientific congresses came to a halt. Scientists went to war, some as volunteers, others as recruits. While some saw military action, others joined scientific boards assigned to bring technological innovations to the front.[12]

The war split also the international community of scholars created by Ernest Rutherford, the father of nuclear physics, at the University of Manchester. Rutherford did his best to keep track of his former students and assistants, now obliged to fight one another. "A. S. Russel—wounded by shrapnel, Andrade—in pretty lively sector—Germans spent 3 shrapnel in trying to get him alone the other day, Moseley now in Dardanelles. Of the Germans, Gustav Rumelin and Heinrich Schmidt have died," wrote the distressed professor. He would soon add Henry Moseley, one of his most talented assistants, to the list of war casualties. But war itself did not destroy the personal ties binding the international community of scholars fathered and nurtured by Rutherford. Two members of his group, the German scientist Hans Geiger and Geiger's former assistant, Ernest Marsden, ended up fighting each other in the same sector of the front in France. But when word reached Geiger that Marsden had been invited to take up an academic position in Wellington, New Zealand, Geiger sent a letter congratulating his former assistant on the appointment.[14]

By cutting international ties, breaking up research groups, and requiring scholars to work on practical matters directly related to the war effort, the global conflict of 1914–18 disrupted and retarded but did not completely end research on the structure of the atom. In Germany, Hans Geiger, wounded on the battlefield, was sent home. There he helped Rutherford's other former assistant, James Chadwick, interned by the Germans at the start of the war, to continue his research on radioactivity. Chadwick's main source of radioactive material was German-produced toothpaste, among whose components was thorium, the radioactive gas first discovered by Rutherford and Soddy at the turn of the century. Geiger managed to find some badly needed equipment, and the work begun by Chadwick in Britain continued in Germany.[15]

Rutherford spent most of the war helping to design an underwater listening device to detect German submarines but managed to return to his studies of the atom in 1917. "I am trying to break up the atom," he wrote to Niels Bohr in December 1917. Using his favorite alpha particles, Rutherford decided to bombard the nuclei of lighter atoms, especially nitrogen. In the same letter to Bohr he wrote: "I have got, I think, the results that will ultimately have great importance." He was perfectly right. Rutherford's was the most consequential "bombardment" of the entire war. He managed to initiate the first artificially induced nuclear reaction and chip off the nucleus a subatomic particle that he would later call a "proton." Rutherford not only smashed the atom but also transformed one element, nitrogen, into another, oxygen. "Way to Transmute Elements Is Found. Dream of Scientists for a Thousand Years Achieved by Dr. Rutherford," read the *New York Times* headline of 1921 that announced his discovery to the world.[16]

Also in 1921, H. G. Wells issued a new edition of *The World Set Free* in which he took account of the hits and misses in the first edition. He reasserted what he considered his major insight: "the thesis that because of the development of scientific knowledge, separate sovereign states and separate sovereign empires are no longer possible in the world, that to attempt to keep on with the old system is to heap disaster upon disaster for mankind and perhaps to destroy our race altogether." But he also admitted that the "dream of highly educated and highly favoured leading and ruling men, voluntarily setting themselves to the task of reshaping the world, has thus far remained a dream."

When it came to more specific predictions, Wells wrote, "As a prophet, the author must confess he has always been inclined to be rather a slow prophet," pointing out that in his book he had predicted 1956 as the year initiating the global conflict that actually began in 1914. The next prediction made in his novel was much closer to the mark. "The problem which was already being mooted by such scientific men as Ramsay, Rutherford, and Soddy, in the very beginning of the twentieth century, the problem of inducing radio-activity in the heavier elements and so tapping the internal energy of atoms, was solved by a wonderful combination of induction, intuition, and luck," wrote Wells in the first edition of his book, "so soon as the year 1933."[17]

One year earlier, in 1932, James Chadwick made a fateful discovery at the Cavendish Laboratory at Cambridge University. He identified a new subatomic particle and measured its mass. The existence of the particle had already been predicted by Rutherford, who called it a neutron. Chadwick published the results of his research in March. Reaction from the international scientific community was almost immediate. In May Dmitrii (Dmytro) Ivanenko, a Ukrainian scholar working in Leningrad, published an article in *Nature* suggesting a proton-neutron model of the atomic nucleus. He was followed a month later by Werner Heisenberg, a German theoretical physicist and a recipient of that year's Nobel Prize in physics, who independently suggested a very similar model. The discovery of the neutron produced a true revolution in thinking about the structure of the nucleus, but even more important in the long run was the identification of the neutron as an ideal bullet for bombarding the nucleus. Unlike Rutherford's alpha particles with which he bombarded atoms, or positively charged protons used for the same purpose by his students, neutrons had no charge, were not repelled by similarly charged particles, and could easily penetrate the nucleus.[18]

The first to realize the potential of the new discovery for unlocking nuclear energy was Leo Szilard, a thirty-five-year-old Hungarian physicist living in Germany who moved to Britain in 1933 to escape the Nazi regime. In the same year he sought to patent the idea of a nuclear chain reaction produced by neutron bombarding of the atomic nucleus. Neutrons, argued Szilard, could not only initiate the splitting of the nucleus. They could also be produced by the very same reaction and set off a self-sustaining chain reaction that would not need additional energy to continue and could itself produce an unlimited amount of energy. It would take a few years for Szilard's idea to be proved in laboratory testing. Once proved, it would change not only science but the world itself.[19]

Chapter 2

FRIGHT

Albert Einstein, sixty years of age, his uncombed hair already solidly gray, spent the summer of 1939 in a rented cottage on Nassau Point, Long Island. His favorite pastime was sailing his 17-foot boat, which he launched almost daily at Horseshoe Cove. He also played violin duets with David Rothman, the owner of the local department store, and read newspapers, which reported ever more disturbing news from Europe: Hitler was poised to seize Danzig, and the British and French were trying to stop him. One day Einstein followed Rothman to the meeting of an organization assisting refugees from Nazi Germany. Invited to speak, Einstein, not pleased with the sudden attention and wanting to leave the meeting as soon as possible, delivered the shortest speech in the history of the organization: "You must organize just as we Jews have organized. Otherwise you will have a big problem."[1]

Einstein knew what he was talking about. He was one of the first Jewish scientists to be targeted by the Nazis after Hitler's ascension to power in 1933. That year he renounced his German citizenship and made the United States his home. The Nazis soon passed a law prohibiting Jews from holding government and university positions: even those who, unlike Einstein, wanted to stay had to leave. As they departed, Jewish scholars were comforted by words of consolation from their German colleagues, but more often than not they found themselves targets of insults from Nazi-indoctrinated students. They left Central Europe for Britain and, in ever greater numbers, for distant and therefore safer America. Einstein, who came to the United States earlier than the others, obtained a position at Princeton. Given his worldwide name recognition, he became a natural attraction for European scientists arriving in the United States and seeking help and advice.[2]

On the hot summer day of July 12, 1939, Einstein welcomed two scientists to his Nassau Point cottage: Leo Szilard, age forty-one, and his fellow physicist, the thirty-six-year-old Eugene Wigner. Both had been born in Budapest when it was joint capital (with Vienna) of Austria-Hungary. Educated in Germany after World War I, they began working in physics at German institutions. Wigner came to the United States in 1929 and accepted an appointment at Princeton in 1930, before the Nazis came to power in Berlin. Szilard, who left Germany for Britain in 1933, was a recent arrival in the New World. Having landed in the United States the previous year, he was working at the Columbia University physics laboratory. Szilard was a former student and assistant of Einstein. Back in the 1920s they had jointly invented a new refrigerator that became known as the Einstein refrigerator. The meeting was in many ways a reunion, but it had a more specific agenda as well.[3]

Szilard and Wigner did not come to Nassau Point for Einstein's scientific advice or help with their careers. Their concern was global and their request international in nature. They wanted Einstein to approach Elizabeth of Bavaria, who was queen of Belgium and whom the great scientist knew socially. Belgium was not yet occupied by Nazi Germany—that would happen ten months later, in May 1940—and the two refugees were concerned not about Belgium per se but about its colony in Africa, the Congo. Deposits of unusually pure uranium had been discovered there in 1915 and mined since the 1920s. Szilard and Wigner wanted Einstein to ask Queen Elizabeth to tell the Belgian government not to sell Congo uranium to the Nazis under any circumstances. They believed that the fate of the world might depend on the success of their mission.

Einstein offered his former students iced tea. He was prepared to listen. The guests began describing the latest developments in nuclear physics and their own recent discoveries and calculations. Einstein was riveted.[4]

The chain of events that brought the two émigré scientists to Einstein in July 1939 had begun in December of the previous year in Europe with the discovery of nuclear fission, or a process of splitting heavier atoms into smaller ones, while releasing energy.

Late in the evening of December 19, 1938, the fifty-nine-year-old German chemist Otto Hahn sat down to write a letter to his longtime assistant, Lise Meitner. Hahn's other assistant, Fritz Strassmann, had been conducting experiments at the Kaiser Wilhelm Institute of Chemistry in Berlin, the same institution that gave the world one of its first poison gases. The experiments entailed bombardment of the nucleus, the central positively charged core of the atom described by Ernest Rutherford, with neutrons, discovered in 1932 by his student, James Chadwick. They targeted the nucleus of uranium, the element with the greatest atomic weight among the primordially occurring elements, its atom having 92 protons and 92 neutrons, making it fissionable through the use of fast neutrons.

That evening Hahn realized that something had gone wrong with the decay products of the uranium atoms that he had been bombarding with Strassmann. One of the new elements appeared to be barium, but its source was by no means clear. "It's almost 11 at night," wrote Hahn to Meitner. "Strassmann will return at 11:30 so that I can see about going home. The fact is that there is something strange about the 'radium isotopes' that for the time being we are mentioning only to you." He then added: "Our radium isotopes act like *barium*." Appealing to Meitner, he wrote: "Perhaps you can suggest some fantastic explanation. We understand that it really *can't* break up into barium."[5]

It would have been easiest to discuss the problem by inviting Meitner to the laboratory, but that was no longer possible. The Vienna-born Meitner had been forced to leave the Kaiser Wilhelm Institute five months earlier because she was Jewish. By July 1938, she had exhausted all legal avenues of return to a research position in the state-owned and -run institution. She had to leave the country illegally, bribing a border guard with a ring that Hahn had inherited from his mother and given to Lise to help her get out of Germany. She had found employment at the Nobel Institute in Stockholm, and they were now corresponding instead of speaking in person. Meitner was still helping Hahn and Strassmann to interpret the results of their experiments.

Although Meitner received Hahn's letter, she had no immediate answer to his puzzle. Intrigued by it, she decided to discuss the unexpected results of Hahn's experiments with her nephew, the thirty-four-year-old physicist Otto R. Frisch, an associate of Niels Bohr, the theoretical physicist who had married nuclear physics with quantum theory. Frisch had come to Stockholm from Copenhagen to visit his aunt over the Christmas break. Listening to Meitner, Frisch suggested that Hahn had simply got something wrong.

Meitner disagreed: Hahn was too good a chemist to mistake barium for something else. In the course of a long winter walk, they came up with an unexpected solution to Hahn's problem. What if the relatively light nucleus of barium, with its 56 protons, was simply one of the elements created by the division of the heavy nucleus of uranium, with its 92 protons? The neutrons, they assumed, did not simply crack or destroy the nucleus of uranium. Instead, by penetrating the nucleus, they caused its transformation into smaller nuclei, just as a drop of liquid can be divided into two drops.[6]

The liquid-drop model of the nucleus had first been proposed by Niels Bohr, and after his Christmas vacation in Sweden, Otto Frisch brought the news of Hahn's experiment and its possible explanation to Bohr himself. Bohr was intrigued, and they devised an experiment to test the theory. The essence of the experiment was to measure the level of energy released by the transformation of the nucleus. Assuming that Meitner and Frisch's interpretation of Hahn's experiment was accurate, the two parts of the nucleus moving away from each other would lose many electrons and release a great deal of energy that could be measured with a Geiger counter, a device named after Rutherford's German assistant Hans Geiger back in 1928. Frisch also came up with a name for the transformation of the nucleus that they wanted to study. He called it "fission," a term borrowed from biology, where it is used to define bacterial division.[7]

But Bohr could not wait for the results of Frisch's experimentation with fission. He had to leave for the Fifth Washington Conference on Theoretical Physics, which was to be held at George Washington University in the District of Columbia. As Bohr crossed the ocean by ship, a telegram reached him announcing the success of the experiment. An unusual amount of energy had indeed been released and captured by the Geiger counter. In Washington, Bohr announced the revolutionary discovery to conference participants, a virtual who's who of the new physics. He offered to repeat the Frisch-Meitner experiment in the university laboratory in their presence and with their participation. The results matched those obtained by Frisch. Some scholars tried to replicate the experiment on their own, and once again it worked.[8]

Everyone at the conference knew what that meant. It was no longer a question of the transmutation of chemical elements, as had been the case when Rutherford split the atom in 1917. There was now no doubt that nuclear energy existed and could be released. The only question remaining was whether a self-sustaining nuclear reaction could be produced. Should that prove possible, the world predicted by Wells back in 1914 would finally have arrived. The news spread like wildfire throughout the small community of nuclear physicists.

Leo Szilard did not attend the Washington conference but heard about Bohr's announcement and the subsequent experiments from Enrico

Fermi, his boss at Columbia University and a fellow refugee from European fascism. Fermi, the former director of the physics institute at the University of Rome, had fled Mussolini's Italy, concerned for the safety of his Jewish wife, Laura. Fermi was not optimistic about the possibility of producing a nuclear chain reaction, estimating its chances at 10 percent.

Szilard, on the other hand, believed that the chain reaction was a certainty: the only question was who would be the first to produce it. Indeed, he literally owned the idea. Back in 1933, when Szilard fled Nazi Germany for Britain, he had filed documents to patent the concept of a chain reaction. He maintained that a self-sustaining nuclear reaction could produce unlimited energy. Recalling his thoughts on hearing the details of Bohr's report, Szilard said: "H. G. Wells, here we come!" He was at once excited and terrified. "You know what that means!" he told Edward Teller, a fellow Budapest-born physicist and Jewish refugee from Nazi Germany. "Hitler's success could depend on it."[9]

Obsessed with his chain-reaction idea, Szilard attempted to be the first to realize it. With Fermi's support, he secured permission to use the physics laboratory at Columbia University to conduct his own experiments. He soon proved that a chain reaction was possible, showing that the bombardment of uranium by neutrons produces more neutrons than are consumed by the uranium nuclei. He needed more uranium, at least five tons, to continue his experiments. Fermi, never a believer in chain reactions, was ambivalent about continuing experiments with Szilard, who failed to mobilize support from the US Navy to obtain more uranium. Szilard turned to Eugene Wigner at Princeton.

Wigner was the first to express concern that if Germany overran Belgium, the Nazis would take over the uranium mines in the Congo. He was afraid that the Germans would win the race to secure uranium, produce a chain reaction, and build an atomic bomb. For both men, the cause of preventing the Germans from getting the Congo uranium suddenly took priority over almost everything else. But how to accomplish that? It was then that Einstein's name first came up in their discussions. On the morning of July 12, 1939, Eugene Wigner picked up Szilard at his hotel near the Columbia University campus in upper Manhattan, and together they drove to Nassau Point on Long Island.[10]

Einstein's first reaction on hearing about fission and the possibility of a nuclear chain reaction was surprise. "I never thought of that!" he told his guests. But Einstein's famous equation $E=mc^2$, stating that energy can be calculated on the basis of its mass multiplied by the speed of light squared, provided the theoretical foundations for an almost unlimited release of atomic energy, and he was soon converted to the nuclear fervor of his guests.[11]

"One thing most scientists are really afraid of is making fools of themselves," Szilard would write years later, recalling the events of that day. "Einstein was free of such fear." Indeed, Einstein agreed to associate himself with what was still a highly questionable scientific idea at the time. But the possibility of the Nazis being the first to produce an atomic bomb filled the Jewish refugees from Nazism with dread. Ironically enough, their conversation that day was conducted in German rather than English. The plot to deprive Germany of uranium was conceived in German, the language with which the three men were most comfortable.

Einstein offered to warn a member of the Belgian cabinet rather than Queen Elizabeth about possible German interest in Congo uranium. Eventually it was decided to write a letter to the Belgian ambassador in the United States, and Einstein began dictating the text to Wigner. But Wigner, who had been in America longer than either Einstein or Szilard, suggested that writing to a member of a foreign government without notifying the American authorities would be inappropriate. They drafted another letter, this time to the US Secretary of State, the idea being to send it along with a copy of Einstein's letter to the Belgian ambassador and give the State Department two weeks to present any objections. Szilard and Wigner left Nassau Point in high spirits but uncertain how best to act in order to deny the Germans Belgian uranium.[12]

Although the three scientists were now in America, the United States government entered their consciousness only as an afterthought. A few days after the meeting, Leo Szilard encountered another Jewish refugee from Eastern Europe, the Lithuanian-born economist Alexander Sachs. He had left the Russian Empire for the United States as a child in 1904, the year after the Kishinev pogrom, which triggered mass Jewish emigration from the empire. Sachs needed little convincing about the seriousness of the threat posed by possible Nazi access to uranium. He considered it urgent enough to be brought to the attention of the president of the United States himself.

Sachs was a vice president of Lehman Brothers, a large investment bank whose collapse in 2008 would trigger a global recession. He had been a consultant to President Roosevelt's electoral campaign of 1933, worked for FDR's National Recovery Administration, and had access to the president. When Sachs volunteered to deliver the letter to the White House, Szilard jumped on the opportunity and drafted a new letter from Einstein, now addressed to Roosevelt. The draft was written in German and mailed from New York to Nassau Point for Einstein's revisions and ultimate approval.[13]

In late July Szilard went to Nassau Point for another meeting with Einstein. This time he was accompanied by another Jewish refugee from Central Europe, Edward Teller, the future father of the hydrogen bomb. Einstein was on board with the idea of writing to Roosevelt. Of course, the letter now had to go beyond the uranium mines in the Congo. It had to alert the US administration to the threat of a devastating new weapon and spur it into action. Moreover, Szilard needed support to continue his own experiments at Columbia University. They rewrote the letter and discussed the drafts, which Szilard took back to New York for translation and typing. He came up with two versions, longer and shorter, which were sent to Einstein, who signed them both.[14]

Szilard gave Sachs the longer version of the letter, preferred by Einstein, for delivery to the president. "Some recent work by E. Fermi and L. Szilard, which has been communicated to me in manuscript, leads me to expect that the element uranium may be turned into a new and important source of energy in the immediate future," reads the matter-of-fact beginning of the letter. The authors cut to the chase when they asserted: "This new phenomenon would also lead to the construction of bombs, and it is conceivable—though much less certain—that extremely powerful bombs of this type may thus be constructed. A single bomb of this type, carried by boat and exploded in a port, might very well destroy the whole port together with some of the surrounding territory."[15]

Einstein and Szilard clearly had in mind the December 1917 explosion of the French ship SS *Mont-Blanc* in the harbor of Halifax, Nova Scotia. Loaded with high explosives, it collided with another vessel, caught fire, and produced an explosion equal to almost 3 kilotons of TNT. As many as 2,000 people had been killed, up to 9,000 injured, and a whole section of the city wiped out. That was the largest explosion in world history until World War II, and Franklin Delano Roosevelt, who had served as assistant secretary of the US Navy at the time, knew the extent of the destruction. Whatever FDR might think about nuclear chain reaction, if anything, a reference to the Halifax explosion was bound to attract his attention.[16]

The letter asked the president "for watchfulness and if necessary, quick action on the part of the Administration." That was to include the appointment of a liaison between the nuclear physicists and the authorities who would inform the administration about the physicists' research and help them speed up their work by securing government and private funds and gaining access to requisite laboratories and equipment. The last paragraph stressed the urgency of the task. "I understand that Germany has actually stopped the sale of uranium from the Czechoslovakian mines which she has taken over," wrote Einstein and his ghost writer. They added a sentence on research at the Kaiser Wilhelm Institute in Berlin, where, according to the letter, "some of the American work on uranium is now being repeated."[17]

The initiative that began with the idea of influencing the Belgian government not to sell uranium to the Germans had turned into an appeal to the United States government to jumpstart a government-backed nuclear fission program with the goal of building an atomic bomb. According to Einstein and Szilard, the race to create such a weapon had already begun: the United States had to enter it and beat the competition. The authors did not discuss whether the bomb should be employed as a threat or used preemptively against a similarly armed Germany. "We realized that, should atomic weapons be developed, no two nations would be able to live in peace with each other unless their military forces were controlled by a higher authority," wrote Eugene Wigner years later.[18]

The imagination of Wigner and his friends had been greatly influenced by H. G. Wells's depiction of nuclear war and its aftermath. The higher authority of which Wigner wrote could only be some form of Wells's world government, which would theoretically have made Wigner and Szilard conspirators not only against the German government but against the American government as well. It is hardly likely that their thinking went so far in the summer of 1939: the imperative then was to stop the Germans. A related goal was to obtain funding for Szilard's research, which would give him an opportunity to prove that his concept of chain reaction was workable. Fear was matched by intellectual curiosity and desire to be the first to solve a scientific problem. Szilard was probably the first but definitely not the last nuclear scientist driven by those often contradictory feelings and desires. They would become the fixture and driving force of the early nuclear age.

Einstein signed the letter sometime after August 2, 1939, the date of the final English version. His pacifism was held in abeyance, given the threat of a Nazi bomb. The world was experiencing its last weeks of peace before the outbreak of World War II. The next day, August 3, the Nazi government stepped up its anti-Semitic campaign, withdrawing licenses from Jewish doctors still practicing in Germany. On August 23, Hitler's foreign minister, Joachim von Ribbentrop, flew to Moscow to negotiate with Joseph Stalin and sign the Molotov-Ribbentrop Pact, dividing Eastern Europe between the two dictators and opening the door to the German invasion of Poland. German troops would cross the Polish border nine days later, on September 1. Britain and

France would declare war on Germany on September 3, setting off a new world conflict.[19]

On Wednesday, October 11, after convincing FDR's key aide, General Edwin "Pa" Watson, that he had a cause important enough to bother the president, Alexander Sachs finally made his way to the Oval Office. "What bright idea have you got now?" said Roosevelt by way of greeting. "How much time would you like to explain it?" Sachs had the unwelcome task of selling the president on an idea that was anything but easy to grasp. In spite of the detailed explanation in the physicists' letter, the only item certain to impress Roosevelt was the name of the signatory, Albert Einstein. The letter itself was simply too long and dry to be read by the president in the course of a meeting or to convince him.

Sachs did not believe that leaving the letter with the president would do the trick. He would have to read parts of it aloud and explain the rest. He spent the previous night in nervous meditation but came up with an effective anecdote to start the conversation. It was taken from a book by the nineteenth-century English historian Lord Acton. Sachs also took along a book by Acton's countryman, the Nobel Prize winner in chemistry Francis Aston. Entitled *Background to Modern Science*, the book had a chapter in which Aston eloquently discusses his belief that humankind would manage to release the energy of the atom.

"All I want to do is tell you a story," began Sachs innocently. "During the Napoleonic wars a young American inventor came to the French emperor and offered to build him a fleet of steamships with the help of which Napoleon could, in spite of the uncertain weather, land in England." Napoleon simply could not imagine a fleet without sails and sent the inventor away. Then came the punch line of Sachs's story: the name of the young American inventor was Robert Fulton, the inventor of the steamboat. Had Napoleon listened to Fulton, argued Lord Acton, he could have defeated Britain and changed the course of European history.

Roosevelt liked the story. In fact, listening to it, he ordered his steward to bring a bottle of Napoleon brandy that the Roosevelt family had had in its cellars for a while. The steward poured brandy into the two glasses he brought along, and Roosevelt toasted Sachs. So far, so good: the Napoleon brandy was by no means a token of esteem for the French emperor's memory. Sachs now moved on to the matter that had brought him to the White House—atomic energy. Having previously advised the president on economics, Sachs began with the economic rather than the military implications of the coming atomic revolution. He spoke first about nuclear energy, then about the medicinal uses of radiation, and finally about "bombs of hitherto unenvisaged potency and scope."

Sachs summarized the Szilard-Einstein letter, but, being all too cognizant that it was hardly eloquent enough to persuade Roosevelt on its own, he quoted from the book by Francis Aston that he had brought along: "Personally I think there is no doubt that sub-atomic energy is available all around us, and that one day man will release and control its almost infinite power. We cannot prevent him from doing so and can only hope that he will not use it exclusively in blowing up his next door neighbour." Roosevelt abruptly cut off his guest. "Alex," he told his adviser turned lobbyist for an as yet nonexistent industry, "what you are after is to see that the Nazis don't blow us up." The letter's hint at the Halifax explosion had clearly registered with the president. "Precisely," responded Sachs with relief. His job was done.

When Roosevelt invited "Pa" Watson into the Oval Office, he told his aide, pointing to the letter brought by Sachs: "Pa, this requires action!" The president was prepared to consider building an atomic bomb to ward off the threat of a similar German weapon. The idea of nuclear deterrence as a political strategy was born at that very moment on October 11, 1939.[20]

Chapter 3

NAZIS AND THEIR FRIENDS

As Leo Szilard, Albert Einstein, and other refugees from Nazi-controlled Europe lived in fear of a Nazi atomic bomb, German physicists worried increasingly about the possibility of an American bomb.

No one in Germany would become more concerned about the Americans getting a bomb first than Werner Heisenberg, the key figure in the German nuclear project. An ethnic German with the looks of an Aryan ideal, Heisenberg had graduated from the University of Munich. In the 1920s he spent some time at the University of Copenhagen, working under Niels Bohr, whom he considered one of his teachers. Heisenberg received his Nobel Prize in physics in 1932 for the "creation of quantum mechanics" when he was only thirty years old. He soon got into trouble with the Nazis for teaching his students about non-German and Jewish contributions to physics, but decided to stay in Germany.[1]

The Nazi government regarded the new field of nuclear physics with suspicion. State anti-Semitism was not only driving some of the most talented scientists abroad but also creating havoc among those who stayed in Germany. The whole field was turning into an ideological battleground between proponents of experimental physics, which became known as *Deutsche Physik*, and theoretical physics, labeled *Jüdische Physik*. Heisenberg ended up in the "Jewish" camp. He was investigated by the SS on the orders of Heinrich Himmler but convinced his investigators, all adepts of *Deutsche Physik*, that he was engaged in worthwhile teaching and research. Himmler issued a letter of dispensation, instructing Heisenberg "to separate clearly for your students acknowledgment of scientific research results from the scientist's personal and political views."[2]

Heisenberg would later complain that "public interest in the problems of atomic physics was negligibly small in Germany between the years 1933 and 1939, in comparison with that shown in other countries, notably the United States, Britain, and France." What Heisenberg had in mind was not public interest per se—if one counts ideological campaigns against *Jüdische Physik*, it was excessive by any standard—but government funding of the field. Despite the discovery of nuclear fission by Otto Hahn in December 1938 and significant achievements of other German physicists and chemists in research on the structure of the atom, the Nazis were not rushing to put that research to use.

Heisenberg was particularly envious of the American cyclotrons that would be used as one of the means to enrich uranium. The United States had several; Germany had none. Heisenberg would later attribute the change in the Nazi government's view of nuclear physics to concern about an American atomic bomb. "Almost simultaneously with the outbreak of war, news reached Germany that funds were being allocated by the American military authorities for research in atomic energy," wrote Heisenberg after the end of World War II. It is not clear what information he had in mind, as it is well known that the US government had not allocated funds for building an atomic bomb before the war broke out.[3]

Nevertheless, anyone reading the *New York Times* in the United States or in Germany might well have been concerned by headlines such as the one that ran on April 30, 1939: "Vision Earth Rocked by

Isotope Blast: Scientists Say Bit of Uranium Could Wreck New York." The report detailed a meeting of the American Physical Society held the previous day. Among the speakers were Niels Bohr, who talked about nuclear chain reaction, and Lars Orsagen of Yale University, who described "a new apparatus in which according to his calculations, the isotopes of elements can be separated in gaseous form."[4]

It was the publication of such articles that most concerned Leo Szilard. He did not want to see the Germans alerted to developments in the United States, and he emerged as an early proponent of keeping new research secret. Szilard tried and failed to delay the publication in April 1939 of an article by Frédéric Joliot-Curie and his group of scientists in Paris reporting on the release of multiple neutrons after bombarding an atomic nucleus (a ratio of 3.5 to each neutron used to bombard the nucleus). Those findings, indicating the means of producing a chain reaction, were published on April 22, 1939, and had the effect on German scientists that Szilard had feared: they sprang into action.[5]

Years later, Heisenberg would recall the publication of that paper. Around the same time, Paul Harteck, a physicist in Hamburg, apprised the military authorities in Berlin that in his opinion and that of his colleagues, the new discoveries in nuclear physics would "probably make it possible to produce an explosive many orders of magnitude more powerful than the conventional one." He continued: "That country which first makes use of it has an unsurpassable advantage over others." On April 29, 1939, a week after the publication of Joliot's paper, a group of German scholars met under the auspices of the Ministry of Education to establish the Uranium Club, a committee charged with sharing the information and coordinating the activities of various scholars and institutions working on the fission problem.

As Szilard had feared, the Germans were interested in the weaponization of nuclear research, but for some time it remained a scholarly initiative with little government support. The number of scholars involved was limited, and once Germany heightened its preparations for an attack on Poland in the spring and summer of 1939, many physicists were drafted into the army, effectively dooming the project initiated by the Ministry of Education, which ranked low in the Nazi pecking order. All that changed on September 1, 1939, when officials at the German Army Ordnance Office decided to take over the project after receiving a number of letters from German scientists about the military implications of their research. The office used its power to release physicists drafted into the army from military service.[6]

The first meeting of scholars involved in the Uranium Club took place on September 16, one day before Stalin's Red Army entered the war by attacking Poland from the east. The new German nuclear bomb project was headed by the young and extremely ambitious nuclear physicist Kurt Diebner, who had served since 1939 as a scientific adviser to the German army. He soon became director of the Ordnance Office's Nuclear Research Council and took charge of the Kaiser Wilhelm Institute for Physics in Berlin. By means of the army draft, Diebner "invited" the country's key physicists, including Otto Hahn, to join the project. Among the new recruits was also Heisenberg, who wrote later: "As early as September 1939 a number of nuclear physicists and experts in related fields were assigned to this problem, under the administrative responsibility of Diebner." The task assigned to them, according to Heisenberg, "was to examine the possibilities of the technical exploitation of atomic energy"—that is, the prospects of building an atomic bomb.[7]

By October 1939, the month Einstein's letter was finally delivered to Roosevelt, the Diebner project was already gathering speed, with research into producing a chain reaction and building a bomb going on at numerous German universities. In Leipzig, Heisenberg began theoretical work on the construction of a nuclear reactor required to separate isotopes of uranium and release uranium-235. He considered two possible moderators, graphite and heavy water, to slow down neutrons and enable a chain reaction. Ultimately, he would choose heavy water, as he did not have access to pure graphite.[8]

In December 1939 Heisenberg submitted his first calculations for the building of an atomic bomb, asserting that it would take hundreds of tons of nearly pure uranium-235 to build the critical mass required for initiating a nuclear explosion. That did not sound encouraging, but in the following year Carl-Friedrich von Weizsäcker, a member of the Heisenberg group and the only German physicist mentioned by name in Einstein and Szilard's letter to Roosevelt, came up with the idea that a nuclear reactor could be used to produce a new element, later to become known as plutonium, as a component of the bomb. Heisenberg embraced the idea. He and his group charged ahead, their first task being the building of a reactor. *Jüdische Physik* was now working for *Deutsche Physik*, and the Germans were ahead of anyone else.[9]

In the spring of 1940, as Hitler took Paris, defeated France, and allowed the British troops at Dunkirk to retreat from the continent in disgrace, two more countries initiated their own nuclear programs. Both were allies of Germany: the Soviet Union and the Empire of Japan.

The Japanese program was led by Yoshio Nishina who, like Werner Heisenberg, had studied and worked under Niels Bohr in Copenhagen. Unlike Heisenberg or, for that matter, any physicist in Germany, Nishina was in possession of the most valuable equipment for his research—the cyclotron. In 1936 he built the first Japanese cyclotron, which was the second such machine in the world and the first outside the United States. A larger cyclotron followed in 1937, both housed at the Institute for Physical and Chemical Research or RIKEN in Tokyo.

Like his German colleagues, Nishina was attuned to the huge potential opened by Otto Hahn's discovery of nuclear fission, but unlike them or, for that matter, British and American physicists, he was in no rush to knock on the doors of government agencies, warning them about the possibility of foreign atomic bombs and asking for funds to support his research. It was purely by chance that in the early summer of 1940 he shared a train ride with Lieutenant-General Takeo Yasuda, the director of the Technical Research Institute in the Aeronautical Department of the Imperial Japanese Army. As they discussed the latest news in nuclear physics, Yasuda became interested in the military applications of nuclear fission.

General Yasuda's original concern was not unlike that of Leo Szilard and George Thomson—the general was worried about supplies of uranium ore. Yasuda ordered his subordinate, Lieutenant-Colonel Tatsusaburo Suzuki, to prepare a report on the feasibility of a uranium project in Japan. A twenty-page report was submitted in October 1940. Suzuki argued that Japan had enough uranium deposits at home and overseas to build the bomb. Yasuda shared the report with other government departments but pretty much stopped there. He did not advocate a crash program to build the bomb or try to classify Suzuki's findings and conclusions: his report was widely circulated in the government. Yasuda made his next move only in April 1941, requesting a feasibility study of producing a chain reaction in uranium. The task was assigned to Yoshio Nishina, the physicist who had inspired the general's initial interest in the project. It had been almost a year since their chance meeting on the train.

The Japanese were clearly taking their time. Nor was there any exchange of information on the nuclear project with Japan's ally, Nazi Germany. The Japanese had cyclotrons but no nuclear lobby driven by frightened, concerned, or ambitious scientists. The government also saw no potential threat from an atomic bomb in the hands of enemies or rivals. History would prove it wrong.[10]

The Soviets were more than vigilant when it came to possible threats from abroad. Born of a revolution that the Western powers, including the United States, had sought to defeat by sending troops and supporting anti-Bolshevik forces, the USSR had reasons to believe that it existed in a hostile environment. In the late 1920s, Stalin began a crash program of modernization with an eye to preparing for a future war. He joined Hitler in attacking Poland in 1939, hoping that the Nazis would turn westward after that. Hitler did as Stalin expected, but the rapid fall of France and the evacuation of British troops from Europe in May 1940 left Stalin one on one with the Führer.

May 1940 marked the beginning of the Soviet nuclear project, but the trigger was in New York, not Paris. On May 5, the *New York Times* published another article about nuclear research. Its author, William Laurence, a Lithuanian-born Jewish immigrant to the United States, was as concerned as Leo Szilard about the possibility of a Nazi bomb and closely followed developments in nuclear research. He had published numerous articles on the subject in 1939, but this one was particularly significant, as it appeared on the front page of the *Times*. Laurence reported on experiments with uranium-235 at Columbia University and speculated about nuclear research going on in Germany.

"Every German scientist in the field, physicists, chemists, engineers, it was learned," wrote Laurence, "had been ordered to drop every other research and devote themselves to that work alone." He added that while it was believed that the Americans were still in the lead, no one knew how far German research had advanced. "The main reason why scientists are reluctant to talk about this development, regarded as ushering in the long dreamed of age of atomic power and, therefore, as one of the greatest, if not the greatest discovery in modern science," wrote Laurence, referring to the separation of uranium-235, "is the tremendous implication this discovery bears on the possible outcome of the European war."[11]

The front-page article attracted the attention of George Vernadsky, a scholar of Russian history at Yale University. Vernadsky had left the Russian Empire via the Crimea during the revolution, leaving behind his family, including his father, Vladimir (Volodymyr), a renowned mineralogist and geophysicist who became the founding president of the Ukrainian Academy of Sciences in 1918. While the younger Vernadsky found his way to Yale, his father assumed important positions in the Soviet Academy of Sciences. George Vernadsky knew that his father would be interested in the *Times* article. Back in 1910, Vladimir had pioneered research in radiology and launched the first search for uranium deposits in the Russian Empire. In 1922 he founded the Radium

Institute in Petrograd, soon to be renamed Leningrad. It was there on his watch that the first Soviet cyclotron was built in 1937.[12]

George Vernadsky clipped the Laurence article and sent it to his father in the Soviet Union. Once it arrived, Vladimir Vernadsky sprang into action. He lost no time in writing a memorandum to the Geological and Geographic Section of the Academy of Sciences, urging it to launch a program of uranium exploration. The memorandum was cosigned by Vernadsky's younger colleague Vitalii Khlopin, who had taken over the directorship of the Radium Institute from Vernadsky in 1939. The academy responded to their memo in late June 1940 by creating a special commission to study the issue, but the two scientists were not content.

In July 1940 they wrote a new letter, this time to the Soviet vice premier, Nikolai Bulganin, informing him about the discovery of nuclear fission and proposing that the government fund the completion of work on the new cyclotron and launch a project on isotope separation. By the end of the month a special Commission on the Uranium Problem had been established under the auspices of the Presidium of the Academy of Sciences. It was headed by Khlopin, a specialist in the chemistry of radioactive materials, who was to be assisted by two deputies—Vernadsky, whose area was geology and mineralogy, and Abram Ioffe, the country's leading nuclear physicist.[13]

Abram Ioffe and his students would play a key role in the development of the Soviet nuclear project. Despite the increasing isolation of Stalin's Soviet Union from the rest of the world, Ioffe and his group were plugged into the international network of physicists throughout the 1920s and most of the 1930s. An ethnic Jew from Ukraine, Ioffe received his doctorate in 1902 from the Ludwig Maximilian University in Munich, Werner Heisenberg's alma mater, and in his youth he had worked with no less a figure than Wilhelm Roentgen. In 1917 Ioffe became the founding director of the Petrograd (later Leningrad) Physical-Technical Institute, the center of atomic studies in the Soviet Union. In 1932, Ioffe's student Dmitrii Ivanenko came up with the neutron-proton model of the atom. A year later, excited by Chadwick's discovery of the neutron, Ioffe, Ivanenko, and a rising star of Soviet nuclear research, Igor Kurchatov, organized an all-Union conference on nuclear physics attended by Frédéric Joliot-Curie and a number of other European scholars that gave a boost to Soviet research on the subject.[14]

In 1928, Ioffe organized the Ukrainian Physical-Technical Institute in Kharkiv, then the capital of Ukraine. The institute would play a key role in Soviet nuclear research during the 1930s. Ivanenko worked there as head of the theoretical physics department in 1929–31. Later in the decade the staff was joined by a number of German physicists fleeing the Nazi regime and anti-Semitic laws in Germany. Among them were Friedrich Georg Houtermans and Fritz Lange. It turned out they were running from one disaster to another. Houtermans, a communist, was arrested during Stalin's Great Terror and, after the signing of the Molotov-Ribbentrop Pact, extradited to Nazi Germany, where he would contribute to theoretical work on the production of plutonium. Lange avoided arrest, allegedly because the documents granting him Soviet citizenship were signed by Stalin himself.[15]

In October 1940 two of Lange's students, Viktor Maslov and Vladimir Shpinel, submitted a proposal for the building of a "uranium bomb" to the patent office of the Soviet Defense Commissariat. The two Kharkiv scientists wrote: "The problem of producing a uranium explosion comes down to creating a mass of uranium in a quantity considerably greater than critical in a brief period of time. We propose that this be done by filling a vessel with uranium, the vessel to be divided by separators impenetrable to neutrons such that every isolated portion, or section, contain a less than critical amount of uranium. Once such a vessel is filled, the separators are eliminated by means of an explosion, leaving a mass of uranium considerably greater than critical." The idea of producing a nuclear explosion by bringing together two subcritical masses of uranium had already been proposed outside the USSR. The significant contribution of the Kharkiv memo was the suggestion of how to remove the separators between the two masses of uranium, which was similar to the one considered for the first American atomic bomb.

Some suspect that the idea proposed by Maslov and Shpinel really belonged to their professor, Fritz Lange, but was signed by Maslov because, as a member of the Communist Party, he was trusted by the authorities more than the unreliable German émigré. While the Nazis were focused on the race with the United States, and the Americans preoccupied with countering the Nazi efforts, the Soviets were concerned first and foremost with the hostile capitalist West. If the scientists like Otto Frisch would have moral reservations about Britain's use of the bomb, Maslov and Shpinel had no such concerns with regard to the use of the bomb against the capitalist states.

"As regards a uranium explosion, aside from its colossal destructive force (obviously, building a uranium bomb sufficient to destroy such cities as London or Berlin poses no problem), one more exceedingly important feature must be noted. The uranium bomb produces radioactive substances," wrote the two young physicists. The choice of possible targets in Maslov and Shpinel's memo is interesting in its own right. In October 1940, the Soviet Union was allied with Nazi Germany and technically at war with Britain. But for the two Soviet researchers, both Berlin and London were centers of the capitalist world equally hostile to their communist homeland.[16]

Reaction to the Kharkiv memo in Moscow was cool. In January 1941, the idea of using explosives to bring together subcritical masses of uranium was rejected by the experts of the Soviet Defense Commissariat as flawed and unrealistic. In April 1941 Vitalii Khlopin, the director of the Radium Institute and head of the Uranium Problem Commission, rejected the proposal as well. Unknowingly echoing the sentiments of Niels Bohr at the time, he wrote: "The situation with the uranium problem at present is such that the practical exploitation of internal atomic energy produced in the process of splitting atoms through the use of neutrons is a more or less distant aim toward which we should strive, and not a problem of the present day."[17]

With the idea of building a bomb in the imminent future rejected by the academic community, the Soviet government had little incentive to dedicate funds and resources to the project as the prospect of conventional war loomed closer. Once the Germans invaded their Soviet ally on June 22, 1941, smashing one division after another, the Soviets had to fight for their survival, and all thought of nuclear research was abandoned. By September, the Soviets had lost not only all the territorial gains of the 1939 Molotov-Ribbentrop Pact with Germany, including the Baltic states, Moldavia, and western parts of Ukraine and Belarus, but also the capitals of those two republics, first Minsk and then Kyiv. In the same month the Germans encircled Leningrad and began the siege of the city. By early December they would be on the outskirts of Moscow.[18]

In September 1941, as the Wehrmacht drove ever deeper into Soviet territory, the German war machine delivered its first tangible results when it came to Werner Heisenberg's uranium project. Heisenberg received 150 liters (some 40 gallons) of heavy water from Norway, which had been occupied the previous fall by the German army. He was finally in possession of the moderator for the nuclear reactor he had envisioned since the end of 1939 and could start building a prototype.

"The uranium and heavy water were arranged in alternate spherical layers with the neutron source at the center," wrote Heisenberg later. He found that this arrangement worked: his assistants were able to detect the increase in neutron production. "The use of pure uranium metal . . . gave such a decided improvement that no further doubt of a real increase in the number of neutrons was possible," recalled Heisenberg. The requisite chain reaction was there, around the corner: it would be achieved in February and March 1942.[19]

The problem was that by the time Heisenberg obtained his first heavy water in the autumn of 1941, he no longer believed that the Germans could build a bomb before the end of the war. His concern was that the Americans, with their superior economic potential, would do it first. What if they dropped the bomb on Germany? Heisenberg's wife noticed that in the fall of 1941 he was preoccupied with thinking through the possible scenarios. In September he decided to visit his old mentor Niels Bohr, who remained in Denmark, now occupied by Germany. Heisenberg wanted to discuss nuclear weapons with Bohr and decide whether it was morally defensible for scientists to produce them in the course of the war. Perhaps, he thought, a general moratorium on atomic weapons research could be attained.

Bohr, who refused to cooperate with the occupying authorities, met Heisenberg, suspicious that his former associate might have been dispatched by the Nazis to discover what was going on with nuclear research in the United States. Heisenberg shared with Bohr a sketch of the German reactor, but Bohr cut off the discussion. Whether Bohr's suspicion was justified or Heisenberg was indeed trying to explore the prospects of a global scientific solidarity to stop the production of nuclear weapons—a discussion already decades old at the time—there is little doubt that Heisenberg's main concern and the cause of his doubts was not the German but the American bomb.[20]

Chapter 4

TRANSATLANTIC ALLIANCE

"In these past two tragic years war has spread from continent to continent; very many Nations have been conquered and enslaved; great cities have been laid in ruins; millions of human beings have been killed, soldiers and sailors and civilians alike," said Franklin Delano Roosevelt, addressing the US Congress on October 9, 1941. He asked Congress to amend the Neutrality Act of 1939, which had kept the United States out of the war raging in Europe and Asia, and allow the US Navy to arm merchant ships delivering Lend-Lease supplies to Britain.

Congress was receptive, voting 259 to 138 to amend the act. The country was in fact at war, and Roosevelt was leading the national legislature toward an open declaration. In the same message to Congress, he continued: "I say to you solemnly that if Hitler's present military plans are brought to successful fulfillment, we Americans shall be forced to fight in defense of our own homes and our own freedom in a war as costly and as devastating as that which now rages on the Russian front. Hitler has offered a challenge which we as Americans cannot and will not tolerate."[1]

On the same day that Roosevelt sent his message to Congress, he took a much less known but no less decisive step with regard to preparing for American participation in the global war and strengthening his country's alliance with Britain. At 11:30 a.m., in his first appointment of the day, Roosevelt met in his office with Vice President Henry A. Wallace and Dr. Vannevar Bush, the head of the recently created Office of Scientific Research and Development. On the agenda was Bush's report on the development of nuclear fission research and prospects of building an atomic bomb.[2]

The fifty-one-year-old Bostonian Vannevar Bush was an empire builder, and his empire was constructed at the crossroads of science, government, and warfare. A graduate of Harvard and the Massachusetts Institute of Technology, with degrees in engineering, he had served as a vice president of MIT before becoming president of the Carnegie Institution for Science in Washington, DC. His tasks in that office, which he assumed in January 1939, included advising the government on scientific affairs.

On June 12, 1940, two days before Hitler's troops entered Paris, Bush proposed that Roosevelt create the National Defense Research Committee to introduce scientific discoveries to the American defense industry. During World War I, Bush had worked for the National Research Council, created by Woodrow Wilson for the same purpose. Roosevelt, then assistant secretary of the US Navy, was one of the beneficiaries of the council's work. Now, with war raging in Europe and worldwide, Wilson's council was reestablished under a new name and leadership. In June 1941, the month Hitler attacked the Soviet Union, Bush became head of another institution linking science and defense, the newly created Office of Scientific Research and Development. The office, charged with the task of developing new weaponry, received independent funding from Congress.

The atomic bomb, a radically new weapon, thus came under Bush's purview, although Bush himself was at first skeptical about the concept of nuclear fission and any practical results that might follow from it. It was not because of the lack of interest. Bush was well informed about the progress of research in nuclear physics: it was at the very start of his tenure at Carnegie that Niels Bohr brought news of experiments with nuclear fission to Washington in January 1939. The Carnegie Institution became the first to hold a press conference announcing the discovery. But Bush shared Bohr's opinion that research on chain reaction and the unleashing of nuclear energy was a task for the future, and under wartime conditions priority should be given to other scientific matters.[3]

Bush also believed that talking about the bomb too much, especially in public, was counterproductive, as speculation about H. G. Wells's prophecy could create panic. That had already happened in October 1938 after the broadcast of a radio dramatization of Wells's *War of the Worlds*, when listeners assumed that the country was being invaded by Martians. Residents of New Jersey hit the highways to escape a nonexistent invasion of aliens equipped with heat wave weapons. Bush wanted none of that.[4]

Part of Bush's skepticism about an atomic weapon was due to the slow progress of American fission research, triggered by Einstein's letter to Roosevelt. The Advisory Committee on Uranium created by presidential order was as yet unable to establish whether uranium could become a source for bombs of enormous destructive power. As head of the National Defense Research Committee, Bush took the uranium project, known as S-1, under his auspices in June 1940. It was a small operation, its original budget limited to the $6,000 requested by Leo Szilard, Eugene Wigner, and Edward Teller to buy purified graphite for their experiments in producing a chain reaction at Columbia University. The experiments went on, but chain reaction was still just a theory.

Bush considered that with little to show, the project did not deserve much more investment. He wrote to a colleague: "I am puzzled as to what, if anything, ought to be done in this country in connection with it." When a request came to allocate an additional $140,000, Bush authorized only $40,000. It was only in the fall of 1941 that he changed his mind as a result of fierce lobbying efforts, not on the part of Szilard and the émigrés who backed him but of key figures in the American scientific establishment.[5]

One of the most important voices was that of Ernest O. Lawrence, an American-born and American-trained Nobel Prize winner who was especially good at bringing science and technology together. Lawrence received his prize in 1939 for inventing and building the world's first cyclotron, a machine for the acceleration of charged particles, essential for nuclear research, at the University of California at Berkeley. In December 1940 Lawrence's cyclotron made a key contribution to the field of nuclear physics. Glenn T. Seaborg and his associates, working under Lawrence, synthetically produced a new element predicted a few months earlier by Carl Friedrich von Weizsäcker in Germany. The new element was named plutonium after the planet Pluto, discovered in 1930 by an American astronomer, Clyde W. Tombaugh, and then considered the planet farthest from the sun.

Plutonium-239 was uniquely able to be fissile under interaction with slow neutrons and thus an excellent candidate for the production of a chain reaction. The production of plutonium on a mass scale was of course a problem yet to be solved, and a group of its original discoverers continued their work at Berkeley throughout the winter of 1941. Lawrence, now a strong believer in the possibility of chain reaction, pushed for government support and funding of the project. As always, he was thinking big: a few years earlier he had obtained a grant in excess of a million dollars. He wanted Bush to come up with serious money for the uranium project, which was now also becoming a plutonium project.

Lawrence's other impetus for active lobbying of Bush and the academic and government establishment to support the uranium project came from overseas. Unlike Szilard, Lawrence was not apprehensive about a German atomic project but concerned that the British might be first to come up with a nuclear weapon.[6]

Britain, the country of H. G. Wells, Ernest Rutherford, Frederick Soddy, and James Chadwick, was no less surprised by the discovery of nuclear fission than the rest of the world.

George Paget Thomson, a Nobel Prize–winning physicist at Cambridge University, was alerted to the possibilities opened by Hahn's experiment in April 1939 after reading Frédéric Joliot-Curie's article in *Nature*. In that regard, the impetus for the start of the British project did not differ greatly from that of the German initiative, where the same article alerted the British government to the implications of the discovery of nuclear fission. What connected the beginnings of the British project with the American one was that Thomson, like Szilard after him, was concerned to block German access to the Congo uranium mines. When Thomson sounded the alarm, the British government responded. A meeting was arranged with the president of the Belgian mining company working in the Congo and an agreement reached ensuring British access to uranium. The British were assured that neither Germany nor any other country was buying up stocks of the mineral.[7]

Most British physicists, like the government officials and politicians they advised, did not believe that uranium would be needed anytime soon. The discoverer of the neutron, James Chadwick, was not sure whether a chain reaction could be initiated and sustained. Those working on uranium at the Imperial College in London considered the production of a chain reaction in uranium-238 impossible. Thomson and his group experimented with natural uranium and reached the same conclusion in February 1940. Nor did the outbreak of World War II in September 1939 help the British nuclear project; rather, it compelled the government and scientists to focus on projects that might yield almost immediate results at the front lines. Nuclear research was left to scientists who could not obtain security clearance to work on military-related projects.[8]

Otto Frisch, the physicist who was first, together with Lise Meitner, to grasp in December 1938 that the Hahn and Strassmann experiments had produced nuclear fission, and who invented the term "fission," belonged to that category of unreliable scientists. Concerned about the possible German takeover of Denmark, where he had worked with Bohr, Frisch left Copenhagen for Birmingham, where he was employed by the Australian-born nuclear physicist Mark Oliphant. Not trusted to work on military projects such as the building of a new radar, Frisch, together with the German-born physicist Rudolf Peierls, was asked by Oliphant to work out calculations for a possible atomic bomb. Earlier calculations suggested that if a uranium-238 bomb were to work, it would have to weigh anywhere between 12 and 40 tons. While that was less than Heisenberg had come up with in Leipzig, it did not sound at all promising. Almost no one had high hopes for the work undertaken by the two refugees from Germany.[9]

In March 1940, however, Frisch and Peierls produced a report that Mark Oliphant believed he had to bring to government attention without delay, especially in a war that was going badly for Britain. Frisch and Peierls suggested that a bomb could actually be built in fairly short order and that it would be relatively small in size. The previous month, Frisch had seized on Bohr's idea of uranium-235, primarily responsible for fission in Hahn's experiment, to suggest that an atomic bomb would require less than a kilogram (2.2 pounds) of that isotope.

That was a far cry from the tons or even hundreds of tons of uranium generally considered requisite for a bomb to work. The memorandum described the basic principles behind the creation of an atomic weapon: two parts of the bomb, consisting of uranium-235, would be brought together to produce critical mass and start a chain reaction that would release destructive energy equal to 1,000 tons of dynamite. To obtain uranium-235, which constituted only 0.7 percent of naturally occurring uranium, a nuclear plant would have to be built at a cost that the authors of the memorandum did not consider prohibitively high. They called the new weapon a "super-bomb."

While making a disclaimer that they were not competent to discuss the strategic importance of the bomb, Frisch and Peierls dedicated a good part of the memorandum to that very question. They suggested that the super-bomb would be "practically irresistible" in explosive and destructive power. However, the two scientists warned the military leaders that it could not be used to break through enemy fortifications because of radiation fallout. For the same reason, use of the bomb would result in civilian deaths, even if the weapon were used under water to target a naval base—a clear reference to the Halifax harbor explosion of 1917. "This may make it unusable as a weapon for use by this country," wrote the two scientists, whose homelands, Austria for Frisch and Germany for Peierls, would be targeted if the bomb that they envisioned could actually be built and used.

Frisch and Peierls advocated for the super-bomb as the only effective response to the threat of a German bomb. "The most effective reply would be a counterthreat with a similar bomb," wrote the two refugees, suggesting the deterrence potential of the new weapon. They urged that production be started as soon as possible, without waiting for Germany to produce a similar weapon, as the enrichment of uranium would take a minimum of several months. That was a gross underestimate of the complexity of the project and the time required to complete it, but a more realistic assessment of several years would have made Frisch and Peierls's argument even stronger. They had no information about what was going on in Germany but urged the British government to move ahead with experimental research and form "detection squads" to deal with the possible radioactive effects of German super-bombs on British territory.[10]

Impressed by the calculations and conclusions of his junior colleagues, Mark Oliphant convinced government officials of their validity. In April 1940, as Germany invaded Denmark and Winston Churchill was appointed chair of the newly created Military Coordinating Committee, the government approved the creation of a special committee under Professor Thomson to look into the possibility of building the bomb. They named it the MAUD Committee. If misleading the enemy was the main objective, the committee's name achieved it in a most unexpected way. Its members believed that they were using a code name invented for nuclear research by Niels Bohr—he had used the word in a telegram sent to Frisch from Copenhagen on April 9, 1940, the day on which the Germans took the city. In fact, as would become known much later, it was a reference to Bohr's housekeeper, Maud Ray. Meanwhile, the committee began its work, coordinating research by groups of scientists at four universities to check the feasibility of the Frisch-Peierls concept.[11]

In March 1941, the MAUD Committee drafted a "Report on the Use of Uranium for a Bomb," the first of its two reports. Compared with Heisenberg's reports of that year, the committee's document was far more optimistic. The Frisch and Peierls calculations about the need of only a small amount of uranium-235 to produce critical mass were confirmed. A plant would have to be built at a cost of £5 million to separate U-235 from U-238. Chadwick, one of the members of the committee, suggested that an atomic bomb could be built within the next two years, that is, in the course of the ongoing war with Germany. In April, the committee's second report recommended that the government proceed with the bomb project.[12]

The final version of the "Report on the Use of Uranium for a Bomb," accompanied by the committee's report on "The Use of Uranium as a Source of Power," was submitted to the Ministry of Aircraft Production, which oversaw the work of the MAUD Committee, in late July 1941. The ministry's leadership was skeptical, questioning what it regarded as the committee's excessively optimistic projections regarding the complexity of the project and the time needed for its realization. Since the theory on which the whole enterprise was premised had yet to be proved, it was a huge risk for the government to undertake a costly project when Britain was already burdened by the needs of the war. An isotope separation plant would be required, as well as facilities and infrastructure to build the bomb itself. Planning and budgeting for the new industrial stage of the project would have to commence. None of it looked easy.

The British had survived the air war with Germany in the fall of 1940 but remained vulnerable to Luftwaffe bombardment of the island. Could they risk diverting huge resources from the war effort and pour them into the building of plants that might be destroyed by a single air raid? Out of the Luftwaffe's reach were the British colonies, notably Canada, where at the turn of the century the young Ernest Rutherford and Frederick Soddy had started their research on the structure of the atom. But Canada lacked the necessary industrial base. As government discussions continued into August and then September 1941, the British looked more and more toward the United States, where such a base existed.[13]

At Berkeley, Ernest Lawrence was impressed by what he heard from Mark Oliphant, who came to the United States to convince American scholars to join their British colleagues in developing an atomic bomb. Oliphant argued that such a weapon could be built fairly quickly. The estimated cost was $25 million, but Britain lacked the money and manpower to complete the project. Lawrence was prepared to listen. He agreed that it was possible to build the bomb, which might decide the outcome of the war, and it was vital to beat the Germans in the race to produce it. To move ahead he needed government support on a much larger scale than the new science tsar, Vannevar Bush, was prepared to offer the uranium project.[14]

In September 1941, Lawrence went to Chicago to deliver a talk and meet with two heavyweights in the American scientific establishment. One of them was James B. Conant, an award-winning chemist and president of Harvard University, largely responsible for turning it into a modern research institution. Bush had appointed Conant head of the National Defense Research Committee after moving on to the Office of Scientific Research and Development. Conant, a fellow Bostonian, had considerable influence on Bush but, like him, was highly skeptical about the feasibility of an atomic bomb. Another figure Lawrence wanted to meet was Arthur H. Compton, a Nobel Prize winner at the University of Chicago. In April 1941, Bush had asked Compton to chair a committee whose task was to assess the prospects of uranium research. Compton produced his report in May 1941, suggesting that a bomb was possible, but it would take until 1945 to build it.[15]

The three men met for dinner at Compton's house. The host was in Lawrence's camp by now, but Conant seemed as skeptical as ever. Responding to Lawrence's passionate criticism of the American nuclear program, Conant said that the time had come to drop support for the nuclear project altogether, as there was no prospect of building a bomb before the end of the war. This remark invited protests and vehement counterarguments not only from Lawrence but also from Compton. Apparently, Conant was counting on such a response. He was trying to find out whether he could depend on Lawrence if he decided to support the project: Would he be prepared to put aside his own research and focus on building the bomb? Lawrence was caught off guard but, after a moment's hesitation, told Conant: "If you tell me this is my job, I'll do it."[16]

Lawrence could celebrate his first victory: in exchange for his commitment to work on the bomb, he had turned Conant from a skeptic into an agnostic. Conant was sufficiently impressed by Lawrence to suggest to Bush that he launch a new review of the uranium project. Conant included two chemists, specialists whose judgment he trusted most, in the committee. None of them turned out to be more influential in changing Conant's attitude toward the atomic bomb project than his fellow Harvardian, George B. Kistiakowsky, a professor of chemistry and an expert on explosives.

Like Szilard, Wigner, and Teller, Kistiakowsky was a refugee from Europe—not from the Nazis, but from the Bolsheviks. A native of Ukraine, he had grown up in liberal circles of the Kyiv intelligentsia. His father was a leading jurist and sociologist who advocated cultural autonomy if not outright independence of Ukraine from Russia, while his uncle became a minister in an independent Ukrainian government in 1918. George, who fought in the ranks of the Russian White Army against the Bolsheviks, became disillusioned with the Whites, fled Ukraine, and received his education in Germany, before coming to the United States. Kistiakowsky was initially skeptical about making an atomic bomb work by combining two quantities of highly fissionable material to produce critical mass. On reconsideration, he changed his mind and told Conant one day as the two met in the middle of the Harvard Yard: "it can be made to work." Conant, who trusted Kistiakowsky, was now fully on board.[17]

By early October 1941, with Lawrence, Compton, and Conant all of the same mind, Bush was also warming up to the project. On October 3,

Conant received the official MAUD report from the British. It proved to be the decisive argument that persuaded Bush to throw his support behind the building of the atomic bomb. Six days later, on October 9, Bush was in Roosevelt's Oval Office, accompanied by Henry A. Wallace, selling the president on the idea to which he had so recently committed himself. The MAUD report proved that the bomb could be built, Bush told the president, and scientists knew how to get it done, but further research and major resources would be needed to make it possible.

As Bush sought to convince Roosevelt to embark on the project, he suggested the British date of 1943 instead of the American estimate of 1945 as the year in which the bomb could be produced. Surprisingly, Roosevelt was interested not only in wartime use of the bomb, but also in postwar control over nuclear weapons as an instrument for changing the existing world order. He gave Bush the go-ahead and agreed to establish a government committee to oversee the research and development of the nuclear bomb. It would include Roosevelt himself, Wallace, Secretary of War Henry Stimson, Army Chief of Staff General George Marshall, Bush, and Conant.[18]

The committee's immediate task was to find out exactly what it would take to build the bomb. Bush tasked his scientists to come up with the answers. In early November they reported that the bomb was a possibility if they obtained enough uranium-235. Estimates of the critical mass needed for the explosion ranged from 2 to 100 kilograms (4.4 to 220 pounds). Anywhere between $50 and $100 million would be required to accomplish the task—a far cry from the $6,000 allocated to Szilard and his group in 1939 on the basis of the letter signed by Einstein, or Oliphant's estimate of $25 million. Bush reported his estimates to Roosevelt on November 27, 1941.

The Japanese attack of December 7, 1941, on Pearl Harbor provoked Hitler to declare war on the United States four days later. The Americans and British were now allies in the war against Nazi Germany, making their cooperation on building the bomb entirely rational. But given the president's immediate preoccupation with launching the war effort, it took him almost two months to respond to Bush's report. Roosevelt's message, written on January 19, 1942, was as cryptic as it was all-important for the future of the project. It read: "V[annevar] B[ush] OK—returned—I think you had best keep this in your own safe. *FDR.*" With that stroke of the pen, the United States officially entered the nuclear race.[19]

Chapter 5

MANHATTAN PROJECT

In Washington the Pearl Harbor attack had removed all moral doubts concerning the development of the atomic bomb. "Are our Prime Minister and the American president and the respective General Staffs willing to sanction the destruction of Berlin and the country round, if ever they are told it could be accomplished at a single blow?" was the question Vannevar Bush and James Conant discussed in the summer of 1941 with Charles Galton Darwin, the director of the British Central Scientific Office in Washington, the UK scientific liaison mission in DC.[1]

After Pearl Harbor that was not their concern anymore. Beating the Germans in the race to produce a bomb became the overriding imperative. "There are still plenty of competent scientists left in Germany," wrote Conant to Bush in the late spring of 1942, noting intelligence reports about the German seizure of heavy water in Norway and increased espionage activity with regard to nuclear research. As far as Conant was concerned, the Americans had to make haste, as the bomb could decide the outcome of the war. "If the possession of the new weapon in sufficient quantities would be a determining factor in the war, then the question of who has it first is critical," wrote Conant. "Three months' delay might be fatal. For example, the employment of a dozen bombs on England might be sufficient to enable an invasion to take place."[2]

But how was the bomb to be built? There were three ways to enrich uranium-239 in order to produce fuel for the bomb: gaseous diffusion, electromagnetic separation, or centrifuge separation. Alternatively, plutonium could be produced as fuel in a nuclear reactor, which had yet to be built. None of the alternatives was cheap or scientifically and technologically easy. But Conant refused to choose between four unproven technologies. He proposed that all four projects go forward simultaneously, the goal being to build six bombs by mid-1944, if not earlier.[3]

Roosevelt approved Conant's proposal on June 17, 1942, asking Bush only one question: "Do you have the money?" The implication was that the government would provide all requisite funds. Roosevelt was giving the full go-ahead not just to explore the means of building the bomb but to actually produce it. The Soviets had given up on the idea of developing a bomb anytime soon, while the Germans were still struggling to figure out how the task could be accomplished. The British believed that they knew how to build the bomb but lacked resources and could not protect their installations from the air. The Americans now decided to throw their enormous resources behind the production of the bomb. They were the only ones prepared and affluent enough to take the scientific, financial and, ultimately, military risk.

In June 1942, Vannevar Bush made a key decision in the American race for the bomb, entrusting the administration of the project to the Army Corps of Engineers. Its expertise in managing huge industrial projects was already established; its military culture and discipline would ensure the level of secrecy needed for the undertaking; and its vast budget would accommodate and conceal the huge expenditures that the endeavor required and the president was prepared to approve. In September 1942 Bush found a man to lead the effort. He was Colonel Leslie R. Groves, to be promoted to brigadier general that month and to lieutenant general by the end of the year.[4]

Forty-six years old at the time of his appointment, Groves had just ended his service as deputy to the Army's Chief of Construction, where he had overseen multi-billion-dollar budgets, built army barracks, depots, and munitions plants, and completed the construction of the Pentagon to house the headquarters of the US Army. He was ready to move on. His next assignment brought him to Manhattan, the headquarters of the program code-named "Development of Substitute Materials." Under him it would become known as the Manhattan Engineer District or Manhattan Project. "If you do the job right," said his superior, the commanding officer of the Army Service Forces, General Brehon S. Somervell, "it will win the war." Groves took the assignment, knowing little about the project he was taking over and nothing at all about nuclear physics. It did not matter. He knew how to get things done, and that was the main quality Bush needed.

General Groves had a reputation for driving his subordinates to the limit. Colonel Dennis D. Nichols found himself under Groves's command in the Manhattan Engineer District. According to him, Groves was "the biggest S.O.B. I have ever worked for." Nichols continued: "He is most demanding. He is most critical. He is always a driver, never a praiser. He is abrasive and sarcastic. He disregards all normal organizational channels. He is extremely intelligent. He has the guts to make timely, difficult decisions. He is the most egotistical man I know."[5]

The unrelenting Groves used to drive his military subordinates crazy. In the fall of 1942 came the turn of the scientists. On his reconnaissance tour in October, Groves visited the two main scientific centers of the project, the University of Berkeley with its cyclotron, overseen by Ernest Lawrence, and Chicago, a new center on the nuclear research map of the country. Chicago was the home of Arthur Compton, who had helped Lawrence the previous autumn to persuade Conant to take another look at the bomb project. In October 1941 Bush had put

Compton in charge of the scientific part of the project. By the following spring, Compton had consolidated research on nuclear fission and chain reaction at his home institution, the University of Chicago.

The Columbia University group, led by two immigrants, Enrico Fermi and Leo Szilard, was brought from New York to Chicago to continue their experiments. Their main task was to build a nuclear reactor that would breed plutonium—the discovery and preoccupation of Lawrence and his group. While Lawrence himself stayed at Berkeley, his key plutonium man, Glenn Seaborg, the discoverer of that new element, had moved to Chicago to work with the others under Compton in a unit given the nondescript name of the Metallurgical Laboratory for security reasons.

Upon visiting Chicago, Groves told the scientists that they would get all the resources they needed but had to speed up their work. They cooperated. In late November 1942, Fermi and his team built the first working nuclear reactor—the all-important element of the plutonium project. It entered the history books as Chicago Pile-1, built on a rackets court underneath Stagg Field at the University of Chicago. The wooden frame held 771,000 pounds of graphite, arranged in fifty-seven layers, 80,950 pounds of uranium oxide, and 12,400 pounds of metal uranium. The reactor went critical on December 2, 1942, securing a place of honor for that date in the history of nuclear physics.

They began the experiment in mid-morning and continued after a lunch break, with Fermi in charge. According to an eyewitness, Fermi was as "cool as a cucumber," showing no signs of stress or uncertainty. After lunch, Fermi ordered one of his men to pull out a cadmium rod, which had delayed a nuclear reaction from the outset, by another 12 inches. "At first you could hear the sound of the neutron counter, clickety-clack, clickety-clack. Then the clicks came more and more rapidly, and after a while they began to merge into a roar; the counter couldn't follow anymore," recalled a participant in the launch. Those gathered heard Fermi announce: "the pile has gone critical."

Fermi ordered the reactor to be shut down after five minutes, when it had produced barely enough electricity to power a lightbulb.

The cost of material alone for building the reactor that had produced such a puny amount of energy was close to a million dollars. It did not matter. Everyone celebrated the achievement. Eugene Wigner, a participant in Leo Szilard's first meeting with Albert Einstein in July 1939, opened a bottle of Italian wine. Szilard congratulated Fermi, shaking his hand. He allegedly told him that the occasion "would go down as a black day in the history of mankind." Compton, meanwhile, called Conant to give him a coded message: "The Italian navigator has just landed in the new world."[6]

General Groves celebrated his first achievement in the nuclear field. Nine days later he would be promoted from brigadier general to lieutenant general. As always, he was advancing on a number of fronts. The task of obtaining fissile material for the bomb was approached from two directions, employing plutonium and uranium.

The plutonium project required the construction of a new nuclear reactor of which the Chicago Pile was just a prototype. The alternative required the transformation of uranium ore, which largely consisted of uranium-138, into uranium-235, which amounted to only 0.7 percent of naturally occurring uranium. By November 1942 it had become clear that the previously suggested centrifugal process of enriching uranium was impractical, as it would have required tens of thousands of centrifuges running simultaneously around the clock. There remained two alternative methods, electromagnetic separation and gaseous diffusion. Another method, thermal diffusion, was later added to the list.

The construction of electromagnetic separation and gaseous diffusion plants began in the spring of 1943 in Tennessee on a 56,000-acre site acquired by the US Department of War the previous year for the Manhattan Project. The Tennessee undertaking became known as the Clinton Engineer Works and gave birth to the new town of Oak Ridge, which would count 75,000 men, women, and children as citizens by the summer of 1945. Three uranium enrichment plants and one experimental nuclear reactor would employ almost the entire workforce of the town.[7]

The electromagnetic separation plant known as Y-12 used calutrons designed by Ernest Lawrence. The calutron, a hybrid of a spectrometer and a cyclotron, utilized electromagnetic separation to deflect positively charged particles according to their mass. Construction of the calutrons required approximately 5,000 tons of copper, which was in high demand during the war. It was replaced with 6,000 tons of silver, an amount that would more than double in the future. The plant produced its first uranium-235 in March 1944, but the level of enrichment was insufficient to build a bomb, as it reached only 15 percent of the concentration of U-235 in the entire mass of uranium. The enrichment process had to be repeated again and again to obtain the 89 percent of uranium-235 in the bomb fuel sufficient for a nuclear chain reaction.

The gaseous diffusion plant, known as K-25, made use of ideas developed by Harold Uri and his associates at Columbia University. It housed cascades of boxes with membranes and pumps that sent uranium hexafluoride gas through the 600 cascades on the assumption that every successive box would collect heavier molecules of the gas, allowing the lighter ones to pass through and be captured at the end of the cascade, where they would turn into atoms of U-235. The plant, built at a cost of almost half a billion dollars, produced its first enriched uranium in early 1945; by summer, it was able to produce a mass with a uranium-235 count of 7 percent.

In the summer of 1944, Groves approved the construction of a third, much cheaper plant, to be built at a cost of $3.5 million in a few short months. S-50, as the new plant became known, used the thermal diffusion method, in which heavier molecules of gas gravitated toward the cold end of the temperature gradient to enrich the uranium ore from 0.7 percent content of uranium-235 to close to 0.9 percent.

By the spring of 1945 all three plants were in full operation, with the slightly enriched uranium from thermal diffusion plant S-50 delivered to gaseous diffusion plant K-25 and from there to electromagnetic separation plant Y-12. It was a costly and laborious process, but by July

1945 Oak Ridge had managed to produce 110 pounds of weapons-grade, 89-percent-enriched uranium-235, making possible the construction of an atomic bomb. The combined effort was enormous, with the Oak Ridge plants consuming up to 10 percent of the entire country's electricity at one point. In 1945 the price of 110 pounds of gold was slightly more than $61,000. By contrast, the cost of producing 110 pounds of Oak Ridge uranium-235 was well above a billion dollars. It was an undertaking that only one country in the world could afford at the time.[8]

The plutonium project, which became possible with the success of the nuclear chain reaction in Chicago Pile-1 in December 1942, yielded a much more effective way of producing fissionable material for the atomic bomb. Enrico Fermi and his group were encouraged by their initial results, but there was a great difference between what they had constructed beneath the stands of a disused football stadium at the University of Chicago and a reactor capable of producing plutonium. Accordingly, they decided to build a fully operational but still experimental reactor at a pilot plant in Oak Ridge.

The reaction to be produced in the reactor was supposed to transmute part of the uranium-238 used as fuel into uranium-239, which would then decay into neptunium-239, which in turn would decay into plutonium-239 for use in the bomb. It worked as planned, and the Oak Ridge X-10 graphite reactor, also known as the Clinton Pile, went critical in November 1943 under Fermi's watchful eye. Within a month the plant had produced its first 500 milligrams of plutonium, which was extracted from the used uranium-238 fuel by a specially constructed separation plant.

Now Fermi and his team could expedite the construction of plutonium-breeding reactors at another Manhattan Project site on the Columbia River near Hanford in Washington State. The area, first surveyed by the Army Corps of Engineers in December 1942, had an abundance of electricity from the nearby Grand Coulee Dam, water from the Columbia River, government-owned land, and farmers poor enough to take compensation and leave their farms without much protest to clear the way for the huge construction project. Work was undertaken in

April 1943 by the DuPont Company, known for its chemical engineering and explosives production. DuPont took on the construction of the plutonium plant for a symbolic profit of one dollar in an attempt to shed its negative image of war profiteering acquired in World War I.

The success of the experimental reactor at Oak Ridge suggested that work on the Columbia River site could proceed and accelerate. The Hanford reactors used in Washington State were water-cooled, as opposed to the air-cooled ones used in the Clinton Pile, but otherwise the design of the reactors was basically similar. Construction of the first of six planned 250 MWt (megawatts thermal) reactors began in October 1943. The first of them, Reactor B, went critical in September 1944 and, after overcoming a number of problems, produced its first plutonium in November of that year. Two more reactors went critical in December 1944 and February 1945. Also in February, the first 80 grams of 95-percent pure plutonium became available for the construction of the bomb.[9]

By April 1945 shipments of plutonium were being made weekly, making available sufficient quantities of the element to build numerous bombs. All three plants in Tennessee had produced only enough uranium-235 for one bomb, but that no longer mattered. With enormous effort, by the summer of 1945 the United States had obtained enough fuel to explode an atomic bomb, and by that time its scientists, engineers, and military knew how the process worked. Nevertheless, the design of plutonium and uranium bombs still presented an enormous challenge. Those facing it would have to overcome the difficulties and deal with numerous problems not only of a technical but also of an ethical and moral character.

General Groves had chosen J. Robert Oppenheimer, a professor of physics at the University of California, Berkeley, the same school that employed Ernest O. Lawrence, to lead the effort of building the atomic bomb. Oppenheimer was a theoretical physicist whom Lawrence had invited to the university in the 1930s to help him deal with the theoretical aspect of the work involved in building and running cyclotrons. Unlike Lawrence or Compton, Oppenheimer had never won a Nobel Prize, had no experience in experimental physics, and was an unknown quantity in administration. But Groves, seemingly like everyone who had ever met Oppenheimer, was impressed by his intellect and convinced others to accept his choice.[10]

Thirty-eight years old at the time of his recruitment by Groves, the tall, thin Oppenheimer was known around the Berkeley campus for his good taste in suits: his father was a wealthy textile importer in New York. Oppenheimer's love of art and music came from his mother, a painter. He was a man of extraordinary intellect, charisma, self-admitted eccentricity, wit, occasional acerbity, and, last but not least, left-leaning views. He hosted fundraisers in support of the Republican side in the Spanish Civil War, and in 1940 he married the twenty-nine-year-old botanist Katherine ("Kitty") Puening Harrison, a one-time Communist Party member. It was her fourth marriage—her second husband had been killed in the Spanish conflict.[11]

Like almost everyone in his circle of friends, relatives, and colleagues, Oppenheimer dreaded the rise of Nazism in Germany. The outbreak of World War II, especially the Japanese attack on Pearl Harbor, made him think about ways in which he might serve the war effort. Most of Oppenheimer's colleagues were recruited to work on war-related projects, radar systems in particular, but he was passed over partly because of his specialization in theoretical physics, and partly because of his leftist background and communist associations. But Oppenheimer wanted to be involved on the home front, and the nuclear project presented an opening.[12]

At his first meeting with Groves in October 1942, Oppenheimer suggested the need for a laboratory whose sole task would be the construction of a bomb. Groves accepted the idea but wanted to place the laboratory as far away from enemy eyes and ears as possible. That meant avoiding universities and major cities and keeping talkative, eccentric, and politically unreliable scientists away from their families and friends. Oppenheimer was receptive. An outdoor enthusiast, he spent his summers riding horses and biking in New Mexico, where the Army Corps of Engineers had found the right spot for a laboratory in an isolated, sparsely populated region of that state where land was cheap. Oppenheimer, who knew the area quite well, suggested an exact location for the new top-secret installation—a boys' school on a hill called Los Alamos, 35 miles (56 kilometers) northwest of Santa Fe.

Groves bought the land and began the construction of housing and laboratory facilities. The isolation of Los Alamos was both a plus and a minus. While it helped to establish a rigid security regime, few scientists, many of them dedicated urbanites, wanted to end up in the middle of nowhere. Convincing Oppenheimer's colleagues to abandon their own projects, laboratories, and families and move to a New Mexico desert was the first test of Oppenheimer's leadership abilities. He succeeded admirably. Szilard refused to come, but most other top-notch scholars, many of them freed from other military projects once the radar system had been developed, came to Los Alamos. Oppenheimer's charm and reputation were part of the reason.[13]

"The prospect of coming to Los Alamos aroused great misgivings," recalled Oppenheimer later. "It was to be a military post; men were asked to sign up more or less for the duration; restrictions on travel and on the freedom of families to move about to be severe; and no one could be sure of the extent to which the necessary technical freedom of action could actually be maintained by the laboratory. The notion of disappearing into the New Mexico desert for an indeterminate period and under quasi-military auspices disturbed a good many scientists, and the families of many more. But there was another side to it. Almost everyone realized that this was a great undertaking. Almost everyone knew that if it were completed successfully and rapidly enough, it might determine the outcome of the war."[14]

But not everyone who agreed or wanted to come to Los Alamos was allowed to do so. The Army Counterintelligence Corps tried to keep people with a communist past, sympathies, or connections away from the project. Among them was the Odesa-born George Gamow, who had defected from the USSR while on an academic trip to the West in 1933. A brilliant theoretical physicist, close to the initiators of the uranium project, Szilard and Wigner, he was never invited to join the Manhattan Project because he had served in the Soviet military as a commissioned officer with the rank of lieutenant. Counterintelligence officers may always have been paranoid but were not always wrong. Decades later, General Pavel Sudoplatov, a retired Soviet spymaster, suggested that Gamow had advised the Soviets on their own atomic bomb project and proposed the names of left-leaning American scientists who could be recruited for Soviet intelligence.[15]

Army counterintelligence kept a close eye on left-leaning scientists who made it to Los Alamos. Among them was Oppenheimer himself, whose telephone conversations, contacts, and correspondence were diligently monitored. They also tried to remove some already there, including the young physicist Robert Serber, who wrote the "Los Alamos Primer," a text intended to bring new recruits to the laboratory up to speed in nuclear physics. It was only Oppenheimer's personal intervention, apparently backed by Groves, that saved Serber and his wife, Charlotte, who was more active in leftist causes than Robert, from being sent away. Serber went to work on the Alberta Project, which designed the process of the actual dropping of the bomb (called a "gadget" for secrecy) and came up with code names for the first bombs produced at Los Alamos, including "Fat Man" and "Little Boy."[16]

Security concerns and interference of counterintelligence officers aside, the Los Alamos scientists were besieged with numerous scientific and technical problems. Their hope of producing the first plutonium bomb before the end of the war encountered many obstacles.

The chain reaction was supposed to be set in motion by "gun fission" process: shooting one piece of fuel into another and thereby creating critical mass. It looked good on paper so long as calculations were based on the super-pure plutonium produced by the cyclotron, but cyclotrons could never produce enough plutonium for a bomb. Once the scientists began to receive the first reactor-produced plutonium from the Clinton Pile in April 1944, they realized that aside from plutonium-239 it also contained plutonium-240, which made plutonium-239 unstable. A chain reaction might therefore begin too early, setting off an explosion at almost any moment.[17]

Oppenheimer had to abandon the gun-type fission principle for creating supercritical mass and look for other options. An alternative suggested at the time was "implosion." Conventional explosives would be used to crush a subcritical mass of plutonium into a smaller sphere and thereby create supercriticality. The new design on which Oppenheimer decided in July 1944 required new engineering decisions and new people. Among those summoned to Los Alamos to work on a new implosion bomb was an explosives expert, the Harvard professor George Kistiakowsky, who had convinced James B. Conant in 1940 that an atomic bomb could actually work.

Recruiting Kistiakowsky was no easy task. Conant had first appointed him to the advisory Uranium Committee and then put him in charge of the explosives branch of the National Defense Research Committee. In late 1943, Kistiakowsky was preparing to go to Europe as part of the Alsos Mission, an effort to uncover German scientific secrets. But Oppenheimer and Groves needed Kistiakowsky at Los Alamos and convinced Conant to let him go. "I went to war work in 1940 because I had a very intense rejection of Hitler and fascism," recalled Kistiakowsky. "And so I said all right, this is war time, and although I'm a civilian, I obeyed the orders of my boss Conant." At Los Alamos Kistiakowsky took command of the group working on the design of the explosive lenses that would direct the power of traditional explosives toward the small sphere of plutonium, turning it from subcritical to critical and causing a nuclear fission reaction.[18]

Oppenheimer and his teams of theoretical physicists, chemists, and engineers would have to overcome numerous other scientific and engineering problems to get the design of the first bombs right. Most of them were resolved by April, when the Oak Ridge plants had produced enough uranium-235 to begin the assembly of a uranium bomb. In late May, plutonium was brought from Hanford to Los Alamos, and near-critical experiments began there. The uranium bomb, called "Little Boy," used the gun-type fission model and was simpler in design. Few doubted that it would work. But there was uncertainty about the design of the plutonium bomb, dubbed "Fat Man," which had yet to be tested. Besides, while they had barely enough uranium for one Little Boy and could not waste a uranium bomb for a trial, they had enough plutonium for a number of bombs and could spare one Fat Man to make sure that subsequent plutonium bombs would actually work. By early July 1945 they were finally ready for a test.[19]

The Manhattan Project, which had cost American taxpayers close to two billion dollars, was nearing completion. It was undertaken in secret, accomplished in record time, and demonstrated the ability of the United States to attract international talent, accumulate and channel enormous resources, solve major scientific, technological, and engineering problems, and produce a weapon that few of its builders had believed feasible a short time previously. But as one set of problems was relegated to the past, another came into view.

The bomb had not yet been tested, but if it worked, who was to decide whether it should be used, and for what purpose? Was it the scientists who had been the first to alert and lobby the government and turned the theoretical possibility of an atomic bomb into reality? The US government, which had taken the wartime risk of marshaling enormous human and financial resources to build the bomb? Or maybe the British government, which was the first to apprehend the threat posed by the German bomb, created a committee that proved the feasibility of the bomb, and then shared that knowledge with the Americans? Each of those parties was in a position to lay claim to a bomb that did not yet exist.

Chapter 6

UNEQUAL PARTNERS

The American atomic bomb project began as an international effort, with British rather than Americans leading the charge. The work of British scientists helped to convince Ernest Lawrence, Arthur Compton, Vannevar Bush, and ultimately President Roosevelt to explore the feasibility of building the bomb. As they considered the next step, key figures in the American scientific and political establishment foresaw the possibility of a joint project with London to build the atomic bomb.

On October 11, 1941, a mere two days after Roosevelt received Vannevar Bush, he wrote to Prime Minister Winston Churchill. "It appears desirable that we should correspond or converse concerning the subject that is under study by your MAUD Committee, and by Dr. Bush's organization in this country, in order that any extended effort can be coordinated or even jointly conducted," wrote Roosevelt. He probably had in mind more than research. A few months earlier, in discussions with Charles Galton Darwin, Bush had been open to building a joint

American-British uranium enrichment plant in the British dominion of Canada or in the United States.[1]

Churchill took his time and responded to Roosevelt's offer only in December 1941. He wrote: "I need not assure you in our readiness to collaborate with the United States Administration in this matter." But the British government showed little initiative to promote cooperation. When the two leaders met in Washington that month during the Christmas break to discuss the war in which both countries were now involved, atomic cooperation was not on the agenda, or at least it was never mentioned in the written record of the conference. Churchill, who showed a good deal of interest in cooperating with the United States on other matters, was not pushing for a nuclear alliance.[2]

The main reason for his lack of interest was the belief of many British scientists that they were far ahead of the Americans and could build the bomb themselves. Prominent among them was the discoverer of the neutron, James Chadwick. "We must take every step to prevent them from learning that we hope to proceed to manufacturing stage," he wrote in the fall of 1941. Mark Oliphant, on the other hand, continued his support for the American-British cooperation. He wrote: "The Americans will undoubtedly go right ahead with both projects [reactor and bomb] and there is little doubt that they with their tremendous resources will achieve both before we have fairly begun." Chadwick played down such concerns, writing back: "We are some way ahead and we shall remain ahead."[3]

Chadwick's position prevailed. In November 1941 Sir John Anderson, Lord Chancellor and cabinet member overseeing the British nuclear project, refused to share the latest information on British research with visiting American scientists, citing lax security in the United States and the possibility of leaking important information to the enemy. But security concerns faded as the British lost confidence in their ability to proceed to what Chadwick called the "manufacturing stage" by the summer of 1942. Besieged on all fronts, Britain simply lacked the resources to build a bomb on its own before the end of the war. The British authorities decided to join forces with the Americans as equals before it was too late.

"We must face the fact that . . . [our] pioneering work . . . is a dwindling asset and that, unless we capitalise it quickly, we shall be outstripped," wrote Sir John Anderson to Churchill in late July 1942. "We now have a real contribution to make to a 'merger.' Soon we shall have little or none." Anderson did not plan for a joint bomb. Rather, he believed that collaboration with the Americans would benefit British research in the long run. After the war, argued the Lord Chancellor, the British "would be able to take up work again, not where we left off, but where the combined effort had by then brought it."[4]

Churchill agreed with Anderson's argument, and the Lord Chancellor wrote to Bush suggesting a joint effort to build a British plant for separating uranium-235 by means of gaseous diffusion—one of the three methods of enriching uranium already pursued by the Americans. Anderson cited the lack of resources in the United Kingdom to build the plant there and proposed adding it to the existing American program. In exchange for the British expertise, he asked for British representation on the committee overseeing the entire bomb project. But the Americans, Bush in particular, who had supported the idea of a joint plant one year earlier, no longer believed that they needed British help at all. Plans for a gaseous diffusion plant based on ideas developed by Harold Uri at Columbia University were already in place, and a project to build such a plant in Oak Ridge would be approved in November 1942.[5]

Now that the Americans believed that they were ahead, it was their turn to worry that the sharing of scientific and engineering secrets with the British might result in the leaking of such information to Germany. But their main concern was to avoid sharing information acquired by such tremendous effort with any other power, especially if the new weapon might help to ensure postwar American supremacy. After all, the main global rival of the United States in the prewar world had not been Germany, Japan, or the Soviet Union, but the United Kingdom.

After listening to a report from his secretary of war, Henry L. Stimson, who was the top government official overseeing the Manhattan Project, Roosevelt agreed with his recommendation to limit the flow of information on the bomb project to the British. No data whatever about the design of the bomb were to be shared. The Military Policy Committee in charge of the Manhattan Project adopted a resolution in that regard in December 1942. But neither Bush nor his second in command, James Conant, was completely averse to cooperation with the British. They were prepared to share American know-how in areas where the British could help with the bomb project, or where American expertise might help the British in the war effort outside the field of nuclear science and technology.[6]

Churchill was informed about what seemed to the British a sudden turn of American policy in January 1943. Sir John Anderson expressed his alarm in a telegram to Casablanca, where Churchill was meeting with Roosevelt. At the top of the prime minister's agenda was obtaining FDR's agreement to postpone the opening of the Second Front in France and invade Sicily instead, as well as to secure British lines of communication with India via the Mediterranean. Not eager to add another request to his long wish list, Churchill turned for help not to FDR but to Harry Hopkins, the president's closest adviser and Bush's political patron. Hopkins promised to help, but the next few months showed no change in the American position.

London had missed Washington's nuclear train, which passed its station in late 1941 and early 1942. The desperate British officials began looking for other ways to influence American behavior. What if they threatened to derail the American nuclear project unless they were allowed on board? They discussed and abandoned plans to cut off the Americans from uranium ore and heavy water production facilities in Canada. If they did so, the Americans could retaliate by withholding their supplies to the Canadian facilities, on which the British relied for their own nuclear project. They had to look for a different solution.

Churchill raised the issue of renewing British-American cooperation with FDR during his visit to Washington in May 1943. In a separate meeting Churchill's scientific adviser, Lord Cherwell, threatened Hopkins and Bush with an inevitable reduction of the British contribution to the war effort if they did not get American cooperation in the nuclear project. The British needed the bomb after the war to counter the Soviet threat and were prepared to divert resources from the ongoing war to achieve that goal, went the argument. The threat had little immediate impact. Although Roosevelt assured Churchill that he was ready to cooperate on the nuclear project, no instructions were issued to Bush to change his policy of noncooperation.

In July 1943, Churchill tried his charm and oratorical skills on the two custodians of the American nuclear secrets, Secretary Stimson and Vannevar Bush, who paid a visit to London. He sought to alleviate worries about possible British-American rivalry in the postwar world. Britain was not interested in the commercial use of nuclear energy, Churchill assured his guests, so it was not going to turn access to American technology into economic competition with the United States. But that technology was required for security purposes. "It would never do to have Germany or Russia win the race for something which might be used for international blackmail," Churchill told the visitors. He then added: "Russia might be in a position to accomplish this result unless we worked together."

The British were visibly unhappy, and the Americans were not prepared to complicate their relations with the island nation and world empire at a time when the battle for Europe was approaching its decisive stage. In the same month that Stimson and Bush visited London, FDR yielded to Churchill's pressure and followed Hopkins's advice to lift restrictions on scientific exchanges with Britain. Feeling bound by his promise to Churchill at their Washington summit in May, the president instructed Bush to renew the "full exchange of information" with America's British ally. Bush met with Sir John Anderson to discuss the exact conditions. The British would gain access to the results of American theoretical research on the bomb but were barred from receiving any information on the design, construction, and operation of plants producing enriched uranium and plutonium unless they had something to add to American knowledge in those areas.

The deal was approved by FDR and Churchill at their Quebec meeting in August 1943. The two leaders agreed to create a British-American

Combined Policy Committee to oversee the project to build the bomb, pledging not to use it against each other or against third parties without mutual consent, and not to inform others about the project. The Soviet Union, whose possible blackmail Churchill wanted to prevent after the war, was of course the main "other." American-British partnership in the nuclear project was reborn, now with the United States as the senior partner. Cooperation on building the bomb was included in the deal, but not information about the production of its fuel. There would be consultation on the use of the bomb but no joint custody. Such was the first international agreement on what would later become known as the nonproliferation of nuclear weapons.[7]

The details of American-British scientific cooperation were hammered out in December 1943. British scientists were given access to the research undertaken at Los Alamos and, selectively, to the results of American theoretical research on uranium and plutonium. In those areas the Americans still had to overcome major scientific problems, and British participation was considered essential to speed up the project. The Americans wanted the bomb for the current war and were prepared to pay the price.

London lost little time in sending its scientists to the United States to engage in a combination of scientific cooperation and industrial espionage. "These men should learn the details of manufacture so that we are in a position to start up on a full scale when required," wrote Lord Cherwell to Churchill with regard to the British scientists commandeered to the United States. The Americans knew what was going on. Bush and Conant stuck to their original position that information would be exchanged only on a need-to-know basis. "There was no reason for exchanging information unless it could be demonstrated that the recipient would indeed help the war effort by the knowledge thus gained."[8]

The British scientists arrived in Los Alamos in the spring of 1944. They would make a major contribution not only to the design of the first two American atomic bombs but also to nuclear proliferation, passing on the information they obtained to Britain and, in the case of at least one of the new arrivals, the German-British theoretical physicist Klaus Fuchs, to the Soviets as well.

The leaders of the Manhattan Project got their first insight into German thinking about the atomic bomb in December 1943 with the arrival in the United States of Niels Bohr, the legendary theoretical physicist highly respected by the American scientific community. Back in early 1939, Bohr had been the first to bring news about the German discovery of nuclear fission to the United States. Now he was reporting on what had happened in Germany since then.[9]

Admittedly, Bohr's information was dated, since it came from his meeting with Werner Heisenberg in Copenhagen in the autumn of 1941. In late September 1943 Bohr was forced to leave Denmark for Sweden because he was apprised that the Nazis were going to arrest him: his mother was Jewish. Bohr took with him the drawing of the German reactor that Heisenberg had left with him. He brought it first to Britain and then to the United States. Bohr's meeting with Heisenberg had convinced him that the Germans were working on a nuclear bomb.

Bohr was right. Heisenberg continued his work on a plutonium-producing uranium reactor or "uranium machine," as German physicists called it at the time. But the project as a whole was significantly affected by the worsening German position on the eastern front. On December 5, 1941, the Red Army managed to stop the German advance on Moscow (German commanders could already see the outskirts of the city through their binoculars) and forced the half-frozen German armies to retreat. Hitler, not believing that the campaign would last into winter, had not supplied his troops with winter clothing. The blitzkrieg was over. The Germans now faced lengthy warfare in the east and a Britain strengthened by American entrance into the war in the west. They had to take a close look at their resources and reassess their priorities. Only projects that could result in the production of new weapons before the end of the war remained under Army control and received priority funding. The uranium project did not make the cut.

In February 1942, the Army Ordnance Office issued a report on "The Production of Energy from Uranium," which stated that plutonium was required to build an atomic bomb. It estimated that between 22 and 220 kilograms (48.5 to 485 pounds) of the element would be needed. An isotope separating plant or plants and a uranium reactor would have to be built in order to produce such amounts of plutonium. In the same month, a one-day conference was convened to discuss the bomb. Among the invitees were Hermann Goering, Heinrich Himmler, Martin Bormann, the Army and Navy commanders, and the all-important minister of armaments, Albert Speer. The speakers included Otto Hahn and Werner Heisenberg, the latter talking about the prospects of building a bomb with either uranium-235 or plutonium. The problem was that none of the Nazi dignitaries came. In the previous month, the uranium project had been de facto demoted by the Army Ordnance Office, which moved it to the jurisdiction of the Ministry of Science, Education, and National Culture.[10]

In February 1942, the month in which the Ordnance Office released its report and the top officials of the Third Reich ignored the scientific conference, Heisenberg and his group in Leipzig produced the first experimental verification of the multiplication of neutrons in a nuclear reactor. With supplies of heavy water received from Norway the previous autumn, Heisenberg created a prototype of the first nuclear reactor using not uranium and graphite but uranium and heavy water. In February he found out that it worked: his assistants were able to detect an increase in the production of neutrons. They managed to do so before the Fermi team achieved similar results in the United States. A chain reaction was there, around the corner, and Heisenberg knew how to produce it. His team also discovered plutonium, a byproduct of the chain reaction.[11]

Even with their diminished status and lack of funding, Heisenberg and his colleagues were either abreast of their British and American rivals or even ahead of them in the race to build a functioning reactor.

In order to obtain funding for the project, however, Heisenberg would have had to tell top government officials that he could deliver a bomb before the end of the war. "We wouldn't have had the moral courage to recommend to the Government in the spring of 1942 that they should employ 120,000 men just for building the thing up," he would say after the war.[12]

Albert Speer, the armaments minister, met with Heisenberg, Otto Hahn, and other nuclear scientists in the late spring of 1942 after being informed about the existence of the nuclear project by General Friedrich Fromm, the commander in chief of the Replacement Army (Wehrmacht reserves and supplies). Fromm presented the atomic bomb project as perhaps the only chance to win the war. Speer also heard that the project was underfunded.

At the meeting, which took place on June 4, Heisenberg complained about the neglect of the project by the Ministry of Education but was reluctant to ask for a dramatic increase in funding. He told Speer that the theoretical and scientific issues had already been resolved: the bomb could be built, but creating an industrial base for its production would take at least two years, even with maximum government support. Speer offered assistance from his ministry. When Heisenberg and his colleagues returned a couple of weeks later with their wish list, which included the allocation of several hundred thousand marks, Speer offered them one or two million. They declined: their laboratory needs had already been met, and they were not ready to proceed to the production stage. Accepting more money would have meant taking responsibility to produce the bomb before the end of the war.

Among the skeptics was also Adolf Hitler. Speer discussed the atomic bomb with Hitler on a number of occasions and reported to the Führer the results of his meeting with the scientists. Hitler, always suspicious of theoretical physics and physicists as Jewish rather than German, was convinced that there would be no atomic bomb in his lifetime. "The scientists in their unworldly urge to lay bare all the secrets under heaven might someday set the globe on fire," Speer remembered Hitler joking about the matter. "But undoubtedly a good deal of time would pass before that came about. . . . he would certainly not live to see it."[13]

Speer's attempts to save the nuclear project from oblivion resulted in the allocation of additional funds, the release of nuclear scientists from military service, and the placement of the project under the command of Reich Field Marshal Hermann Goering, the supreme commander of the Luftwaffe and head of the Reich Research Council. Goering, who maintained his positions as president of the Reichstag and minister president of Prussia, had neither time nor the requisite scientific insight and imagination to prioritize nuclear research: among his ideas remembered by Speer was building locomotives of concrete, given the wartime lack of steel. Until the end of the war, the German nuclear project, known as *Uranverein*, would be conducted by a loose network of research centers, each guided by its own research agenda and priorities. The ultimate cost of the project amounted to eight million German marks, or roughly two million US dollars, one-thousandth of the funding of the Manhattan Project.[14]

Until early 1942, both the Germans and their American competitors and adversaries were more or less on the same level in the research and development of nuclear weapons. From that point, the two projects took opposite directions: while the American project intensified, the German one slowed down significantly. Different scientific approaches—German reliance on heavy water as opposed to American experimentation with graphite to slow down neutrons—affected this divergence, but other factors were also involved. In the final analysis, the two events outside the field of nuclear physics that determined the development of the two projects both resulted from the war and happened in December 1941. The Japanese attack on Pearl Harbor brought the United States into World War II and gave the nuclear bomb project a priority it would probably not have attained otherwise. The Soviet counteroffensive that began near Moscow in the same month dramatically changed calculations in Berlin and led German officials to downgrade their uranium project.

In September 1944 Roosevelt and Churchill met again, this time in Roosevelt's home town of Hyde Park, New York, and once more the nuclear bomb was on their agenda. They committed their countries to a wartime and postwar nuclear alliance. The two leaders agreed that once the bomb became available, it would be used after "mature consideration" but without warning. Only after the first explosion would the enemy be warned that "bombardment will be repeated until they surrender."[15]

The most astonishing aspect of the agreement was not that the scenario of nuclear bombing, discussed in September 1944, would be followed almost to the letter eleven months later, with Roosevelt dead and Churchill out of office, but that Germany was no longer on the target list for the atomic bomb. Instead, Roosevelt and Churchill agreed to bomb Japan. Ironically, American spies were still working on operations to kidnap or kill Heisenberg in order to "deny the enemy his brain," as one US officer suggested. In February 1944 American bombers completely destroyed Otto Hahn's institute of chemical research. As late as December 1944, an assassin was sent to Switzerland, where Heisenberg gave a lecture, to kill him if it should become clear from discussions with him that the Germans were building the bomb. The assassin, a former major-league catcher, concluded that they were not doing so and spared Heisenberg.[16]

Meanwhile, as of September 1944, Roosevelt and Churchill were showing no concern whatever about the German atomic bomb: their scientists, unlike their spies, did not believe that the Germans were close to producing one. Nor did the two leaders believe that an American atomic bomb would be ready before the end of the war in Europe. In that regard they were close to Hitler and Heisenberg, who concluded that no one would manage to produce a bomb before the end of hostilities. By the time of the Hyde Park meeting, Allied forces were already on the approaches to Germany, taking over not only most of France but also Belgium. It was hard to imagine that with the Red Army starting its advance into East Prussia, Germany could sustain its war effort until an American bomb was ready.

The anti-Hitler bomb on which scientists were working night and day at Los Alamos was now out of the picture as far as the politicians were concerned. In Roosevelt and Churchill's plans it became the anti-Japan bomb, and in Churchill's long-term view it remained an anti-Soviet bomb. But the two Allied leaders were not the only ones claiming possession of the yet unbuilt bomb or at least trying to influence its future use. By the time they met in the fall of 1944, the idea of international control over nuclear weapons, inspired by H. G. Wells's novel of 1913, *The World Set Free*, and formulated by no less an authority than Niels Bohr, was gaining currency among the scientists working on the Manhattan Project.

Both Roosevelt and Churchill had personal meetings with Bohr. In May 1944 Bohr proposed to Churchill that he take the USSR into his confidence on the Manhattan Project. Thinking about the bomb's postwar future, Bohr believed that not including the Soviets would produce a nuclear arms race. Churchill was not receptive. He had spent so much time and effort lobbying the Americans to allow the British to join the project that he could not countenance such a proposal. The meeting did not go well. To remedy the situation, Bohr sent Churchill a semi-apologetic letter assuring the prime minister that he was not trying to infringe on his prerogatives. "The responsibility for handling the situation rests, of course, with the statesmen alone," wrote Bohr, to no avail. Churchill later claimed with regard to Bohr that he "did not like the man," while Bohr suggested that they "did not even speak the same language."[17]

Bohr had little choice but to start knocking on American doors. He suggested to Oppenheimer that involving the Soviets would accelerate the project. Oppenheimer seemed receptive to the idea but suggested that Bohr first discuss it with Roosevelt. In July 1944 Bohr sent the president a memorandum in which he dropped the idea of speeding up the project and focused instead on the postwar situation. He wrote: "The terrifying prospect of a future competition between nations about a weapon of such formidable character can only be avoided through a universal agreement in true confidence." The two met in August in an encounter arranged with the help of Supreme Court Justice Felix

Frankfurter. Roosevelt appeared sympathetic to Bohr's argument but sent the scientist back to discuss the matter once again with Churchill.[18]

Meeting in September at Hyde Park, Roosevelt and Churchill discussed Niels Bohr's ideas and decided to turn them down. "The suggestion that the world should be informed regarding Tube Alloys [the British code term for the bomb project], with view to an international agreement regarding its control and use, is not accepted," wrote Churchill in an aide-memoire summarizing his exchange with Roosevelt. The last of the three entries in the short memo left no doubt whose suggestion Churchill had in mind. It read: "Enquiries should be made regarding the activities of Professor Bohr and steps taken to ensure that he is not responsible for leakage of information, particularly to the Russians."[19]

In a memo to Frederick Lindemann, his main scientific adviser, Churchill mentioned with extreme disfavor Bohr's successful attempt to meet President Roosevelt with the assistance of Felix Frankfurter, to whom he had allegedly disclosed the details of the project. But he was even more concerned about Bohr's association with the Soviet Union. "He says he is in close correspondence with a Russian professor, an old friend of his in Russia to whom he has written about the matter and may be writing still," wrote Churchill about Bohr. The Russian professor in question was Petr Kapitsa, a leading Soviet physicist, who had spent ten years at the Cavendish Laboratory in Cambridge, working under Ernest Rutherford. "The Russian professor has urged him to go to Russia in order to discuss matters," continued Churchill. "What is all this about? It seems to me Bohr ought to be confined or at any rate made to see that he is very near the edge of mortal crimes."[20]

Bohr, who had quite a following among British and American scientists, had shaken the belief of a good many of them in the bona fides of the cause for which they had been working. Also among the doubters was Joseph Rotblat, a thirty-five-year-old scientist from Poland who had come to Los Alamos in February 1944 as a member of the British scientific mission led by Chadwick. There he became friendly with Bohr: both hated listening to American radio news programs, as they were constantly interrupted by commercials, and Bohr shared with Rotblat "his worry about the dire consequences of a nuclear arms race between East and West." Rotblat, on his part, was taken aback by General Groves's remark that the bomb was being built to "subdue the Soviets."

When by the end of 1944 Rotblat concluded that the Germans were not building a bomb of their own, he decided to leave the project, Los Alamos, and the United States. But an American counterintelligence officer accused him of unauthorized secret meetings with a suspicious person in Santa Fe and of planning to be parachuted into the Soviet-occupied part of Poland after returning to Britain. Rotblat denied the charges and was allowed to leave. He sailed back to Britain on Christmas Eve 1944.[21]

Rotblat was the only scientist who left Los Alamos but not alone in having doubts about the purpose of his work. In March 1945, the thirty-year-old Robert R. Wilson, who headed the experimental physics division at Los Alamos, decided to call a meeting to discuss the moral implications of what he and others were doing at Los Alamos. Oppenheimer tried unsuccessfully to convince Wilson not to hold the meeting, which was advertised as "The Impact of the Gadget on Civilization" and attended by some twenty scientists. Oppenheimer spoke and convinced Wilson and others to carry on. He turned Bohr's idea of the need to inform the world about the bomb's destructive potential on its head, urging those in attendance that they were duty-bound to produce the weapon and demonstrate its destructive potential. Otherwise, the nuclear bomb would remain a dreadful secret to be loosed on an unsuspecting world in the next war.[22]

Work at Los Alamos continued around the clock. Roosevelt was building an anti-Japanese bomb, Churchill an anti-Soviet one, and the scientists the bomb envisioned by H. G. Wells, terrifying enough to end all wars.

Chapter 7

AMERICAN BOMB

What was to be expected when it came to the military use of an atomic bomb? In H. G. Wells's *The World Set Free*, atomic bombs were basically "dirty bombs," black spheres two feet in diameter dropped from airplanes, and they did more damage by spreading radiation ("exploding indefinitely") than by the destructive power of explosion itself. In 1924 Winston Churchill imagined a bomb the size of an orange that could destroy an entire block of buildings.[1]

On August 1939, Einstein and Szilard warned President Roosevelt of "extremely powerful bombs" that, if brought to a port by a vessel, could destroy the entire port and its environs. In March 1940, the authors of the original British report on the bomb, Otto Frisch and Rudolf Peierls, estimated the blast as equivalent to 1,000 tons of dynamite, which could destroy the center of a big city and, additionally, emit powerful and dangerous radiation that would prevent occupation of the area for at least a few days. They considered the expected harm to civilians so serious as to preclude Britain's use of the bomb on land and suggested a "depth charge near a naval base" instead.[2]

James Chadwick, the discoverer of the neutron and primary author of the 1941 MAUD Report, also thought of the atomic bomb as a weapon for the destruction of ports. He imagined that it would create an explosion at least equal to the one of December 6, 1917, in Halifax harbor, though the devastating power of the atomic bomb was expected to be even greater.[3]

The test of the first American atomic bomb took place at the Alamogordo US Army Air Force Bombing and Gunnery Range in the New Mexico desert on July 16, 1945. It entered the annals of history as "Trinity," a code name proposed by Robert Oppenheimer. The newly constructed plutonium bomb was full of untested devices, from Kistiakowsky's lenses to switches that were supposed to ignite conventional explosives in thirty-two different locations in a split second. The "gadget" was placed atop a specially constructed 100-foot tower. The systems were checked, last-minute fixes carried out, and some parts held together by nothing more than duct tape. Everything was ready, but it was anyone's guess whether the bomb would explode and, if so, to what effect.

General Leslie Groves was upset when he heard that some scientists working on the bomb were placing bets on whether the explosion would lead to the destruction of the entire universe or not. Groves's main concern was whether the bomb would explode, whatever the consequences. Failure would mean that billions of taxpayer dollars had been spent in pursuit of a mirage. Oppenheimer, who was growing visibly more nervous as the test approached, also hoped for success. The test was scheduled for the morning of July 16, subject to favorable weather conditions. There was a lightning storm in the area on the evening before. But weather and technical issues related to the bomb were not the only concerns on Oppenheimer's mind.

"Oppenheimer was really terribly worried about the fact that the thing was so complicated and so many people knew exactly how it was put together that it would be easy to sabotage," recalled Don Hornig, a young physicist who devised the switch that helped to detonate the explosives. Hornig knew what he was talking about. He spent the night before the test on top of the bomb tower during the lightning storm. "All I had was a telephone," remembered the physicist turned night guard. "I was not equipped to defend myself, I don't know what I was supposed to do. There were no instructions." He read a book and tried to dismiss the thought that a bolt of lightning might explode the bomb. Hornig had been posted on the tower for one reason only: to deter colleagues who regretted having helped to devise the "gadget" from disarming it.[4]

Concern about possible sabotage was real. As preparations for the Trinity test gathered speed in early June 1945, a revolt of nuclear scientists against the use of the bomb in the war with Japan took place at the Chicago University Metallurgical Lab, the birthplace of the first nuclear reactor and the first chain reaction. Arthur Compton, in charge of the laboratory but not involved in the revolt, tried to warn people around Groves about the coming academic "explosion." In a memo of June 4, 1945, he wrote: "The scientists have a very strong feeling of responsibility to society regarding the new powers they have released." He continued: "The fact that the control has been taken out of their hands makes it necessary for them to plead the need for careful consideration and wise action to someone with authority to act."[5]

And plead they did. On June 11, seven scientists from the Metallurgical Lab, led by the Nobel Prize–winner James Franck and Leo Szilard, sent a "report" to the Interim Committee, a special body created with the approval of Roosevelt's successor, President Harry Truman, to advise him on the use of the atomic bomb. Franck, Szilard, and their colleagues were of two opinions about the test. It could be conducted as planned to warn the world about the destructive power of the atomic bomb. But they also saw reasons not to carry it out: keeping the bomb a secret would allow the United States to "substantially increase the lead which we have established during the present war, and our position in the armament race or in any later attempt at international agreement will thus be strengthened." Divided on the test, the signatories were unanimous in their support for international control of nuclear weapons and opposition to the use of the bomb against Japan.[6]

The Chicago appeal was addressed to the Scientific Panel of the Interim Committee, a group that included Oppenheimer, Fermi, Lawrence, and the head of the Chicago Metallurgical Laboratory, Compton. It had no effect. On June 16, Oppenheimer sent the Interim Committee a statement in which the members of the Scientific Panel, admitting that scholarly opinions differed, voiced their support for the use of the atomic bomb in the ongoing Pacific war. They wanted to save American lives and concluded that the bomb was the only way to do that: "We can propose no technical demonstration likely to bring an end to the war." Edward Teller, who had driven Szilard to Einstein's summer house in August 1939, used H. G. Wells's argument about the end of all wars in one of his letters to Szilard: "Our only hope is in getting the facts of our results before the people. This might help to convince everybody that the next war would be fatal. For this purpose actual combat-use might even be the best thing."[7]

On July 16, Groves was up at 1:00 a.m. He and Oppenheimer repeatedly checked the weather to determine whether the stars were visible through the clouds. The rain finally stopped by 4:00 a.m., and they scheduled the test for 5:30 a.m.

Twenty minutes before the test, Groves left the scientists and went to the main observation point, where Vannevar Bush and James Conant were waiting for him. As the loudspeaker from the control station started the countdown, Groves and those with him lay on the ground, their heads turned away from the site of the coming blast and their eyes covered. They waited. Conant told Groves that he had never "imagined that seconds could be so long." Then the whole observation station was suddenly illuminated with a bright burst of light, followed some forty seconds later by a blast wave. Conant, lying on one side of Groves, shook his hand. So did Bush on the other side. The experiment had worked.

Groves later wrote that all of them were struck by the "feeling that the faith of those who had been responsible for the initiation and the carrying on of this Herculean project had been justified."[8]

There was relief, jubilation, and more scientific experimentation in the shelter that housed Oppenheimer and the top scientists. When the bomb exploded, those present saw a clear expression of relief on Oppenheimer's face in the sudden burst of light. Less than a minute later, the incoming blast wave knocked down those who were standing. As they got back on their feet, there was jubilation. George Kistiakowsky, whose explosive lenses had worked against all odds, now seemed ready to explode with emotion himself. He hugged Oppenheimer. But Enrico Fermi, at work on a different experiment, seemed unaffected by the general euphoria. Immediately before the blast, he had started to throw small pieces of paper in front of himself. Once the shock wave reached the shelter, the pieces of paper were blown a distance of 8.2 feet. Fermi estimated that the power of the blast was equal to 10,000 tons of TNT. He was considerably off the mark. A later measurement estimated the yield at 20,000 tons, almost seven times greater than the largest previous explosion, that of December 1917 in Halifax harbor, estimated at the equivalent of 3,000 tons of TNT. A new era had arrived, and not everyone was jubilant.[9]

News of the successful Trinity blast put Leo Szilard into overdrive in his efforts to stop the government from using the bomb against Japan. On July 17, the day after the test, 155 scientists at the Chicago Laboratory and the Oak Ridge plants in Tennessee signed a letter to Harry Truman emphasizing his moral responsibility for unleashing "new means of destruction" and of a forthcoming nuclear arms race. The scientists argued that the bomb should not be used "unless the terms which will be imposed on Japan have been made public in detail and Japan knowing these terms has refused to surrender."[10]

July 16, 1945, the day of the successful test at Alamogordo, was Truman's first full day in Berlin, where he had arrived the previous afternoon to take part in the Potsdam Conference. Intended to put the final touches on the World War II peace settlement, the conference would go on with some interruptions for sixteen days, from July 17 to August 2.[11]

Truman spent July 16 meeting with Winston Churchill in the morning and touring bombed-out Berlin in the afternoon. The next day, he met Stalin, already informed about the nuclear explosion. Truman was uncertain about the implications of the successful test for American wartime policy. He was pleased by the Soviet leader's assurances about his country's forthcoming participation in the Pacific war—the item at the top of Truman's agenda. "He'll be in the Jap War on August 15th," noted Truman in his diary on the evening of July 15. "Fini Japs when that comes about."[12]

It took Churchill to explain to Truman that the bomb was changing calculations with regard not only to American but also to Soviet prospects in the war. The two had lunch on July 18 and discussed the implications of the news from Alamogordo. Truman emerged from the meeting with the idea that the bomb could end the war even before the Soviet Union entered it. "Believe Japs will fold up before Russia comes in," he wrote in his diary that day. "I am sure they will when Manhattan appears over their homeland. I shall inform Stalin about it at an opportune time." The Soviets were no longer needed in the war on Japan—a view that would become the basis of the new American line at the Potsdam Conference.[13]

No one contributed more to the formulation and implementation of that line than the new US secretary of state, James Byrnes. Throughout the first days of the conference, Byrnes watched with worry the unfolding battle between Churchill and Stalin over the composition of the East European governments and Soviet Mediterranean ambitions, demonstrated in an attempt to establish a base in the Dardanelles and a claim to a former Italian colony in North Africa. Stalin's behavior in Eastern Europe, which Churchill claimed was in violation of the Yalta agreements, concerned Byrnes, who wanted to avoid a repetition of the same situation in the Far East. The Yalta agreements had promised Stalin the Japanese Kuriles, a Manchurian sphere of influence, and warm-water seaports in northeastern China, but only in exchange for Soviet participation in the war. If the war ended before Stalin could join it, then the agreements would lose their validity. The bomb could save Japan and parts of China from turning into an analogue of Soviet-controlled Eastern Europe. One could hardly imagine a better political use of the new weapon.[14]

Byrnes decided to delay or even prevent Soviet entry into the war by urging the Chiang Kai-shek government in China to stall negotiations with Moscow on the concession to the USSR agreed at Yalta by the Americans but not yet approved by the Chinese. On July 23, he told an American diplomat that "somebody had made an awful mistake in bringing about a situation where Russia was permitted to come out of a war with the power she will have." Four days later he confided to Secretary of the Navy James Forrestal that he wanted to end the war with Japan before the Soviets got into it, as it would be difficult to get them out of Dairen and Port Arthur—the two Chinese ports conceded by Roosevelt to Stalin at Yalta—afterward.[15]

The bomb, originally conceived as anti-Nazi and then reclassified as anti-Japanese, was now also becoming anti-Soviet, not only for the British but also for the Americans. At Byrnes's urging, the Interim Committee had rejected the idea of international control over atomic weapons, but scientists now insisted on informing the international community about the existence of the bomb. That applied even to Oppenheimer who supported the use of the bomb against Japan. Back in June, he and his Scientific Panel had urged the administration to inform not only the USSR but also China and France about the bomb and its coming use, as well as to promise cooperation with them.[16]

Truman decided to inform Stalin after all. After the plenary session on July 24 he walked around the big circular table, engaged Stalin in conversation, and spoke to him with only Stalin's interpreter, Vladimir Pavlov, present. "I casually mentioned to Stalin that we had a new weapon of unusual destructive force," recalled Truman. "The Russian premier showed no special interest. All he said was that he was glad to hear it and hoped we would make good use of it against the Japanese."

Churchill, who watched the conversation from a distance, remembered later that Stalin seemed to be delighted. When Churchill asked Truman later how it had gone, the response was: "He never asked a question."[17]

Truman did not mention the atomic bomb by name, and it remained anyone's guess whether Stalin understood what he was talking about. Since Stalin asked no questions, Truman was glad to be spared any further explanations. The American president had formally done his duty as an ally, while Stalin showed no indignation about not having been informed earlier, nor did he ask to join the project. The Soviet leader knew about the US work on the bomb from his spies but had difficulty imagining the destructive power of the new weapon and its full political implications.

On July 26, two days after his brief chat with Stalin, Truman issued the Potsdam Declaration, an ultimatum to the Japanese government urging it so surrender or face "prompt and utter destruction." The signatories, apart from Truman himself, included Churchill and Chiang Kai-shek, the head of the Chinese nationalist government, which was politically, economically, and militarily supported by the United States. The declaration did not mention the atomic bomb by name, following the pattern established by Truman in his conversation with Stalin, but spared no effort describing the devastation to come. "The might that now converges on Japan is immeasurably greater than that which, when applied to the resisting Nazis, necessarily laid waste to the lands, the industry and the method of life of the whole German people," read the text. "The full application of our military power, backed by our resolve, will mean the inevitable and complete destruction of the Japanese armed forces and just as inevitably the utter devastation of the Japanese homeland."[18]

The Potsdam Declaration was released to the media on the evening of July 26 and received by the Japanese on the morning of July 27. It is not clear whether the Japanese understood that the Americans had an atomic bomb in mind. Their own nuclear program, begun in 1940, was as far from producing a bomb as that of Germany. The Imperial Japanese Army had authorized the builder of the Japanese cyclotrons,

Yoshio Nishina, to start working on the bomb project back in April 1941. Five projects, including the enrichment of uranium by the thermal gaseous diffusion process and the construction of the bomb itself, began simultaneously. But Nishina's conclusions were quite pessimistic: the bomb was a theoretical rather than a practical possibility.

The Japanese navy, always in competition with the army, launched its own probe into the building of an atomic bomb. The naval group, which also included Nishina, came up with an only slightly more optimistic finding: the bomb could be built, but the project would take at least ten years to complete. The enterprise was abandoned, and an American air raid destroyed the thermal gaseous diffusion facility in April 1945.[19]

The Japanese government originally was inclined to remain silent and await a response from the Soviets, whom they had asked to serve as intermediaries in proposed negotiations with the United States, but the military wanted to reject the ultimatum right away. They won the day. On July 28 Admiral Kantaro Suzuki, who had assumed the office of prime minister in April 1945, stated to journalists that his government considered the Potsdam Declaration irrelevant and saw "no other recourse but to ignore it entirely and resolutely fight for the successful conclusion of the war."

This flat rejection of the ultimatum was hardly a surprise to Truman and his advisers: if anything, it came as a relief. They had satisfied the demands of their domestic critics by informing Stalin about the bomb and issuing a public ultimatum to the Japanese, giving them an opportunity to surrender. Now they could go ahead and use the bomb with a clear conscience. In fact, the order to drop the bomb was issued on July 25, one day before the release of the Potsdam Declaration and three days before its rejection by the Japanese.[20]

On the day Truman gave the order to Secretary of War Henry Stimson to use the bomb, he noted in his diary: "He and I are in accord. The target will be a purely military one and we will issue a warning statement asking the Japs to surrender and save lives. I'm sure they will not do that, but we will have given them the chance. It is certainly a good thing for the world that Hitler's crowd or Stalin's did not discover this atomic bomb. It seems to be the most terrible thing ever discovered, but it can be made the most useful." The same diary entry had a note on the timing of the delivery of the bomb: "This weapon is to be used against Japan between now and August 10th."[21]

The first atomic bomb was dropped on Japan on the morning of August 6, 1945. It was code-named "Little Boy" and contained 64 kilograms (141 pounds) of uranium-235—almost the whole supply of the isotope produced at the Oak Ridge plants in the previous year. Unlike the plutonium bomb exploded at Alamogordo in July, it had never been tested before. Nevertheless, it exploded as planned at approximately 8:15 a.m. Japanese time over the courtyard of a Hiroshima hospital—not exactly the military plant imagined by President Truman.[22]

What to target with the atomic bomb had first been discussed by the Interim Committee back in May. On May 31, the committee agreed to Henry Stimson's proposition not to bomb a big war armaments plant surrounded by workers' dwellings or other similar targets, and aim for the greatest possible psychological effect. Hiroshima, which was later chosen as a target, fit the committee's requirements and was considered ideal by the planners. It was a densely populated city, housing a military base untouched by previous bombings and thus a perfect choice to estimate the bomb's scale of devastation and produce the greatest possible psychological effect. An additional benefit of choosing Hiroshima was that the city had no POW camps.[23]

At 8:15:15 Tokyo time the *Enola Gay*, a Boeing B-29 Superfortress piloted by Colonel Paul Tibbets Jr. and named after his mother, flew over Hiroshima, the bay doors were opened, and once the bomb cleared its security hook, it began its descent from a height of 31,060 feet. "Bomb away!" announced the bombardier. The B-29 was soon illuminated with a bright light followed by a shockwave that reminded the pilots of being hit by flak. The bomb's yield is now estimated at having been close to 15 kilotons of TNT. When the pilots looked down, all they could see was a black cloud assuming a mushroom shape, with numerous fires licking the mountains around the city.[24]

The bomb instantly killed more than 3,200 Japanese officers and troops exercising on the parade grounds close to the epicenter. Altogether more than 60,000 people died on the spot, crushed by the strength of the explosion and the shockwave that knocked down the walls of all buildings in a one-mile radius from the epicenter, damaging and burning the rest of the city. Like the American pilots above them, the Japanese on the ground away from the area directly hit by the bomb first experienced a burst of bright light, followed by the shockwave.

Kikue Shiota, a young woman who lived with her siblings in a family house a little bit over one mile away from the epicenter, remembered the explosion as a "blinding light that flashed as if a thousand magnesium bulbs had been turned on all at once." The walls around her collapsed, and it was only through the efforts of her father that Shiota and her sixteen-year-old sister were able to get from under the rubble. Still in shock, they started to walk the streets of the city in search of their other siblings, a fourteen-year-old sister and ten-year-old brother. When Shiota finally spotted the boy, she could barely recognize him. "All the skin on his face was peeling off and dangling," remembered Shiota. "He was limping feebly, all the skin from his legs burned and dragging behind him like a heap of rags." They would never locate the younger sister, Mitsue, just a scrap of her school uniform stuck in the overheated asphalt.[25]

Shiota was lucky to survive and live to tell her and her family's story. Many survivors never got that chance. Before the end of the year another 60,000 would die of injuries and radiation poisoning sustained during the explosion. Among those was Shiota's mother, who died one month after the explosion. Gone were more than half the population of the city, estimated to have been 225,000 on the morning of August 6, 1945.[26]

President Truman received a report about the explosion as he crossed the Atlantic on board the cruiser USS *Augusta*, en route from

Plymouth, England, to Newport News, Virginia. He and Byrnes were lunching with crew members when, close to noon, the ship's captain delivered a message from the Navy Department. It informed the president that "the Japanese port of Hiroshima had been bombed a few hours before, under perfect weather conditions and with no opposition. The results of the bombing were reported to be even more successful than previous tests of the new weapon had led us to hope for." According to Truman's travel log, "The president was excited and pleased by this news." He shook the captain's hand and told him: "This is the greatest thing in history." After Stimson confirmed the news, Truman called the crew to attention and declared to them that "a terrifically powerful new weapon, which used an explosive 20,000 times as powerful as a ton of TNT," had been used successfully against Japan. The crew applauded. Everyone expected a speedy end to the war.[27]

As the *Augusta*'s crew cheered the news delivered by Truman, the ship's radio picked up broadcasts from Washington, first news bulletins announcing the dropping of the bomb, and then a statement preapproved by the president. It revealed that the new weapon used against Japan was an "atomic bomb," mentioned the Japanese rejection of the Potsdam Declaration, and threatened more of the same if the Japanese continued their resistance. "If they do not now accept our terms, they may expect a rain of ruin from the air, the like of which has never been seen on this earth," read the text. "Behind this air attack will follow sea and land forces in such number and power as they have not yet seen and with the fighting skill of which they are already well aware."[28]

The Japanese government was silent. Shocked by the power of the new weapon, the civilian and military leaders could not agree on a course of action. The last hope of avoiding surrender without going down in the flames of nuclear Armageddon seemed to be the Soviet Union. The Japanese ambassador in Moscow requested a meeting with the Soviet foreign commissar, Viacheslav Molotov, but when they met on August 8 Molotov told the ambassador that the USSR had declared war on Japan. In the early hours of August 9, Red Army forces numbering more than 1.5 million officers and men crossed the border into

Manchuria, going after the Japanese troops there. Attacked by the Americans from the air and by the Soviets on land, the Japanese had no further recourse.

On August 9, the day of the Soviet advance into Manchuria, an American bomber dropped a plutonium bomb of the type tested at Alamogordo the previous month on the city of Nagasaki. The Fat Man, 128 inches in length and 60 inches in diameter, had only 12 pounds of plutonium in its belly, but that was enough to produce a blast equivalent to 21 kilotons of TNT—more powerful than the one delivered by Little Boy three days earlier. It caused less damage because clouds obstructed the pilot's view, and the bomb was dropped a few miles away from its intended target. Moreover, the surrounding hills limited the impact of the shockwave. An estimated 40,000 people died instantly, and 25,000 were injured by the blast and radiation. Fires started by the explosion were the main cause of death, accounting for 90 percent of the casualties, as opposed to 60 percent in the case of Hiroshima.[29]

The Japanese government was split in its reaction, with the military arguing that the Americans could not have more than a couple of atomic bombs ready and hoped to survive future atomic attacks. Although the Soviet offensive brought a new element into their calculations, the military still were not ready to surrender. The government's impasse was broken by Emperor Hirohito himself. Having barely escaped a military coup, he decided to accept the Potsdam ultimatum with one caveat: in a telegram received by the US government on the morning of August 10, surrender was conditioned on the maintenance of "the prerogatives of His Majesty as a Sovereign ruler." Truman decided to accept the deal, and the American response was broadcast on August 11, but there was no immediate response from Japan. The Japanese government was still debating conditions.[30]

Finally, on August 14, a few days before another plutonium bomb was to become available to the US Air Force, Japanese radio signaled that the emperor would accept the Potsdam Declaration. Hirohito addressed his nation the next day in a radio broadcast, referring directly to the atomic bomb. "The enemy has begun to employ a new and most cruel bomb, the power of which to do damage is indeed incalculable, taking the toll of many innocent lives," stated the emperor. Historians still argue whether the Soviet surprise attack was a greater factor in the Japanese surrender than the dropping of the atomic bombs. The emperor's statement points to the bombs as decisive. Most scholars agree.[31]

As had been hoped, the bomb indeed helped to end the war and save American lives. It did not, however, prevent the Soviet entry into the war, rather speeding it up, nor did it stop the USSR from claiming its Yalta loot in the Far East. The bomb had two targets, Japanese and Soviet: it hit the first and missed the second.

Chapter 8

STOLEN SECRET

The atomic bombing of Hiroshima and Nagasaki made a sea change in Joseph Stalin's thinking about the unprecedented weapon. What Albert Speer said about Hitler, who "was unable to grasp the revolutionary nature of nuclear physics," could be applied to Stalin as well. Before he learned of the devastation wrought by the atomic bomb in Japan, the Soviet leader could not grasp the nature of the revolution created by the arrival of the nuclear age in the military and diplomatic realms.[1]

Much has been made of Stalin's lack of reaction to Truman's words about the nuclear test at Potsdam. "Not a muscle twitched on his face," wrote the Soviet interpreter, Vladimir Pavlov, recalling Stalin's encounter with Truman. "He turned and left. Truman, puzzled by such a reaction, stood for a few seconds as if rooted to the spot, gazing at Stalin as he walked away." The usual explanations range from Stalin's incomprehension of the news to a deliberate attempt on his part to remain impassive as he confronted a now nuclear-armed rival. But there is good reason to believe that Stalin's reaction indicated first and foremost his lack of appreciation of the bomb's importance.[2]

Marshal Georgii Zhukov, who met with Stalin later that day, remembered him laughing the matter off. When Stalin, speaking in Zhukov's presence, mentioned his conversation with Truman to Viacheslav Molotov, the Soviet foreign commissar, the latter commented: "They're trying to build themselves up." Stalin laughed and responded: "So let them build themselves up; I'll have to talk to Kurchatov about speeding up our work." Molotov seems to have regarded the bomb as little more than a diplomatic ploy on the part of the US government. Stalin probably agreed. Contrary to widespread opinion, he did not hasten to call Igor Kurchatov, the young nuclear physicist in charge of the rather dormant Soviet nuclear project. Nor did he instruct his military commanders to adjust or speed up their preparations for war in the Pacific. Stalin's first documented conversation about the Soviet nuclear project with government officials responsible for the armaments industry would take place only after the American bombing of Hiroshima and Nagasaki.[3]

Averell Harriman, the American ambassador to the USSR, met with Stalin in his Kremlin office on August 8, 1945, two days after the dropping of the American bomb on Hiroshima. It was 8:20 p.m. on August 9 in Moscow and 2:20 a.m. on August 9 in the Far East. The Soviet invasion of Manchuria had begun about two hours earlier, and less than nine hours remained until the explosion of Fat Man over Nagasaki. According to the minutes of the meeting compiled by George Kennan, a counselor at the American embassy who accompanied Harriman, both issues were discussed: Stalin raised the question of the Soviet offensive and Harriman that of the American bomb.

The Generalissimo, as Kennan referred to Stalin in his report, informed the ambassador that the Soviet attack had taken the Japanese by surprise, and the operation was proceeding better than expected. Four columns, one composed of cavalry units, were attacking the Japanese and preparing the ground for a massive invasion that would include the island of Sakhalin. After trying unsuccessfully to elicit Stalin's opinion about the likely Japanese reaction to the invasion, Harriman asked a similar question regarding the bomb: "what he thought of the effects of the news of the atomic bomb would have on the Japanese." Stalin expressed concern that the Japanese might use the bomb as "a pretext to replace the present government with one which would be qualified to undertake a surrender." He clearly did not want that to happen anytime soon.

With the bomb now a subject of the conversation, Harriman did his best to explain away the fact that the Western allies had kept Stalin in the dark about the atomic project for so long. "The Ambassador observed that it was a good thing we had invented this and not the Germans," wrote Kennan. "For long, he said, no one had dared think it would be a success. It was only a few days before the President had told Stalin about it in Berlin that we had learned definitely that it would work successfully." Stalin played along: "Soviet scientists said that it was a very difficult problem to work out." When Harriman mentioned cooperation on the bomb project with the British, stressing Churchill's role in its development, Stalin, the member of the Big Three who had been left out of the loop, once again showed no apparent concern. He told Harriman: "Churchill was a great innovator, persistent and courageous. He too had pushed the development of the tank at one time."[4]

The Soviet memoir literature suggests that Stalin's demeanor—playing the gracious host unconcerned either about the American bomb or about the Western allies' having concealed its development from him—was a mere sham. According to a rumor picked up by Yakov Terletsky, a Soviet physicist turned atomic spy, from his colleagues at secret police headquarters, "Stalin had a tremendous blow-up for the first time since the war began, losing his temper, banging his fists on the table and stamping his feet." The outburst took place after news about the bombing of Hiroshima reached Stalin's country house in Kuntsevo near Moscow. Stalin's daughter, Svetlana, visited her father that day with her newborn child, Stalin's grandson, whom he had never seen before. But the grandfather and his aides were too busy dealing with the news about Hiroshima to pay any attention to Svetlana.[5]

On August 8, two days after the Kuntsevo episode, Stalin accepted Harriman's explanation of why the bomb had been kept secret from him, but he and Molotov, who was present at the meeting, could hardly conceal their disdain for American dissembling on the matter. While agreeing with Harriman that "if the Allies could keep it [the bomb] and apply it for peaceful purposes it would be a great thing," Stalin added: "that would mean the end of the war and aggressors. But the secret would have to be well kept." Molotov entered the discussion, saying: "You Americans can keep a secret when you want to." Harriman detected "something like a smirk on his face" and concluded that the American atomic bomb project was anything but a secret to the Soviet leadership.[6]

The meeting, which might have ended with a diplomatic explosion, concluded in seemingly amicable fashion. Stalin decided not to make an issue of having been excluded from the Grand Alliance on one of the most important issues of the war. Either he did not want to spoil relations with the Americans when the question of his sphere of influence in the Far East was on the line or he wanted to feign lack of concern about the powerful new American weapon. But the bomb was now definitely on Stalin's radar as it had never been before. A few days later he would make a key decision to revive his own atomic bomb program.

Ever since the spring of 1941, when the Soviet academic establishment decided that the tapping of nuclear energy for military or other purposes was a task for the distant future, Soviet nuclear research had been confined to the laboratory.

The German invasion of the Soviet Union on June 22, 1941, delivered a new blow to the project, as scientific research that might result in the improvement of conventional weapons took priority over theoretical physics. The Academy of Sciences had dissolved the Uranium Commission, and scientists working on the uranium problem were drafted into the army. Viktor Maslov, who worked at the Ukrainian Physico-Technical Institute in Kharkiv, was among the authors of the Soviet design for an atomic bomb. Having volunteered for military service, he was mortally wounded at the front and died in a hospital in December 1942. About the same time, Soviet scientists apprised the government that nothing had been published in the West about chain reaction since the discovery of nuclear fission in 1938, and Soviet intelligence began to pick up more and more information about research on the atomic bomb conducted in Britain.[7]

The first detailed information on the British nuclear project came from the Cambridge Five, a group of Englishmen recruited by Soviet intelligence in the 1930s ostensibly to help the Soviet Union fight fascism. Now they were stealing nuclear secrets for the Kremlin. The stolen documents contained a summary of discussions conducted by the MAUD Committee at its meeting on September 16, 1941. The committee concluded that an atomic bomb could be built within two years. Lavrentii Beria, the Soviet security tsar, was suspicious of the intelligence report: was it not a ploy to fool the Soviets into pouring their meager resources into a project that the British knew to be a dead end?

Further information about the development of the British project was received in the following months, and by the fall of 1942 there was enough evidence to suggest that the British and the Americans were serious about building a bomb. The Germans also appeared to be working on one. On the Soviet-German front, north of the Sea of Azov, Red Army officers found a notebook on the corpse of a German officer with calculations on the construction of a German atomic bomb. Soviet physicists began pushing for an assessment of information on the atomic project collected by the intelligence services and resumption of work on uranium, lobbying the military to recall from the Red Army key scientists working on the problem.[8]

In late September 1942, almost exactly a year after first intelligence information on the MAUD project began to arrive in Moscow, the State Defense Council, the de facto wartime government of the USSR, adopted a resolution proposed by Viacheslav Molotov to resume Soviet research on uranium. The resolution was endorsed personally by Stalin, who was informed about the American project for the first time and met with Soviet scientists to discuss the matter. The resolution called for the creation of a special laboratory to work on the uranium nucleus, tasking Soviet industry to build centrifuges and provide other equipment required for the thermodiffusional separation of uranium. The State Defense Council requested a report on "possibilities of building a uranium bomb or providing uranium fuel" by April 1, 1943.[9]

The request went to Igor Kurchatov, the forty-year-old student of the founder of Soviet nuclear physics, Abram Ioffe, who had spent years pushing for aggressive research on nuclear issues but was ignored and overruled by his senior and more powerful colleagues in the Academy of Sciences. In March 1943 Kurchatov was appointed to head secret laboratory no. 2, a nascent Soviet atomic bomb project with a staff of ten scientists besides Kurchatov himself. With this appointment, the young physicists led by Kurchatov eclipsed the mineralogists and chemists of Vladimir Vernadsky, the founder of the Soviet Radium Institute, who considered the release of nuclear energy a long-term goal, not an immediate task of Soviet science. In the world of nuclear physics, the Leningrad group represented by Kurchatov pushed aside their Moscow and Kharkiv colleagues. The leader of the latter, Aleksandr Leipunsky, the former director of the Ukrainian Physico-Technical Institute, did not believe that the country should spend enormous resources needed in wartime to chase the mirage of a nuclear bomb. Kurchatov had no such doubts.[10]

Viacheslav Molotov, who oversaw the atomic project during the first years of the war, recalled later that he had chosen Kurchatov from a list of scientists prepared for him by the secret police—those whom the Soviet government trusted. Kurchatov had to be able to keep secrets not only about the Soviet atomic program but also about those of the British and the Americans. In the spring and early summer of 1943, Kurchatov was all but buried under mountains of intelligence data coming not only from Britain but also from the United States. He regularly sent his assessments of the information to the government. Kurchatov would file his most comprehensive report in July 1943, concluding that uranium fuel and a uranium bomb could indeed be produced.[11]

Kurchatov's conviction was based primarily on the work of American and British rather than Soviet scientists. After describing in impressive detail British work on heavy-water reactors and Fermi's graphite reactor in the United States, Kurchatov concluded: "there is every reason: 1) to maintain that positive conclusions about a chain reaction in the combinations of 'uranium and heavy water' and 'uranium and carbon' produced in America and England are more correct than our previous negative conclusions, and 2) to consider that reactors employing such combinations can be made to work." He reported that his laboratory was ready to build its first reactor using 1 tonne (metric ton) of uranium and 10 tonnes of graphite to verify the American results.

Kurchatov's next conclusion was that an atomic bomb could be built using plutonium-239. "Subsequently," he wrote, "additional material was received from America not only confirming the proposed schema but also containing a series of new and most important data on the question." On the basis of the American material, Kurchatov believed that it would be easier to build a plutonium than a uranium bomb. He concluded his report with the following statement: "At the moment, of course, it is impossible to assert with perfect certainty that a quantity of metallic eco-osmium of acceptable size will solve the problem of building an atomic bomb. But neither can the same be asserted with perfect certainty about a quantity of metallic uranium-235. Theoretical expectations, however, are somewhat more favorable to plutonium than to uranium-235."[12]

It was now up to the Soviet leadership to decide what to do with Kurchatov's conclusions. The next stage in catching up with the Americans would mean a significant allocation of funds, as well as administrative and technical resources, that the Soviet Union simply lacked in the middle of the war with Germany. Given the needs of the war effort and the largely theoretical rather than practical prospects for building the bomb, the Soviet leadership refused to prioritize the atomic project over all other military-related scientific endeavors. Without official approval, it had no chance of succeeding. The Soviets found themselves in the same position as their former allies and now adversaries, the

Germans, and their past and future adversaries, the Japanese. They decided not to chase the mirage of the atomic bomb at a time when conventional bombs were needed at the front.[13]

Kurchatov kept working on building a new cyclotron and obtaining materials for his graphite reactor. He even organized a bomb group to design the future weapon. But without appropriate resources, progress on all those efforts was exceptionally slow. Sometimes he had nothing to show for his work but an assessment of American results supplied by Soviet intelligence. "In the course of the past month I have been occupied with preliminary study of new and quite extensive (300 pages of text) materials pertaining to uranium problems," wrote Kurchatov in September 1944 to the People's Commissar of Internal Affairs Lavrentii Beria, who was also responsible for foreign intelligence. "That study showed once again that abroad there has been a concentration of scientific, engineering, and technical resources unprecedented in the history of world science to deal with the problem, and that they have already achieved extremely valuable results. At home, however, despite great progress in the development of work on uranium in 1943–44, the state of affairs remains completely unsatisfactory."[14]

All the way into the summer of 1945, the frustrated Kurchatov continued to run his underfunded and largely neglected uranium project. But then, all of a sudden, the American attack on Hiroshima and Nagasaki produced a most dramatic change of political fortune. In mid-August 1945, Stalin called Kurchatov and People's Commissar of Munitions Boris Vannikov into his office to discuss the atomic project. "A single demand of you, comrades," Stalin told his guests, "provide us with atomic weapons in the shortest possible time." He continued: "You know that Hiroshima has shaken the whole world. The equilibrium has been destroyed. Provide the bomb: it will remove a great danger from us."[15]

Vannikov's memoirs leave little doubt that Stalin's approach to the nuclear project was influenced by the American experience. "Stalin dwelled briefly on the atomic policy of the USA and then spoke about organizing work on the exploitation of atomic energy and the building of an atomic bomb here in the USSR," recalled Vannikov. Stalin asked Vannikov what he thought of entrusting the project to the Ministry of the Interior, which had administrative, logistical, and intelligence capabilities. The idea had come from Lavrentii Beria, but Stalin appeared to have his doubts. After listening to Vannikov, who favored a civilian committee to oversee the project, Stalin decided to keep Beria in charge but raise the profile of the project by creating a special committee. He summoned Beria and told him that the committee "should be under the control of the Central Committee, and that its work should be kept strictly secret.... The committee should be endowed with extraordinary powers."[16]

On August 20, Stalin signed a decree creating a Special Committee of the State Defense Council, which had been the country's wartime government. The committee's task was to provide "leadership of all work on the exploitation of energy within the uranium atom." Beria, the Soviet security tsar and Stalin's deputy in the State Defense Council, became its head, with Vannikov chairing the committee's technical and scientific panel. The decree also created the First Chief Directorate within the Soviet government to execute the bomb project and ordered other branches of government to provide it with all necessary resources, while prohibiting them from interfering in any way with the directorate's work.

The last, thirteenth point of Stalin's decree read: "To commission Comrade Beria to take measures to organize foreign intelligence work with regard to obtaining fuller technical and economic information about the uranium industry and atomic bombs." Unlike Churchill, Stalin never discussed a partnership in the nuclear project with Truman. He knew how to take advantage of American knowhow without burdening himself with the formalities of partnership. Kurchatov's reports leave no doubt that the Soviets already knew a good deal about the American achievements. Now they simply wanted to know more.[17]

From then on, Stalin would pay close attention to his atomic project. The pressures of the exhausting war with Germany were finally over, while the American bomb was turning the scientists' dream into a military and political reality that threated Stalin's postwar plans. Kurchatov would receive more attention from the top Soviet leadership than his American counterpart, Robert Oppenheimer, had ever been accorded in his work on the bomb. The Soviet atomic weapon would become Stalin's bomb to a degree that the American bomb had never become Roosevelt's. In fact, Roosevelt never got the bomb he built, while Truman never built the bomb he got. Stalin would build his bomb and have it too. Never conceived as an anti-German or anti-Japanese bomb, it was always imagined as and would remain an anti-American weapon. The fact that it had been built with the help of stolen American scientific and technological knowledge was an ironic bond between the atomic projects of the two increasingly adversarial countries.

Secretary of War Henry Stimson, the Truman administration's top official responsible for making the bomb, believed that the United States needed to engage not only the United Kingdom but also the USSR in an international agreement on the control of atomic energy. He had abandoned his earlier hopes that letting the USSR in on the atomic energy secret might help democratize the communist dictatorship but still insisted on an agreement as a way to prevent a nuclear arms race.

Stimson wrote to Truman on September 11 to warn him against creating an Anglo-Saxon atomic club that would include the United Kingdom but exclude the USSR. "Such a condition will almost certainly stimulate feverish activity on the part of the Soviet toward the development of this bomb in what will in effect be a secret armament race of a rather desperate character," wrote Stimson. "There is evidence to indicate that such activity may have already commenced." He continued: "If we fail to approach them now and merely continue to negotiate with them, having this weapon rather ostentatiously on our hip, their suspicions and their distrust of our purposes and motives will increase. It will inspire them to greater efforts in an all out effort to solve the problem."[18]

But Stimson, already on his way out, would leave office a few weeks after sending this memo to Truman. His views, which proved prophetic, and his hopes, admittedly not very realistic with regard to influencing Soviet policy, were largely ignored. The administration's new star, Secretary of State James Byrnes, was bullish on the bomb. The atomic bombs dropped on Japanese cities did not prevent Soviet entry into the war in the Far East, as Byrnes had hoped at Potsdam, but they speeded up the Japanese surrender and thus limited Soviet ambitions in the region to the territories promised to Stalin at Yalta. Now Byrnes believed that the bomb could strengthen his hand in dealing with the Soviets in Eastern Europe.[19]

The first round of atomic diplomacy between the United States and the Soviet Union took place in London in September 1945, slightly more than a month after the explosion of the American bombs over Hiroshima and Nagasaki. The scene was a meeting of the Council of foreign ministers of the United States, the United Kingdom, the USSR, France, and China—a platform created by the Potsdam Conference to draw up peace treaties with Germany's former allies in Europe, including Romania, Bulgaria, Hungary, and Finland—all now under Soviet occupation.[20]

The American atomic bomb indeed influenced the negotiations, but not in the way Byrnes had hoped. Stalin's plenipotentiary, Viacheslav Molotov, kept mentioning the bomb, leaving little doubt that it bothered the Soviets, but he did so to mock rather than to appease the Americans. At a reception he asked Byrnes whether he had "an atomic bomb in his side pocket." Byrnes responded in kind: "If you don't cut out all this stalling and let us get down to work, I'm going to pull an atomic bomb out of my hip pocket and let you have it." Everyone laughed. At a banquet, Molotov raised a toast: "Here's to the Atom Bomb! We've got it!" At another dinner he remarked: "Of course we all have to pay great attention to what Mr. Byrnes says, because the United States are the only people making the atomic bomb."[21]

Molotov had an atomic chip on his shoulder, but he refused to budge on any of the issues under discussion. If anything, the Soviets were determined now more than ever to strengthen their hold on Eastern Europe and refused to make concessions on the composition of the Hungarian and Bulgarian governments. When Stimson suggested that unless the Soviets were brought into the atomic club, they would have no trust in the United States, he was underestimating the negative impact of the bomb on his Soviet allies. Not only had they never trusted the Americans, but now they felt threatened. Molotov later recalled his thoughts in the first years after the war: "The bombs dropped on Japan were not aimed at Japan but rather at the Soviet Union." He continued: "They said, bear in mind you don't have an atomic bomb and we do, and this is what the consequences will be if you make a wrong move."[22]

On November 6, 1945, delivering a report in Moscow on the anniversary of the Bolshevik revolution, Molotov warned Washington against nuclear pressure and indicated a possible threat to the alliance. He declared that "the discovery of atomic energy should provoke neither enthusiasm for exploiting that discovery in the play of forces in foreign policy nor carelessness with regard to the future of peace-loving peoples." Molotov made two points. First, the nuclear weapon had not yet been tested as an instrument for preventing war or maintaining peace, and second, he warned Washington, "There can be no technical secrets of great significance that might remain the property of any one country or any limited group of countries."[23]

Molotov sounded as if he was in full agreement with Leo Szilard and his colleagues at the Metallurgical Laboratory in Chicago, who argued that sooner or later other countries would get the bomb too. But Molotov was not predicting the future. He was simply describing the state of affairs at the time: the American secret was no secret to the Soviets.

Chapter 9

UNITED NATIONS

The dropping of atomic bombs on Hiroshima and Nagasaki was met variously by the American public with surprise, shock, horror, and joy but, more than anything else, with approval. "Exploding with a blinding flash visible for miles, and a terrific roar, the new secret bomb that harnesses the power of the universe to destroy men and material was loosed against Nagasaki at 11:58 o'clock yesterday morning," wrote the dean of nuclear journalism, Bill Laurence, in the *New York Times* on August 10, 1945. After Trinity this was the second nuclear explosion that Laurence had witnessed, and one could feel a sense of excitement in his prose. The Gallup poll conducted between August 10 and 15 showed overwhelming American support for the bombing: 85 percent of respondents were in favor, with only 10 percent opposed; 5 percent had no opinion.[1]

The lyrics of the 1945 song "When the Atom Bomb Fell" by Karl and Harty, a popular country music duo, celebrated the bombing as an answer to the prayers of American soldiers fighting the Japanese. "And a great a ball of light filled the Japanese with fright / They must have thought it was their judgment day," they sang. "Atom bomb dancers" were welcomed with applause in Los Angeles theaters, adults ordered "Atomic Cocktails," and their children looked excitedly for "Atomic Bomb Rings" in their General Mills cereal boxes.[2]

Some were shocked rather than excited, among them the thirty-year-old journalist Norman Cousins, editor of the *Saturday Review of Literature*, a weekly with approximately 50,000 readers. On August 18, three days after the Japanese government offered its surrender and the US government accepted it, Cousins published an editorial, "Modern Man Is Obsolete." It became a manifesto of the 10 percent of Americans who did not approve the nuclear bombing of Japan.

Instead of feeling triumphant and looking forward joyfully to the coming victory, Cousins wrote about fear. "Whatever elation there is in the world today because of final victory in the war is severely tempered by fear," read the opening line of the essay. "It is a primitive fear, the fear of the unknown, the fear of the forces man can neither channel nor comprehend. This fear is not new; in its classical format it is the fear of irrational death. But overnight it has become intensified, magnified. It has burst out of the subconscious and into the conscious, filling the mind with primordial apprehensions." Cousins wondered how much time remained before man "employs the means he has already devised for the ultimate self-destruction—extinction?"[3]

The response to Cousins's editorial among those concerned about the way in which victory had been achieved was overwhelming: the *Saturday Review of Literature* received 35,000 requests for reprints. "You have written a sparkling message which I hope starts a conflagration of constructive thinking and conduct among the peoples everywhere," wrote one reader. It was clearly the editor's purpose to affect "modern man's" thinking and conduct. Cousins not only addressed the problem of humankind's possible self-destruction but offered a solution: for him, it was the creation of world government.[4]

An effective writer who expressed the concerns and fears of tens of thousands of his readers, Cousins was piggybacking on an idea expressed at the dawn of the nuclear age by its major prophet, H. G.

Wells, who back in 1913 predicted the advent of a new atomic age and provided an antidote for the specter of nuclear war—world government. Cousins's contribution to the decades-long debate was not the idea itself but the timing of his declaration. Unlike the scientific fantasy of 1913, the atomic bomb had become a reality in 1945. The urgent question was whether the idea of world government could make a similar transition.

Those in favor included not only traditional supporters of the world government idea but also American nuclear scientists who had played a key role in the creation of the bomb and were now dismayed at the possibility of its further use. Some of them believed that moral justification for their research had expired with the bombing of Hiroshima. The bombing of Nagasaki or any future atomic bombing struck them as completely unjustified. Albert Einstein, a godfather of the American atomic bomb, would become chairman of the Emergency Committee of Nuclear Scientists, which opposed the use of nuclear weapons. He would tell its members in May 1946 that "a world authority and eventual world state was not just 'desirable' in the name of brotherhood, they are 'necessary' for survival."[5]

In October 1945, Cousins attended a conference of proponents of world government, including such a prominent figure as the former US Supreme Court Justice Owen J. Roberts. Participants put their hopes in the United Nations, the world peace organization that was inaugurated in April and acquired official status in October. The General Assembly would meet for the first time on January 10, 1946, in London. Expectations ran high that the UN would establish international control over atomic weapons. Indeed, the assembly's very first resolution dealt with the subject but fell far short of the expectations of Cousins and his supporters.[6]

Prime Minister Clement Attlee, who replaced Churchill as leader of the British government after general elections in the summer of 1945, had high hopes for the first meeting of the UN General Assembly, which his country was going to host. But he was also deeply concerned that any productive work of the session could be derailed by the nuclear issue.

On October 16, Attlee wrote to Truman proposing a meeting to discuss the international control of nuclear weapons. "It is our view here that the meeting of foreign ministers was overshadowed by the [atomic bomb] problem," wrote Attlee, referring to a tense set of meetings between his foreign secretary, Ernest Bevin, US Secretary of State James Byrnes, and the Soviet foreign commissar, Viacheslav Molotov. He continued: "the prospective conference of the United Nations will be jeopardized unless we have some clearness on our own attitude to this problem." Attlee proposed a tripartite meeting that would include, apart from Truman and Attlee, also Prime Minister William Lyon Mackenzie King of Canada, whose scientists had made a major contribution to the Manhattan Project. Truman invited both to visit Washington.[7]

Attlee was a proponent of transferring American and British knowledge about atomic weapons to the custody of the United Nations. In his thinking about the atomic bomb, Attlee was very much under the influence of Sir John Anderson, Viscount Waverley, a key figure in Churchill's war cabinet whom Attlee appointed to advise the Labour government on nuclear issues. Anderson in turn was much influenced by his friend Niels Bohr and the international movement for control of nuclear armaments that Bohr had helped to create among scientists. Back in September, Anderson had drafted a letter to Truman proposing that the president, as an "act of faith," transfer knowledge about the production of the atomic bomb to the United Nations in order to avoid an arms race. The letter was never sent to Washington, partly because of the objections of Winston Churchill, now the leader of the opposition in the British parliament.

Churchill, with whom Attlee had informally shared the draft, did not believe in international treaties and obligations as a solution. He believed instead in nuclear deterrence. "Nothing will give foundation [for peace] except the supreme resolve of all nations who possess or may possess the weapon to use it at once unitedly against any nation that uses it in war. For this purpose the greater the power of the US and GB in the next few years the better are the hopes. The US therefore should not share their knowledge and advantage except in return for a system of inspection of this and all other weapon-preparations in every country, which they are satisfied after trial is genuine." Churchill thought that the United States would have two to three years of nuclear monopoly and should use that time to set up a peace system guaranteed by the threat of using the bomb that it possessed.[8]

In Churchill's view, Attlee's suggestion that the United States would turn over its theoretical knowledge and technical knowhow to the UN had no chance of realization. Indeed, on October 3, 1945, Truman sent a message to Congress that was very different from Attlee's suggestions. He proposed an international agreement controlling atomic weapons as an alternative to an arms race but immediately limited the items that it would cover. Truman assured Congress that there would be no disclosure of manufacturing processes leading to the production of the atomic bomb. Negotiations would focus rather on "the terms under which international collaboration and exchange of scientific information might safely proceed."[9]

On November 15, 1945, Truman welcomed Attlee and Mackenzie King to Washington. They issued a joint declaration that addressed relations between the nuclear haves and have-nots. Its text followed the principle of separating theoretical knowledge of nuclear physics from technological processes, as Truman had specified in his message to Congress. The declaration read: "We believe that the fruits of scientific research should be made available to all nations, and that freedom of investigation and free interchange of ideas are essential to the progress of knowledge." Then came the caveat: "We are not convinced that the spreading of the specialized information regarding the practical application of atomic energy, before it is possible to devise effective, reciprocal, and enforceable safeguards acceptable to all nations, would contribute to a constructive solution of the problem of the atomic bomb."[10]

The three leaders brought the United Nations into the picture, but not as a future custodian of nuclear weapons. Instead, they entrusted the UN with the practical execution of scientific exchange and nuclear disarmament. "In order to attain the most effective means of entirely eliminating the use of atomic energy for destructive purposes and promoting its widest use for industrial and humanitarian purposes, we are of the opinion that at the earliest practicable date a commission should be set up under the United Nations Organization to prepare recommendations for submission to the organization." The basic principles of the declaration had been proposed a few weeks earlier by Vannevar Bush, the head of the wartime American nuclear project. Britain and Canada had little choice but to adopt the American position.[11]

The Washington Declaration became the first international document dealing with the control of nuclear issues. Yet the Western allies knew perfectly well that no UN agreement would work without the Soviet Union. Would Stalin get on board? That question headed the agenda of James Byrnes when he embarked on his trip to Moscow in December 1945 to attend the next round of foreign ministers' talks, in which Viacheslav Molotov and the British foreign secretary, Ernest Bevin, would take part.[12]

The creation of a UN Atomic Energy Commission, the panel of diplomats proposed by Truman, Attlee, and King, was first on the eight-point American agenda for the meeting. The Soviets did not object, but Molotov put the issue of the commission at the very end of the agenda. They finally reached it on December 22, the sixth day of the conference. Molotov agreed to co-sponsor the American proposal but wanted changes. The commission, he argued, should be subordinate to the Security Council and not the General Assembly, as suggested by the Americans. Byrnes promised to study the Soviet proposal.[13]

The Soviets had apparently decided that their influence was strongest in the Security Council, not in the Assembly, where they had few clients or allies, while the United States had plenty. The Soviet position was akin to that of the British, who also believed that they would have more influence if the commission reported to the Security Council, of which Britain was a permanent member. Ernest Bevin had suggested the idea a week earlier, but Byrnes responded to him that "this procedure would enable the Russians to use the veto, which would be obviated if the Commission reported to the Assembly." Now, with the Soviets asking for the same arrangement as the British and the success of the whole project depending on the subordination of the commission, Byrnes had to look for a compromise solution.[14]

They eventually agreed that the commission would be "accountable" to the Security Council but not run by it. That formula was attained with the help of Stalin, who took a direct interest in the question and met with the American and British delegations. A further compromise was reached on the distribution of the commission's reports. Reports to the General Assembly were to be published unless the Security Council ruled otherwise. Thus the Americans sold the idea of the commission to the Soviets, while the Soviets managed to establish a significant degree of influence, if not control, over the work of the international body that could allow legal access to American nuclear knowledge and possibly tie America's hands diplomatically.[15]

James Byrnes celebrated his first achievement in nuclear diplomacy, but it came at a high price. Instead of helping to strengthen America's hand in dealing with growing Soviet assertiveness in Eastern Europe and Iran, agreement on the Atomic Energy Commission came at the expense of American interests in those regions. Byrnes agreed to recognize the communist-controlled governments in Hungary and Romania, as well as the continuing presence of Soviet troops in Iran.[16]

Byrnes returned to Washington to a frosty reception from President Truman, who was unhappy to see his secretary of state making foreign policy without consulting the White House. Truman had not been shown the draft communiqué of the Moscow Conference, and his relations with Byrnes would never recover from the blow. Nevertheless, the atomic deal that Byrnes helped to negotiate in Moscow would live on. The Soviets kept their word, and the UN Atomic Energy Commission became a reality on January 24, 1946, with the affirmative vote of the General Assembly.[17]

The UN Assembly charged the Atomic Energy Commission with the preparation of proposals for the consideration of the Security Council on the control of atomic energy. The exchange of scientific information and the inspection regime—two key elements of any successful international effort toward achieving the first two goals—also came under the commission's mandate. Its members included representatives of the permanent members of the Security Council: China, France, the United Kingdom, the United States, and the Soviet Union. The platform for debates, disagreements, and cooperation on the uses of atomic energy had been created.[18]

The UN Atomic Energy Commission came into existence before either President Truman or the commission's early promoter, James Byrnes, developed a clear idea of what kind of nuclear policy they wanted the commission to promote.

To determine what the country wanted from the UN Commission and what American policy on the international control of atomic energy should be, in January 1946 Truman and Byrnes put together a committee headed by Undersecretary of State Dean Acheson. Acheson, for his part, brought in a group of consultants chaired by David Lilienthal, the head of the Tennessee Valley Authority, and including no less a figure than Robert Oppenheimer, who became a key drafter of the committee's report. Its main premises were based on ideas discussed by Oppenheimer and fellow scientists during the war. The new report was very much a continuation of two earlier reports drafted by scientists working on the Manhattan Project.[19]

The Acheson-Lilienthal report, drafted by some of the same scientists, postulated that the prevention of nuclear war required international control of atomic weapons. The authors proposed the creation of a UN Atomic Development Authority with the ultimate task of establishing such control and overseeing the production of nuclear energy. Under the new regime enforced by the agency, governments would give up their existing atomic bombs or refrain from building them in exchange for scientific information and assistance in developing peaceful nuclear energy.

The key question was how to ensure that nuclear energy facilities would not be diverted to military purposes. "Take the case of a controlled reactor, a power pile, producing plutonium. Assume an international agreement barring use of the plutonium in a bomb but permitting use of the pile for heat or power," wrote the drafters. "No system of inspection, we have concluded, could afford any reasonable security against the diversion of such materials to the purposes of war." Their proposed solution was to vest ownership or control of deposits of fissionable ores in the agency. The known deposits of uranium and thorium were limited, and control of them was now even more important than it had been back in 1939, when Szilard approached Einstein with the idea of writing a letter to the queen of the Netherlands to keep the Germans away from the Congo. The Atomic Development Agency's mandate would not be limited to deposits: it would also have to own or control the entire production cycle of fissionable materials.[20]

According to the Acheson-Lilienthal plan, nuclear disarmament was to be achieved at an unspecified point in the future, with the American government sharing scientific and technological secrets with the USSR in exchange for an agreement banning the production of any new bombs. The atomic bombs already in the American arsenal were to be decommissioned and the fissile materials used for peaceful purposes. President Truman, who received the report in the spring of 1946, liked its basic premises but worried about giving up the American nuclear monopoly without guarantees that other countries, the Soviet Union in particular, would not violate the agreement and build bombs as the United States destroyed its own.[21]

Truman's concerns were addressed in a most unexpected way by a plan prepared by the first American representative to the UN Atomic Energy Commission, the distinguished businessman and philanthropist Bernard Baruch. Truman assured Baruch that he could produce his own proposal on the basis of the Acheson-Lilienthal report. Baruch brought in as drafters people who knew little about nuclear energy and made his first presentations to Wall Street bankers. Although the scientists' monopoly on drafting American nuclear policy was broken, many of their ideas remained in place.[22]

The Atomic Development Authority, the UN body proposed by Baruch, like the Atomic Development Agency proposed by Acheson and Lilienthal, would not only be responsible for control of fissile fuel deposits but also manage nuclear facilities capable of building atomic bombs and have the right to inspect plants producing nuclear energy for peaceful purposes. But if the Acheson-Lilienthal plan sidestepped the question of inspections, the Baruch Plan was preoccupied with it. The proposed authority would report to the Security Council, which would impose sanctions on countries found in violation of the authority's rules. The five permanent members of the Security Council would be open to sanctions like any other member nation of the UN.

Baruch presented his plan to the UN Commission on Atomic Energy on June 14, 1946. He did so in the most dramatic terms, addressing himself to "fellow citizens of the world." "We are here to make a choice between the quick [i.e., living] and the dead," declared Baruch. His rhetorical skills did not help him much with the Soviets, who probably did not understand the archaic expression and were inclined to see, in what became known as the Baruch plan, an attempt to perpetuate the American monopoly on the bomb.[23]

The proposed Baruch Atomic Development Authority was supposed to establish its control over the American nuclear arsenal only after it had made certain that no atomic bomb project was under way anywhere in the world. And what would happen if the USSR ended its project and the United States rejected the authority's claim on its bombs? That, the Soviets believed, was the essence of the Baruch Plan. They were also averse to giving up their veto power on anything, especially such a key issue as nuclear weapons and atomic energy, and wanted no foreigners inspecting their facilities, nuclear or otherwise. They were accelerating their own bomb project, confident of its success, and did not believe that the Americans would give up their bomb.[24]

Instead of outright rejection of the Baruch Plan, the Soviets proposed a plan of their own. Andrei Gromyko, the ambassador to the

United States, who also acted as the Soviet representative to the UN Atomic Energy Commission, presented the plan in late June 1946. It proposed that the United States, the only country that possessed and had used the atomic bomb, divest itself of nuclear weapons before the Baruch Atomic Energy Authority introduced its inspection regime. The Soviet plan called for signing a convention on the abolition of atomic weapons, postponing the introduction of the inspection regime. Harry Truman and his advisers were by no means receptive. Neither the president nor Congress was prepared to give up nuclear weapons without a system of rigid control in place.[25]

With the Baruch Plan tabled at the United Nations, the Truman administration proceeded with policies that made the rejection of the plan by the Soviets almost inevitable. In June and July 1946, the US government invited the UN Atomic Energy Commission to observe the detonation of two "Fat Man"–type plutonium bombs, one exploded above water, the other under water in the Bikini Atoll lagoon of the Marshall Islands, where the US Navy brought ninety-five target vessels. Only nine of them were eventually scrapped: the rest, too contaminated to be of any further use, were scuttled. The Soviets accepted the invitation but treated it as an act of American intimidation, not openness. One of the Soviet observers, Semen Aleksandrov, told an American scientist that the Americans wanted to "frighten the Soviets," but they were "not afraid." According to Aleksandrov, the Soviets had "wonderful planes" to bomb American cities. That was a bluff, but Aleksandrov's message was clear: the Soviets were not about to be intimidated.[26]

The Baruch Plan was presented for formal discussion and vote by the member states of the UN Atomic Energy Commission in December 1946. With the Soviet Union refusing to accept it, there was virtually no chance of success. Speaking to the United Nations in October 1946, the Soviet foreign commissar, Viacheslav Molotov, called on the delegates to approve the USSR's proposed convention banning the manufacture and use of atomic weapons in a manner similar to the ban on other weapons of mass annihilation, in particular the poisonous gas and bacteriological weapons that had been banned after World War I.

On December 30, 1946, when the UN Atomic Energy Commission voted on the plan, two member states, the USSR and Soviet-controlled Poland, a non-permanent member of the commission, abstained. Ten members voted in favor of the plan, but acceptance required unanimity. It was never achieved. The nuclear arms race was officially on.[27]

Chapter 10

UNION JACK

Clement Attlee replaced Winston Churchill on July 26, 1945, the day after Churchill signed the Potsdam Declaration, which implicitly threatened Japan with the nuclear bomb. He shared his predecessor's conviction that the atomic bomb was the product of a joint venture between the United States and the United Kingdom and was supposed to be in their common custody. He said as much in the letter he sent to Truman on August 8, 1945, two days after the bombing of Hiroshima. Attlee wrote: "I consider, therefore, that you and I, as heads of the Governments which have control over this great force, should without delay, make a joint declaration of our intentions to utilize the existence of this great power, not for our own ends, but as trustees for humanity in the interests of all peoples in order to promote peace and justice in the world."[1]

Soon after the bombing of Nagasaki, Attlee set up a group of cabinet members to assess the future of the British nuclear project. The group, officially known as Gen-75, was dubbed the "Atomic Bomb Committee" by the prime minister. In a memo to the committee, he presented his views on the new situation that had emerged with the arrival of the atomic bomb. He considered it a truly revolutionary development that had "rendered much of our post-war planning out of date" and "all considerations of strategic bases in the Mediterranean or the east Indies . . . obsolete." The arrival of the bomb, wrote Attlee, "offers to a Continental Power such targets as London and the other great cities." In short, he was concerned about a future Soviet nuclear weapon. "The vulnerability of the heart of the empire is the one fact that matters," he wrote.

What was to be done? Attlee did not believe in the prohibition of nuclear weapons, citing the failure of attempts to outlaw the use of poisonous gas during the interwar period. "This method is bound to fail and it has failed in the past," wrote the prime minister. "Gas was forbidden but used in the first world war. It was not used in world war 2, but the belligerents were armed with it. We should have used it, if the Germans had landed on our beaches." Though not intended for the public, that was quite an astonishing admission. But if not an international ban, then what? An option suggested by Attlee was using the bomb as a weapon of retaliation. "We recognized, or some of us did before this war that bombing could only be answered by other bombing," he argued. "We were right. Berlin and Magdeburg were the answer to London and Coventry. Both derive from Guernica. The answer to an atomic bomb on London is an atomic bomb on another great city."

Attlee had effectively laid the foundations of British nuclear deterrent policy but was not yet prepared to follow it. "Dueling with swords and inefficient pistols was bearable," continued the prime minister. "Dueling had to go with the advent of weapons of precision." Attlee's vision of the future called for the continuation of the wartime Grand Alliance and the creation of an American-British-Soviet triumvirate to ban the use of force throughout the world and end wars altogether. Although he did not call for world government, he had clearly got the H. G. Wells bug. "This sort of thing in the past was considered a Utopian dream. It has become today the essential condition of the survival of civilization and possibly of life on this planet." Attlee concluded the memo by proposing a first step: "I should on behalf of the Government put the whole of the case to President Truman and propose that he and I and Stalin should forthwith take counsel together."[2]

While Attlee dreamed of using the bomb to end all wars and then establishing joint custody of the bomb secrets, if not of the weapon itself, with the Americans, his advisers thought in more practical terms and were concerned with the future of the British-American nuclear alliance. As Truman, Attlee, and Mackenzie King declared in Washington on November 15 that they were prepared to share scientific information on the peaceful uses of atomic energy with the rest of the world, British diplomats used the opportunity of the high-level visit to negotiate a British-American agreement. It was meant to continue and deepen the wartime collaboration in the nuclear sphere started by FDR and Churchill and authorized by the Quebec Agreement of 1943 and the Hyde Park memorandum of 1944.[3]

The negotiations proved a major disappointment for the British, ending their illusions about joint custody of the bomb. They were constrained to renounce the "consent" clause of the agreement on the use of the atomic weapon. The Americans now promised only to consult with them before dropping the next bomb. In exchange they got the right, as well as American assistance, to develop commercial nuclear energy. The additional price for that right and assistance was an agreement that all raw materials pertaining to nuclear energy would be distributed between the allies on the basis of need, which meant that for the foreseeable future all uranium from the British Empire would go to the Americans alone.

Attlee left Washington believing that he had secured "full cooperation" with the United States not only on the development of nuclear power for peaceful ends but also for the bomb, if the British decided to build their own. But the reality was much more complex. One of the documents signed as a result of the negotiations promised full cooperation in the realm of nuclear energy, but the other limited it to basic scientific research, excluding industrial knowhow.[4]

A major obstacle to Attlee's "full cooperation" was the dramatic change on the American political scene that followed the death of Roosevelt. Unlike his predecessor, Truman did not have full control over the bomb and its secrets, to which he had never aspired. The Congress that took over the supervision of the US nuclear program after the war felt that, given the nation's investment in the project and its $2 billion price tag, the Americans alone owned the project. In September 1945, an estimated 90 percent of congressmen and 70 percent of the public opposed sharing the secret of the bomb with any nation. The British and their supporters in the ranks of the American administration were facing a high tide of nuclear nationalism. While most Americans polled in opinion surveys did not believe that their country could maintain the nuclear monopoly for long, they were in no rush to give it up.[5]

The British encountered a formidable opponent of nuclear cooperation in no less a figure than General Leslie Groves, the former head of the Manhattan Project. He began to raise the alarm about excessive British demands for exchange of technological information as early as January 1946. Many suspected that it was he who leaked news about Soviet nuclear spies in the United States to the media. The information came from Igor Gouzenko, a cipher clerk at the Soviet embassy in Ottawa, who defected to the Canadians and told them everything he knew about Soviet spying on the bomb in Canada and the United States. The leak produced a major public scandal, presenting the Soviets as a threat and the Canadians and British as unreliable partners. Groves wanted tighter security controls to protect nuclear secrets from enemies and allies alike, as well as military representation on the government commission that was supposed to take control of the American nuclear program on behalf of Congress.

Groves succeeded on both fronts. In April 1946 he managed to obtain the backing of President Truman himself, who dismissed Attlee's complaints about lack of cooperation by stating that the agreement he had signed in November 1945 made no provision for American assistance in the construction of British atomic energy plants. Truman also added an all-important caveat regarding his understanding of the agreement: exchanges of information could take place only on a "mutually advantageous basis." In July 1946 Congress passed the McMahon

Act, named after Brian McMahon, a senator from Connecticut and chair of the Senate Atomic Energy Committee. The act created a Joint Congressional Committee on Atomic Energy (JCAE), which was obliged to work in close contact with a military liaison group. More importantly for the troubled American-British cooperation, it prevented American authorities from sharing any information pertinent to the nuclear program with any country, including Britain, unless the exchange might benefit the United States.[6]

The McMahon Act became law on August 1, 1946, with Truman attaching his signature to the document. The British were caught by surprise. One of their diplomats tried to appeal to American lawmakers by showing them copies of the 1944 Quebec Agreement, but that was already old hat. Although Roosevelt and Churchill had agreed in that document on the creation of a uranium condominium, joint research, and joint custody of atomic bombs, the wartime consensus was gone. The US Congress that took possession of the bomb was in nonproliferation mode, making little distinction between former partners and new nuclear rivals. The British were on their own, whether they liked it or not.[7]

With Attlee's original ideas about a Soviet-American-British triumvirate and the abolition of war long gone, and his hopes for continuing nuclear cooperation with the United States dashed, his only remaining option was the concept of nuclear deterrence. Britain would have to build its own bomb. But that realization came only later and in stages, as did the key decision to proceed from nuclear research to production of the bomb. The original calculation was that an independent British nuclear program would force the Americans to reconsider their stand and return to the cooperation of the war era. As that hope waned, the goal of the program became the building of a British bomb.[8]

The idea that Britain should develop its own atomic bomb was gaining supporters among the British elite in the course of 1945. The country should "be prepared to use it ourselves in retaliation," stated a committee of British scientists in June 1945. In October, the same opinion was voiced by the British chiefs of staff, who urged that the production of bombs begin as soon as possible. In November 1945, Churchill, already in opposition, expressed the opinion of the political class when he declared in the House of Commons: "This I take it is already agreed, we should make atomic bombs." The Baruch Plan and its subsequent failure in the United Nations encouraged similar sentiments in the government. "Let's forget about Baroosh," allegedly stated Attlee's second in command, Foreign Secretary Ernest Bevin, referring to the Baruch Plan, "and get on with making fissile."[9]

The British began to reestablish their own nuclear enterprise—the original one had been very much rolled into the Manhattan Project—one step at a time. The job went to John Cockcroft, a forty-seven-year-old physicist and former member of the MAUD committee. Cockcroft had spent most of the war working on radar systems, but in May 1944 he took control of the Montreal laboratory that worked on the design of heavy water reactors and helped build one at the Chalk River Laboratories northwest of Ottawa. In September 1945 the reactor became operational—the first heavy water reactor in the world and the first operational reactor of any kind outside the United States. Called back to the United Kingdom, Cockcroft was asked to replicate and build on his Canadian success.[10]

In December 1945, Attlee's Gen-75 ministerial committee decided to approve the construction of an industrial-type graphite reactor modeled on the American reactors at Hanford in Washington State. There was no doubt in the mind of those present that they were building a reactor to produce plutonium, which was required to produce a bomb. "How many piles should be built depended in part upon the output of bombs which the Government thought necessary," Attlee told the gathering. Concerned about the cost of the undertaking, they decided on one pile. "It was pointed out . . . that the construction of two piles would make heavy demands upon the capacity of chemical engineering and heavy electrical industries, both of which were of great importance to the revival of our export trade," reads the protocol of the meeting.[11]

The decision was a victory of British scientists working in North America over their counterparts who had stayed in Britain and mastered gaseous diffusion technology for the enrichment of uranium. The "Americans" among the British scientists knew that a plutonium bomb could be produced more quickly than a uranium one. The ministerial committee went along. The view of the skeptics, who claimed that producing a handful of bombs would not protect Britain but provoke an attack, was ignored. The production target was set at fifteen atomic bombs per year and the price tag for the construction of the reactor at £20 million (over £1 billion in 2024 prices)—an enormous amount by the standards of postwar Britain.[12]

In March 1946 Air Chief Marshal Lord Charles Portal, one of the planners of British nocturnal terror bombing of Germany, assumed the newly created position of Controller of Atomic Energy within the Ministry of Supply. That made him the equivalent of General Groves in the British atomic project, with the important difference that he had much more power and a much higher public profile than his American counterpart. Portal used that profile to hedge his and his country's bets for producing a bomb by pushing for the construction not only of uranium piles but also of a gaseous diffusion plant for the enrichment of uranium. In the atmosphere of uncertainty about the primary goal of the British nuclear program, the uranium enrichment facility idea was supported by the argument that uranium was needed for the industrial production of nuclear energy. The cost of a uranium facility was double that of the graphite pile—approaching £40 million, or over £2.1 billion in 2024 prices.[13]

Portal's request was not met with surprise, as there was a growing consensus that Britain needed a bomb of its own. But there was the as yet unanswered question of whether Britain could afford the construction not just of reactors that theoretically could be used to produce commercial electricity but also the bomb, which would yield nothing productive. It was one thing to commit the country to the research program, another to make a long-term commitment to the production of the bomb, of which the reactor was just the first and a relatively simple step. The construction of plants to enrich uranium and produce plutonium, and eventually to build bombs, was complex not only in scientific and technical but also in financial terms. And there lay the greatest weakness of a wholly independent British project: the government had no money.

An influential group of members of the new Labour cabinet that swept into power in the summer of 1945 opposed the atomic bomb project as too costly for the troubled British treasury. The austerity measures introduced by the Labour government resulted in food rationing more severe than that of the war years. The country emerged from the war all but broke, its gold reserves depleted and two-thirds of its foreign trade gone because of the loss of ships to German submarines. The debt was approaching 250 percent of GDP. For Britain to survive, the government had to replace American Lend-Lease with a multibillion-dollar loan from the very same Americans, who insisted in return on convertibility of the British pound and liquidation of imperial preferences or free-trade agreements within the British Empire—steps that made speedy economic recovery all but impossible. There was simply no money in the treasury to build the atomic bomb.[14]

But Portal was pushing for a new gaseous diffusion plant, and the government had to decide whether it was prepared to part with anywhere between £30 and £40 million to build it. The matter came before the Gen-75 Committee on October 25, 1946. Hugh Dalton, the chancellor of the Exchequer, and Sir Stafford Cripps, the president of the Board of Trade, were skeptical, to say the least. "We must consider seriously whether we could afford to divert from civilian consumption the restoration of our balance of payments, the economic resources required for a project on this scale," reads the protocol of the meeting. Attlee's financial and economic experts were telling him that there was no money to build the plant, and he was considering whether to kill the proposal. The rest of the nascent British nuclear program might well go with it.[15]

The man who swayed the committee's opinion was the British foreign secretary, Ernest Bevin. As minister of labor in Churchill's wartime coalition government, he had been a key figure in mobilizing British labor resources for the war effort. Now he was entrusted with maintaining the country's international standing in the midst of an enormous economic downfall. Bevin was prepared to sacrifice the British Empire but not British security in Europe, where he was working hard to counteract growing Soviet influence. The Stalin regime had already established its control over Eastern Europe, and many considered it likely that a takeover of Western Europe was in the making. Bevin was considered anticommunist and pro-American, but whatever his position in relations with Moscow and Washington, British interests came first, as did his commitment to Britain's great-power status.

At the fateful afternoon meeting on Downing Street, Bevin objected to Attlee's proposal to scrap the project because of its high cost. "No, Prime Minister, that won't do at all," he told Attlee. "We've got to have this thing. I don't mind it for myself, but I don't want any other Foreign Secretary of this country to be talked at or to by the Secretary of State of the United States as I have just been in my discussion with Mr. Byrnes. We've got to have this thing over here, whatever it costs. . . . We've got to have the bloody Union Jack flying on top of it." Bevin was referring to his recent discussions with James Byrnes about sharing nuclear know-how. Like Churchill before him, Bevin needed the atomic bomb to deter the Soviets. But unlike Churchill, he did not want to be dependent on the increasingly uncooperative Americans.[16]

Bevin's argument carried the day. "Our prestige in the world, as well as chances of securing American cooperation would both suffer if we did not exploit to the full a discovery in which we had played a leading part at the outset," reads the protocol of the meeting. The gaseous diffusion plant project was provisionally approved. The committee agreed to allocate half a million pounds to the project and review its progress after that money was spent. The decision had a broader implication. An attempt by the economic ministers to kill an important component of the atomic bomb project was rebuffed, and the program itself was saved. The Attlee government was seriously setting out to build its own atomic bomb. Attlee himself was slowly but surely returning to the idea that he had first presented in his memorandum of August 28, 1945, to the newly created Gen-75 Committee: "The answer to an atomic bomb on London is an atomic bomb on another great city."[17]

The Royal Air Force made its requisition for the first atomic bomb in August 1946. The formal decision to build the bomb was made by Attlee's government in January 1947. The initiator of the decision was Lord Portal. In the last days of 1946 he penned a memorandum that reached Attlee on December 31. "I submit that a decision is required about the development of Atomic weapons in this country," began the two-page memo. Portal explained that he and the ministry were "charged solely with the production of 'fissile material,' i.e. of the 'filling' that would go into any bomb that it was decided to develop." That, he argued, was not enough: "Apart altogether from producing the 'filling,' the development of the bomb mechanism is a complex problem of nuclear physics and precision engineering on which some years of research and development would be necessary." He asked the ministers for direction.[18]

The meeting of ministers called by Attlee on January 8, 1947, conspicuously did not include those responsible for the treasury and the economy. The Gen-75 Committee was now gone, replaced with the Gen-163 Committee, which excluded ministers opposed to building the gaseous diffusion plant. Not only the composition of the new committee but also the timing of its meeting could not have been better for Portal and the British bomb project. A few days earlier, news had reached London that the Soviet Union and Poland, with their two votes, had killed the Baruch Plan in the UN Atomic Energy Commission.

The cabinet secretary, Sir Edward Bridges, provided the ministers with a memo criticizing the Baruch Plan for its provision that research projects of the Atomic Energy Authority would be carried out by scientists seconded by their national governments. There was no way of preventing those scientists from sharing their secrets with their governments, argued Bridges, with the clear approval of Attlee. Under the circumstances, Attlee saw no alternative to building a British atomic bomb. The British government now considered itself free of the limitations imposed by the Baruch Plan.[19]

The Gen-163 meeting began with a presentation by Lord Portal, who suggested that approximately three years would be needed to build the bomb. Ernest Bevin was next to speak. He reported on the discussions about the Baruch Plan at the UN, stressing that it would be important not to allow an international authority to hamper the development of atomic energy in Britain. "We could not afford to acquiesce in an American monopoly of this new development," argued Bevin. "Other countries also may well develop atomic weapons. Unless therefore an effective international system could be developed under which the production and use of the weapon would be prohibited, we must develop it ourselves."

The ministers of defense and supply agreed with Bevin. Unsurprisingly, given the absence of the ministers of finance and trade, discussion was limited, and the protocol took up only one full page of text. Indeed, the decision had already been made implicitly months earlier. The British bomb project, as much anti-Soviet as it was non-American, was now officially born. While this first meeting of the Gen-163 Committee was also its last, it produced one of the most consequential decisions in Britain's postwar history.[20]

The secrecy surrounding the British nuclear project was akin to that of the American and Soviet projects. But there were substantial differences as well. While the Americans sought to conceal their efforts from the Germans and the Soviets from the Americans, the British government or, more precisely, one faction of it was trying to hide its nuclear program from the rest of its own government, parliament, and people. Given Britain's dire postwar financial circumstances, as well as the ideological rivalries and moral reservations about nuclear weapons that divided its parliament, it was anyone's guess whether a British atomic bomb program would have proceeded at all if Attlee had not kept it secret.[21]

Lord Portal, the British counterpart of General Leslie Groves, found his own Robert Oppenheimer in the person of W. G. Penney. A thirty-seven-year-old mathematician, Penney had been seconded to Los Alamos to calculate the height at which the atomic bomb had to be detonated and the damage it could cause. He was present at the Trinity test and gave advice on the choice of Japanese targets for the bombing. He was also present at the Bikini Atoll tests in mid-1946. The passing of the McMahon Act later that year sent Penney and other British scientists at Los Alamos packing. In mid-1947 Penney was back in Britain, talking to Portal. "We're going to make an atomic bomb," Portal told his future top scientist. "The Prime Minister has asked me to coordinate the work. They want you to lead it."[22]

Penney, who assumed the newly created position of chief superintendent of armament research, set himself up in the Royal Arsenal at Woolwich in southeastern London, where he directed the work of the High Explosive Research Unit. He took upon himself the role of director of the armament division responsible for the construction of the bomb. The role of director of research went to the reactor expert John Cockcroft at Harwell. Penney's first task was to figure out how much British scientists knew about the work done on the American bomb. What they knew went into a document entitled "Plutonium Weapon – General Description."

The British were important participants in the research and bomb-building enterprise at Los Alamos. In those areas, the transition from the joint American-British project to the British one was rather smooth: they were replicating the American "Fat Man"–type plutonium bomb detonated over Nagasaki. The production of plutonium—the part of the Manhattan Project to which the British had had no direct access—was trickier. Penney entrusted that task to Christopher Hinton, a forty-six-year-old Ministry of Supply insider who had spent the war constructing and running British ordnance factories. His task was building the reactors and enrichment plants needed for the production of plutonium and weapons-grade uranium.

Hinton did not start entirely from scratch. In August 1947, a few months after he took charge of the project, Cockcroft launched his first research graphite reactor—GLEEP, or Graphite Low Energy Experimental Pile—at Harwell. It went operational about a year and a half later. Construction soon began of two more reactors under Cockcroft's supervision at Windscale on the northwestern coast of England in Cumberland. They were plutonium-breeding reactors whose design was based on the X-10 Graphite Reactor built by the Americans at Oak Ridge. The British introduced one major change, using air instead of water to cool the active zone of the reactor. A water-cooling system was considered less safe in densely populated Britain.[23]

The Americans knew what was going on at Windscale and, although formally prohibited from exchanging any scientific information with the British about the reactors, they refused to leave their cousins completely on their own. A group of Americans arrived at Windscale in 1948 and advised their British colleagues on "Wigner growth," an effect named after Eugene Wigner, who had driven Leo Szilard to his first meeting with Albert Einstein in 1939. "Wigner growth" was the tendency of graphite in a reactor to expand in response to a fission reaction. Edward Teller, another of Leo Szilard's companions in meetings with Einstein, warned the British about another of Wigner's discoveries known as "Wigner energy," a buildup of energy in graphite that could ignite it.

Very often, American expertise reached the British after they had already made mistakes in the construction of reactors, so corrections had to be made on the fly. That was the case with adjustments to correct problems caused by the Wigner phenomena. When Cockcroft returned from a trip to the United States with the idea of adding filters to the coolant discharge chimney in order to prevent the escape of radioactive particles, the chimneys of British reactors were all but built. Since it was too late to install filters at the bottom of chimneys, they had to be placed on top, adding tremendously to the cost of the whole enterprise. Nevertheless, American advice saved the British from more than one potential pitfall. The filters, which became known as "Cockcroft's folly," would prevent a major ecological disaster in Britain in 1957, when a fire broke out because of the buildup of Wigner energy in one of the reactor piles.[24]

The British were now only a few steps away from an atomic bomb of their own. Unlike the American and Soviet bombs, the British one was conceived with the idea of status in mind. Churchill wanted a joint bomb with the Americans to counter the Soviets. But it was not the case with Attlee. The Labour prime minister was a reluctant Cold Warrior who originally wanted to share secrets about atomic physics with the Soviets. His ministers insisted that the British military refrain from planning a future war, including the use of atomic weapons, with a focus on the Soviet Union as Britain's potential enemy. Although the Soviet threat became more pronounced over time, the decision to build the British bomb was due mainly to the United Kingdom's desire to maintain its great-power status and freedom of action in world politics, especially vis-à-vis the United States.[25]

With American help, the construction of the Windscale reactors and the design of the British nuclear bomb moved ahead. Plans were made for the production of the first plutonium in 1950, with construction and testing of the bomb to follow soon thereafter. The British knew that they were in competition with the Soviets, but never in their wildest dreams did they imagine that the Soviets were in fact ahead of them. When news came in September 1949 that the Soviets had exploded their first atomic bomb, the British were taken completely by surprise.[26]

Chapter 11

STALIN'S BOMB

From the very beginning of its work in August 1945, the Soviet Special Committee charged with building the Soviet atomic bomb and chaired by the country's security tsar, Lavrentii Beria, was blessed with an abundance of secret American and British documents showered on it by Stalin's intelligence services. Igor Kurchatov's very first presentations to the Special Committee were already based on an analysis of materials supplied by Soviet agents working at Los Alamos. On October 18 Minister of State Security Vsevolod Merkulov sent Beria a package of documents including a detailed twenty-two-page description of the construction and assembly of the atomic bomb.[1]

But not everyone in the Soviet scientific establishment believed in the benefits that security police patronage brought to the project. Petr Kapitsa, a former student of the founder of Soviet nuclear physics, Abram Ioffe, and a star of Soviet nuclear science, had difficult relations with the secret police and serious misgivings about Beria's management of the project. Kapitsa, who had defended his doctoral dissertation under Ernest Rutherford, worked at the Cambridge University Cavendish Laboratory from 1921 to 1934. Detained on one of his visits home, he was prohibited to travel abroad. A renowned physicist and friend of Niels Bohr, with whom he corresponded during the war, Kapitsa was appointed to a technical panel of the Soviet Special Committee with responsibility for theoretical work on the enrichment of uranium. There he clashed with Beria and his appointees over the role that scientists should play in the project and its overall direction.[2]

Between October 1945 and January 1946 Kapitsa wrote personally to Stalin, voicing his unhappiness with the treatment of scientists by Beria and munitions boss Vannikov. He wrote: "Comrade Beria's basic weakness is that the conductor should not only wave the baton but also understand the score. Beria is weak in that regard.... Comrade Vannikov and others on the Technical Council remind me of the story about the citizen who did not believe the doctors and drank every brand of mineral water at the Yessentuki spa in order, hoping that one of them would help." Kapitsa argued for enhancing the role of scientists in running the project and changing its overall direction from emulating the Americans to looking for original solutions to scientific and engineering problems. "We should build the atomic bomb and the jet engine in our own way," urged Kapitsa. Pointing out that the USSR simply did not have the kind of money spent by the Americans on their atomic project, he maintained that cheaper and better solutions had to be found. He also proposed a two-year plan to raise the overall quality of Soviet science.[3]

Kapitsa insisted on resigning from the Special Committee, and Stalin let him go in December 1945. He would later tell an American scholar: "I refused to work on those weapons because the man who was in charge of the weapons project—the secret police chief Lavrentii Beria—had no respect for scientists. I could not work for such a person." He would be removed from his position at the Academic Institute of Physics in August 1946 and fired from his last remaining job, that of professor at Moscow University, in 1950. Kapitsa's resignation meant that there was no longer anyone in a position of relative power in the Soviet atomic project to question reliance on stolen information and thus on American models for constructing reactors, enriching uranium, producing plutonium, and eventually building the atomic bomb. If American scholars suspected of lacking loyalty or refusing to fall in line with the overall direction of the Manhattan Project were simply left alone, Kapitsa was removed from the academic world altogether. And he was lucky to suffer no further consequences. Rumor had it that Stalin personally protected Kapitsa from Beria's rage, keeping him out of prison.[4]

On January 25, 1946, the day after the UN General Assembly created its Atomic Energy Commission, Stalin met with Kapitsa's young colleague in the atomic project, Igor Kurchatov, to discuss the state of affairs. The meeting began around 7:30 p.m.—the elderly dictator preferred to work at night. Also present were the two government leaders of the project: the previous one, Viacheslav Molotov, and the new one, Lavrentii Beria. Stalin raised the question of "the expediency of Kapitsa's work." By that time, the decision had already been made to let Kapitsa go. Stalin wondered for whom Kapitsa was working and whether his activity was directed "to the benefit of the Motherland or not." Judging by Kurchatov's notes, Kapitsa, who had insisted on indigenous Soviet ways of solving scientific and technological problems related to the bomb project, was being accused of lacking patriotism.

On the wall of Stalin's office Kurchatov noted a portrait of Lenin along with those of Russian imperial military commanders: the leader's secular iconostasis blended symbols of the imperial past with those of communist ideology. Recording his impressions of the meeting for posterity, Kurchatov scribbled: "Comrade Stalin's great love of Russia and V. I. Lenin, of whom he spoke in connection with his great hope for the development of science in our country." Russian imperial nationalism and communism in its Leninist version now belonged to the new Soviet credo. The bomb was supposed to be Russian and communist at the same time.[5]

In the following month George Kennan, the American chargé d'affaires in Moscow, would stress the importance of Russian history in explaining Soviet aggressiveness in Europe. Kennan's "Long Telegram" to Washington, which detailed his understanding of the imperial roots of Soviet foreign policy, would set the American government on a path to the Cold War. As the Stalin regime blended Russian expansionism with its communist vision of world revolution, Kennan remained unaware that the project to build a Soviet atomic bomb was under way. If Britain was building a bomb to preserve its great-power status as its empire crumbled, the Soviet Union was building one in hopes of ensuring the existence and growth of its communist empire.[6]

Stalin's meeting with Kurchatov that January day of 1946 was equivalent to the approval that President Roosevelt had given Vannevar Bush in June 1942 to proceed to the industrial level of nuclear fuel production. His question to Bush, "Do you have the money?" suggested that money was not an issue. Bush would get everything he asked for. Stalin demonstrated the same attitude toward his own atomic project. "As regards his views on the future development of the project," wrote Kurchatov in his notes, "Comrade Stalin said that it was not worth engaging in small projects but that work should proceed on a large scale, with Russian sweep; that the broadest assistance, involving every measure, would be provided in that regard. Comrade Stalin said that there was no need to seek cheaper methods... that work should proceed swiftly and in rough basic forms."[7]

In his letter of November 1945 to Stalin, Petr Kapitsa had argued that the first step toward making the bomb was to "improve our scientific institutes and the welfare of our scientific workers." Stalin took Kapitsa's advice most seriously. In January 1946 he told Kurchatov: "[O]ur scientists are very modest and never point out that they are living poorly: that is really bad, and even though... our state has suffered greatly indeed, it is always possible to see to it that [a few thousand?] people live comfortably, with country houses of their own so that they can relax, that they have cars."[8]

Stalin meant business. Funding for the sciences tripled in 1946 as compared with the previous year. In March 1946 the salaries of scientists and scholars were increased across the board by official decree, whether they worked on the atomic project or not. The salary of the president of the Academy of Sciences more than quadrupled, jumping from 7,000 rubles to 30,000, and those of candidates of science in the Academy doubled from roughly 1,000 rubles to 2,000. A university professor now received 5,500 rubles per month. At the time, average salaries in the USSR did not exceed 600 rubles per month. A family sedan, the Moskvich, cost about 10,000 rubles, while the luxurious Pobeda was priced at 16,000. Stalin made sure that his scientists could get both a car and a country house, or *dacha*. In return, they were supposed to produce the bomb.[9]

The leading scientists working on the bomb project received more than 10,000 rubles monthly, or almost twice the salary of a university professor. Lavrentii Beria made sure that special salaries were also introduced for engineers and workers employed at enterprises directly related to bomb construction. Bonuses that exceeded regular salaries tens of times were offered to ensure that scientists and engineers were motivated to meet their targets on time. The overall goal was to build the bomb by 1948.

Stalin demonstrated the same resolve to push ahead with the atomic project as had Roosevelt four years earlier. Despite the parallels between the Soviet and American atomic projects, which were initiated by scientists and embraced and eventually owned by the state, the differences between them were more pronounced than the similarities. The Soviet atomic project was run very differently from the American one. The Manhattan Project bore only a distant relation to the country's top official, was overseen by academics doubling as bureaucrats, and run by the army, which struck a partnership with the scientists and managed to keep the security services at bay. The Soviet project was closely monitored by the supreme leader, run by the secret police chief, and was unrelated to the army. The scientists working on it, although being well paid, found themselves in a position inferior to that of their American counterparts.

In large part, the differences between the two projects were rooted in the contrasting political and social characteristics of the rival countries. The Soviet Union was essentially a police state run by a dictator who wanted to sideline the army and its generals, believing that the war had made them too powerful and dangerous to his rule. The Soviet economy was run exclusively by the government bureaucracy, and scientists had no employment other than that given and taken by the state. But there were also other reasons for the power of the secret police and the weakness of the scientists. When it came to scientific and engineering solutions, Soviet intelligence operatives and not scientists were in the driver's seat.

Beria and Kurchatov were not just emulating the American approach but doing so with the benefit of insights that only the stolen American documents could provide. Kurchatov's key task was to check the information through his own research and experiments to make certain that the Americans had not got it wrong and, second, that US intelligence had not fed the Soviets false data. It was a difficult and challenging job in its own right, but nothing like what the Americans had had to go through to acquire the knowledge in the first place.[10]

The Soviet bomb, as pointed out above, was at once Russian and communist. It was the second characteristic that accounted for the success achieved by Soviet intelligence in recruiting atomic spies in Britain and the United States. Most of them were engaged before or during the war, when the Soviet Union appeared to be the best hope of stopping Nazi Germany, its nuclear project was in its infancy, and the Soviet bomb was not yet perceived as either anti-American or anti-British.

The very first sign that the Soviets obtained the atomic bomb project came from Britain and was supplied by the Cambridge Five, communist-leaning members of the British intellectual elite. They decided to work for the Soviets in the 1930s because they saw the USSR as the only bulwark against the rise of fascism in Europe. The information most probably came from John Cairncross, who worked in 1941 as the private secretary of Lord Maurice Hankey, a member of the committee that supervised the British atomic project. More information about the atomic bomb was received in Moscow through the efforts of Donald Maclean, another member of the Cambridge Five who was working on economic matters in the Foreign Office in 1941. Between 1944 and 1948 he was stationed in Washington, where among other things he acted as secretary of the US-British Combined Policy Committee on atomic energy.[11]

But by far the best-placed and most productive Soviet spy inside the Manhattan Project was Klaus Fuchs, a German physicist, age thirty-two when he came to Los Alamos in August 1944 as a member of the British team assigned to assist the Americans in building the bomb. Fuchs was a committed communist. The son of a Lutheran pastor, he joined the party at the age of twenty, emigrating to Britain after the Nazi takeover in Germany. A brilliant theoretical physicist, he was invited to work on the British nuclear project; in 1942 he contacted Soviet intelligence, offering the information to which he had access. While the bomb existed only in theory at the time, Fuchs did not think that capitalist Britain should have any secrets from its communist ally.

When in 1943 Churchill convinced Roosevelt to resume British-American cooperation on the Tube Alloys project, Fuchs was invited to continue his work on the bomb in the United States. He declined the invitation at first, concerned that he would no longer be in a position to pass information to the Soviets. His concern turned out to be unfounded, as the Soviets managed to keep in touch with him in the United States. They contacted him first in early 1944, when he was working on the diffusion method of separating uranium at Columbia University, and then again after he moved to Los Alamos in the summer of 1944. Fuchs conducted research on a plutonium-based bomb and on blast waves that turned out to be truly pioneering. He would be among the select scientists who attended the Trinity test in July 1945. Fuchs gave his all to the building first of the anti-Hitler bomb and then of the anti-Japanese one but did not consider them exclusively American property and had no qualms about sharing his privileged information with the Soviets.[12]

Fuchs passed the copious data he could get his hands on to Harry Gold, a thirty-three-year-old laboratory chemist who doubled as a courier for Soviet civil intelligence, the People's Commissariat of State Security. The son of immigrants from the Russian Empire, Gold was recruited as a Soviet spy in 1934, soon after Hitler's ascension to power in Germany, and was a member of a Soviet spy network run by one of the founders of the Communist Party of the United States, the Ukrainian-born intelligence operative Jacob Golos. As the Soviets showed ever-increasing interest in the Manhattan Project, Gold was employed as a courier passing secret information between New York City and Los Alamos. Fuchs became the primary client of Harry Gold's courier service at Los Alamos but was not the only one to benefit from it.[13]

David Greenglass, a twenty-two-year-old GI when he was transferred to Los Alamos from Oak Ridge in August 1944, was Harry Gold's other important contact in New Mexico. Like Gold, Greenglass was a son of immigrants from Eastern Europe and a former member of a communist organization. He was recruited by Soviet intelligence on the advice of another communist activist, Julius Rosenberg, who was married to Greenglass's sister, Ethel. In 1953 the two would pay the ultimate price for their role in Soviet nuclear espionage, sentenced to death in one of the most publicized and controversial trials in American history. At Los Alamos Greenglass worked as a technician on molds for the explosive lenses designed by George Kistiakowsky and his crew. He passed numerous documents to the Soviets, including a sketch of the implosion-type atomic bomb. A Soviet intelligence report gives some inkling of the motives inspiring the atomic spies: "They are young, intelligent, capable, and politically developed people, strongly believing in the cause of communism and wishing to do their best to help our country as much as possible."[14]

Communist connections, if not beliefs, played a role in the recruitment of another Los Alamos spy, Theodore Hall. He ended up there in the summer of 1944 after graduating from Harvard at the age of eighteen. Hall was deeply concerned about the American monopoly on the bomb and feared that a fascist government might come to power in the United States. He was never a communist, but the Great Depression proved catastrophic for his family, ruining his father's business. In October 1944 Hall walked into the New York headquarters of the Communist Party to get in touch with Soviet intelligence. He eventually managed to do so and passed numerous documents to the Soviets, including a sketch of the plutonium bomb.[15]

The majority of the atomic spies for the Soviet Union were either dedicated communists or shared communist views. They were eager to help the first communist country in the world, which was also an ancestral homeland for some of them. But it was not communism alone that drove Americans and British sympathizers into the ranks of the Soviet espionage network. The moral doubts of many participants in the Manhattan Project, as well as their concern about the postwar American monopoly on the deadly weapon, also played a role. That was the case with Hall and to some degree with Fuchs, who believed in proliferation for peace: a nuclear monopoly seemed likely to lead to war, which would be rendered impossible if more than one country possessed the weapon.[16]

The sheer amount and quality of the information produced by the atomic spies made it difficult for the Soviet leaders in charge of the atomic project not to emulate the Americans in all they were doing, from science to technology to the designation of certain towns and cities as closed to outsiders and heavily controlled by the security services.[17]

The Soviet equivalent of the Chicago Pile—the F-1 uranium-graphite reactor (the first physical reactor)—was built by Igor Kurchatov in Moscow as a project undertaken by the Special Laboratory no. 2. Unlike the Chicago Pile, built beneath the stands of Stagg Field at the University of Chicago, the Soviet reactor was housed in a concrete pit beneath a specially constructed building. It went operational on December 25, 1946, slightly more than four years after its American counterpart. This was the only Soviet nuclear installation to be built in Moscow. Given security considerations, as well as the experience of World War II, when Stalin temporarily lost most Soviet territory west of Moscow to Hitler, postwar Soviet nuclear facilities were located quite far east of the Soviet capital. The Soviet equivalent of the Washington State–based Hanford Project, a facility with industrial reactors producing plutonium, was built in the southern Urals near the small town of Kyshtym, becoming known as Cheliabinsk-40 and eventually as the city of Ozersk.[18]

The construction of the Kyshtym industrial complex known today as Maiak (Lighthouse) began in 1946 under the overall supervision of the forty-eight-year-old Interior Ministry General Yakov Rapoport, one of the creators of the Gulag system of forced-labor camps. Rapoport made a name for himself in the 1930s as the administrator of a large project, involving forced labor, to build a canal from the White Sea to the Baltic Sea. The construction site for the postwar nuclear complex was chosen near a German prisoner-of-war camp. The Gulag work force played an important role in the construction of the Soviet Hanford in the Urals. There were 16,000 Soviet citizens of German nationality interned by the regime during the war and not released after it, as well as 10,000 soldiers and close to 9,000 Gulag prisoners whom Rapaport set to work building the Soviet plutonium-producing site.

In June 1948 the first industrial reactor at Kyshtym went critical. Known as A-1, or Annushka, a Russian diminutive for Anna, the reactor was built by Nikolai Dolezhal. He used the Hanford reactor as a model but turned the dual channels from the horizontal position to vertical, partly to simplify construction. Annushka had numerous problems that necessitated two shutdowns and restarts after its original launch, but by the end of the year it had produced 16.5 kilograms (36 pounds) of plutonium. That plutonium had to be delivered to newly constructed radiochemical works for separation from other isotopes. In February 1949 the "cleansed" plutonium was delivered to a chemical-metallurgical plant, also newly built, for the production of metallic plutonium. The first Soviet weapons-grade plutonium was produced at Kyshtym in April 1949.[19]

The Soviet equivalent of Oak Ridge, a site for the enrichment of uranium, was built in the city of Novouralsk, some 230 kilometers (143 miles) north of Kyshtym, also in the Ural Mountains. It was a plant for gaseous diffusion enrichment of uranium—the method first developed by Harold Uri at Columbia University. Some of the equipment used at the plant was brought from Soviet-occupied Germany. A plant for the production of uranium hexafluoride, used in the gaseous diffusion process developed by Soviet scientists, was built in Kirovo-Chepetsk, a town about 1,000 kilometers (621 miles) northwest of Kyshtym—a settlement almost equidistant from Moscow and Cheliabinsk-40. The first Soviet weapons-grade uranium was produced there before the end of 1949.

As in the United States, the enrichment of uranium turned out to be a lengthier and more cumbersome process than the production of plutonium. While many ideas of what to do and how to proceed came from the United States, Soviet scientists and those brought to the USSR from Germany solved quite a few theoretical and practical problems independently to make the diffusion process work. The Soviets also tried electromagnetic and thermal-diffusion methods of enriching uranium, which had been tested in the United States and found ineffective, but these were eventually abandoned.[20]

The Soviet equivalent of Los Alamos was called Arzamas-16. A location for the center entrusted with the design and actual construction of the Soviet bomb was chosen less than 400 kilometers (249 miles) southeast of Moscow, in a settlement called Sarov, where a seventeenth-century Orthodox monastery had stood. The construction bureau and works around it were supervised and built by enterprises belonging to the Ministry of the Interior. Like all other nuclear project sites, Arzamas-16 was a closed city whose very existence constituted a state secret. The first scientists arrived in Sarov in the spring of 1947, charged with producing an atomic bomb ready for delivery by airplane by the summer of the following year.[21]

With the help of stolen American data, they produced the bomb as soon as they received the first plutonium in August 1949. In that same month the first Soviet atomic bomb, a plutonium-based weapon like the one exploded four years earlier at Alamogordo, was ready for its first test. It took place on August 29 in the Kazakh steppe 170 kilometers (106 miles) west of the city of Semipalatinsk. The bomb's explosive power was registered as equivalent to 22 kilotons of TNT, approximately the same as that of the American "gadget." Like its American predecessor, the Soviet weapon was placed on a tower and exploded there instead of being dropped from an airplane. The damage caused by the explosion was measured by studying its impact not only on buildings and tanks but also on animals, hundreds of which were killed by the blast.[22]

On the following day Lavrentii Beria and Igor Kurchatov, both present at the test site, signed a report to Stalin describing the explosion and its physical, thermal, and radioactive impact on the immediate surroundings and environment. It was apparent from the opening sentences of the confidential report that the country in which the test had been carried out was a dictatorship. "We report to you, Comrade Stalin," read the report, "that through the efforts of a large collective of Soviet scientists, builders, engineers, supervisory personnel and workers, as a result of four years of intensive labor, your instruction to build a Soviet atomic bomb has been fulfilled. The creation of an atomic bomb in our country has been attained thanks to your daily attention, care, and assistance in carrying out that instruction."[23]

The Soviet atomic bomb now in the possession of the USSR was of course a communist one, as envisioned by the Soviet atomic spies in the United States; it was also Russian, as imagined by the Western governments. But first and foremost it was Stalin's bomb. He alone had the power to decide how to use it.

Chapter 12

BRITISH HURRICANE

The success of the Soviet atomic project took the Americans and their British rivals by surprise. In July 1949, a month before the Soviet test, the CIA suggested mid-1953 as the most probable time of arrival of the first Soviet atomic bomb. But on September 3, 1949, a B-29 flying over the North Pacific registered a level of radioactivity in the atmosphere three times higher than normal. Soon clouds with a heightened level of radioactivity were spotted not only over Scandinavia but also off the Pacific coast of Canada and the United States. American scientists eventually figured out that it was the result of an atomic explosion. They knew that it was not an American bomb: it could only be a Soviet one.[1]

President Truman announced the news to the nation and the world on September 23, 1949. "I believe the American people, to the fullest extent consistent with national security, are entitled to be informed of all developments in the field of atomic energy," he began. "We have evidence that within recent weeks an atomic explosion occurred in the U.S.S.R." He did his best to play down the importance of the event, saying, "This probability has always been taken into account by us," and called for "truly effective enforceable international control of atomic energy." The announcement produced a shock wave of its own. "An atomic explosion has occurred in Russia—a fateful portent that the Soviets have broken an American A-bomb monopoly on which the non-Communist world depended so heavily," wrote the *Los Angeles Times*. The newspaper called it "historic news, comparable only in significance to the announcement of the Hiroshima blast of Aug. 6, 1945." It asked the question that was on many minds: "Does it mean World War III?"[2]

Washington tried to calm the public, indicating that Soviet atomic bomb production was still in its infancy, and the United States had a four-year lead. The State Department suggested that the development had no impact on prospects of war and peace, but in Congress many believed that the threat of war had increased dramatically. Democrats on the Hill called for a meeting between Truman and Stalin. Republicans suggested an agreement with the Soviet Union banning first use of nuclear weapons. In the minds of many, war had come one step closer. Truman's announcement followed the approval that day of a $1.3-billion bill of military assistance to NATO countries.[3]

The Soviets responded to Truman's announcement in a most unexpected way. Their State Information Agency, TASS, denied the atomic bomb test. The heightened radiation levels were blamed on explosions conducted for peaceful purposes "with the use of the latest technical apparatus." The first lie was followed by another, suggesting that the USSR had had the bomb for almost two years. "TASS finds it necessary to make a reminder that as early as November 6, 1947, the minister of foreign affairs of the USSR, V. M. Molotov, made a statement about the secret of the atomic bomb, saying that 'the secret has long ceased to exist.' That statement meant that the Soviet Union had already discovered the secret of atomic weapons and has that weapon at its disposal."[4]

The Soviets did not want to provoke a possible American overreaction to the news and trigger an atomic attack on themselves before they acquired a substantial supply of atomic bombs. But the ostensible modesty with which they announced their ascendance to nuclear status did not change the basic fact that the American monopoly on nuclear weapons had been broken. A new era in the history of nuclear arms had arrived.

Truman's shocking announcement of the Soviet nuclear test was accompanied by a joint statement issued by the American, British, and Canadian governments, the Western partners in the Manhattan Project. The statement reminded the world of the November 1945 declaration of Truman, Attlee, and Mackenzie King about the need for international control of atomic weapons. The declaration was presented as an example of the Western leaders' foresight, since it had stated that no one nation could long maintain a monopoly over atomic weapons. The new statement called for the creation of an enforceable system of international control over nuclear weapons.[5]

The three-power statement was more a flashback to the past than an assertion of continuing unity between the three wartime atomic partners. By September 1949, the three governments were on separate tracks with regard to the development of nuclear policy and nuclear arms. In fact, there was no unanimity even on the way in which the statement was to be released. The British tried to persuade the Americans to arrange a leak instead of an official announcement. They were afraid of panic in Britain and a backlash against a government that had failed to produce an atomic bomb before the Soviets did so. London had good reasons for concern.[6]

News of the Soviet bomb was a political embarrassment to the Labour government: How could the country that initiated nuclear arms research and whose scientists had been the first to come up with the idea of the bomb have fallen behind not only the Americans but the Soviets as well? In 1947 the British chiefs of staff estimated that the Soviets would not be able to produce a bomb of their own before January 1952. In 1948 they suggested a target date of January 1951, and in January 1949 the estimate was changed to July 1950, but the Soviet test was carried out almost a year earlier, in August 1949! With no nuclear weapons of its own and none supplied by the Americans, the British government could not avoid painful self-assessment.[7]

Few people were more shocked by the news of the Soviet nuclear test than the British nuclear scientists. Like the German scientists in 1945 who could not believe that the Americans had succeeded where they themselves had failed, their British counterparts could not believe that the Soviets had sped past them. How could that have happened? Had they gained access to American information? That seemed likely but did not explain the Soviet achievement. After all, no one had more access to American knowledge than the British, and yet they had neither an atomic bomb nor even a supply of plutonium. The conclusion was obvious to some if not to all: the Soviets had stolen not just American know-how but plutonium itself. That explanation was advanced by Sir Henry Tizard, the chief scientific adviser to the British Ministry of Defense.[8]

That kind of thinking helped calm British nerves and protect egos, but it did not change the basic fact: the Soviets had the bomb, and the British were now more vulnerable than ever before. They needed to rethink their strategy and, in particular, their relations with Washington, the only source of atomic bombs to protect their island. British-American negotiations, focusing from the British side on obtaining American technical information and from the American side on access to Congo uranium, had been proceeding in desultory fashion since 1946, resulting in temporary agreements with little hope of a real breakthrough.

The British insisted on the right to run their own program while demanding access to American industrial knowhow. The Americans resisted—Congress more than the administration—and the standoff continued until the arrival of the Soviet atomic bomb. It changed the dynamic in more ways than one. The Americans woke up to the reality that they had lost their nuclear monopoly and thus the "secret" they were reluctant to share with the British. They had to become more flexible. The British, for their part, recognized that insisting on the independence of their program was not making them any safer.

In December 1949 a British delegation went to Washington to breathe new life into the ongoing negotiations. The Americans were ready to make a deal. With the atomic monopoly gone, they had to produce as many bombs as quickly as possible to offset the future Soviet nuclear arsenal. Increasing production meant involving the British. First, the Americans needed full access to the uranium produced in the Belgian Congo, which they had had to share with the British, who traditionally had enormous influence over the Belgian government. They also needed the British to stop building reactors that required large quantities of uranium needed for American bombs. Washington wanted London to move its nuclear facilities to Canada, where they could not be overrun by the Soviets in case of war. Moreover, British expertise could be useful to the American atomic project, and duplication of effort looked like a wasteful and now inadmissible luxury.

The offer that the US delegation put on the table for the British to consider was more attractive to the British than their previous proposals. The Americans offered deep integration of the two national programs, with the goal of producing as many bombs as possible in the shortest time frame. The proposal included the temporary assignment of British scientists to the United States, as in World War II, but now with full access to scientific and technological information and the right to take it home.

The British would have the right to maintain their own program in the United Kingdom on condition that they not duplicate research and other work conducted in the USA. The UK would ship to the US all the plutonium produced by its as yet non-operational reactors in exchange for enriched American uranium (American scientists toured the British facilities and decided that work on the two reactors there was too advanced to be halted, but that work on the gaseous diffusion plant could easily be terminated). The United States was prepared to ship twenty American-made atomic bombs to Britain in order to provide it with a nuclear deterrent.[9]

This was a breakthrough that the British had long been looking for. London was satisfied, although it wanted assurances regarding the supply of the bombs and resented the idea that new work, scientific or technical, could not be duplicated in the United Kingdom—a provision that would make the British program incomplete and dependent on its American counterpart for years to come. They addressed those concerns in their counterproposal, which removed restrictions on the duplication of work in the UK, limited the assignment of British scientists to the US to three years, and asked that some of the British share of jointly produced atomic bombs, in excess of the twenty to be deployed in the United Kingdom, be delivered to Canada.

In early January 1950 a deal appeared to be in sight. Wartime collaboration between the two allies was about to be renewed on conditions much more favorable to the British than those stipulated by the 1943 Quebec Agreement. The Soviet threat, like the German threat before it, was bringing the two cousins closer together.[10]

Unexpectedly for everyone, the high hopes of renewed transatlantic cooperation suffered a major blow on January 24, 1950. The news that killed them came not from Moscow or Washington but from London, the capital of the country most interested in reviving nuclear cooperation with the United States.

The newsmaker was the German-born theoretical physicist Klaus Fuchs, the Soviet spy who had worked at Los Alamos and then headed the theoretical physics department at Harwell. Fuchs told an officer of MI5, the British counterintelligence service, that he had shared his knowledge about the Los Alamos project not just with the British but also with the Soviets. His confession was hardly a voluntary admission of guilt or declaration of pride in having been a simultaneous participant in three nuclear projects, British, American, and Soviet. Approached by MI5 agents, Fuchs originally pleaded innocence and denied divulging the secrets. But a few days later he apparently cracked under psychological pressure, returned to the investigators, and confessed to having been a spy for the Soviets.

Fuchs ended up on the MI5 suspect list thanks to the Americans. They were proceeding slowly but steadily with their Venona Project, which analyzed intercepted Soviet diplomatic cables sent from the United States to Moscow during World War II. It took the Americans a while to prioritize the project, crack the code, and gradually start deciphering the messages. The picture that emerged was shocking: the Soviets had been after American atomic secrets even before the Manhattan Project was launched. They appeared to have had sources at Los Alamos, and one of the cables pointed to Fuchs as a likely culprit.[11]

On February 4, 1950, Fuchs was charged with espionage. The Americans, who had been suspicious of the British and their lax security system since Igor Gouzenko's exposure of a Soviet spy ring in Canada in 1945, had every reason to be unhappy with the British once again. But what Fuchs told the interrogators indicated that the British were not the only party to be blamed in the Soviet espionage story. Americans were guilty as well. It turned out that the person through whom Fuchs had been passing information to the Soviets while stationed at Los Alamos was the naturalized American citizen Harry Gold. When the FBI picked up Gold, he identified another Los Alamos spy, David Greenglass. Once the FBI arrested Greenglass in June 1950, he named Ethel—Greenglass's sister—and Julius Rosenberg. The pair was arrested in July 1950. They would be executed in June 1953.[12]

For a while it looked as if the British were no worse than Americans when it came to keeping nuclear secrets. But then, in September 1950, a month after the arrest of the Rosenbergs, another spy scandal exploded in London, destroying whatever was left of British credibility in the eyes of the Americans. The Italian-born British citizen Bruno Pontecorvo, one of Enrico Fermi's early collaborators who was employed at Harwell, disappeared with his wife and three sons in the middle of their Italian vacation. It turned out that Pontecorvo had first flown to Stockholm and was brought to Moscow from there, along with British nuclear secrets.

What prompted Pontecorvo's sudden escape were the heightened security measures introduced at Harwell after Fuchs's confession. Pontecorvo was considered a security risk because of his relatives' membership in the Communist Party, but neither the British nor the Americans had any proof that he was actually a spy. It was suggested that he leave Harwell. He was offered a professorship at the University of Liverpool but decided to escape to the Soviet Union instead. Whether Pontecorvo had been a spy no longer mattered: now that he was in Moscow, all that he knew about the Manhattan Project, in which he had been involved while working in Canada, and all his knowledge of the British nuclear project was assumed to be known by the Soviets as well. Fingers were now pointed at the British alone.[13]

The year 1951 brought more revelations about Soviet penetration of the British-American nuclear project. Once again, it happened on the British side of the partnership. On May 26, 1951, two British civil servants took a ferry from Britain to France—the first leg of their escape to the Soviet Union. The two men, Donald Maclean, head of the American desk at the British Foreign Office, and Guy Burgess, a second secretary at the British embassy in Washington, belonged to the Cambridge Five, the group of left-leaning students at the University of Cambridge recruited by Soviet military intelligence in the 1930s. Maclean and Burgess, suspected of spying for the Soviet Union, were tipped off about the danger of arrest by another member of the Cambridge Five, Kim Philby, who served as the chief British intelligence representative in Washington. The escapees had access to American and British nuclear secrets.[14]

The public announcement about the escape was made on June 7, 1951. The Americans, the Pentagon in particular, had had enough. They now regarded the British as partners who could not be trusted with any secret information whatever. The proposal made to the British by the US chiefs of staff to exchange American bombs for British plutonium was withdrawn. Other parts of the proposed agreement giving British scientists access to American nuclear secrets were also in serious doubt. The talks were not broken off, but the optimism with which they had been pursued now disappeared. The British, facing growing restrictions, were not eager to make a deal either. Cooperation would only slow down their nuclear program without gaining them any useful American information.[15]

Soviet possession of the bomb made the Americans consider a policy of limited proliferation of technological data. But the Soviet espionage exposed by the scandals involving Fuchs, Pontecorvo, and the Cambridge Five put the brakes on plans for the joint British-American nuclear project. The British would have to build a bomb on their own with whatever information they had already gleaned from the Americans.

Winston Churchill, the World War II architect of British nuclear policy, returned to power in October 1951. Clement Attlee and his Labour Party lost the general elections, and the voters or, rather, the British electoral system—Labour still won the popular vote—brought back the Conservatives.

What Churchill did not expect was to inherit a fully developed atomic bomb project from Attlee. The two plutonium-breeding reactors at Windscale were now operational, and the chemical separation plant was scheduled to produce the first weapon-ready plutonium early in the following year. One separation plant for uranium-235 was operational, another under construction. All that came as a complete surprise to Churchill. Despite budget difficulties and economic hardship, Labour had managed, in complete secrecy from the opposition, to allocate enormous amounts of money to the project. All told, the British bomb would cost taxpayers £140 million, amounting to 23 percent of all expenditures of the Ministry of Supply in fiscal 1950–51, twice the percentage allocated in previous years. Those were enormous figures for the cash-strapped British government, but the overall cost was only about half of what the US government had spent on the Manhattan Project.[16]

Churchill spent six years in opposition believing that the Labour government had done the country a major disservice by not insisting on the continuation of the Quebec Agreement, which he had signed with Roosevelt back in 1943, and which had launched the American-British cooperation in the nuclear sphere. He considered the American bomb a

British-American project, and his favored solution to Britain's deterrence problem was to share atomic bombs with the Americans, not to build them at British expense. Churchill knew in what dire economic straits his country had been during the war, and after returning to No. 10 Downing Street he found that conditions were no better in 1951. Because of the Korean War, his predecessor had pushed defense spending from 7 to 14 percent of GDP, exacerbating the country's balance-of-payments crisis.

As Churchill familiarized himself with the progress made by the Labour government on the British nuclear project, he wrote to his former chief scientific adviser, Frederick Lindemann, now serving as Paymaster General, the government's money man: "I have never wished since our decision during the war that England should start the manufacture of atomic bombs." Churchill was in favor of limiting the British effort and expenditures to research and having "the art rather than the article." Thus his plan was to travel to the United States, resume the good relations that he believed to have been spoiled by Attlee's "socialists," never trusted by Washington, and convince Truman to provide the United Kingdom with "a reasonable share of what they have made so largely on our initiative and substantial scientific contribution."[17]

Churchill was clearly stuck in the past. That was what Lindemann took pains to explain to his superior, stating that the Quebec Agreement was long dead, renegotiated by the Labour government in a way that allowed Britain to develop a separate nuclear program. Unlike Churchill, who still cherished his special wartime relationship with Roosevelt, Lindemann well understood the nuclear realities of the postwar world. "If we are unable to make the bombs ourselves and have to rely on the American army for this vital weapon," explained Lindemann, "we shall sink to a rank of second class nation, only permitted to supply auxiliary troops like the native levies, who were allowed small arms but no artillery." Comparing Britain to a colony could not sit well with Churchill, and it took him a while to reply to Lindemann's letter, but when he did, he no longer insisted on asking the Americans to share their bombs with the British. Instead, he approved preparations for the test of a British atomic bomb scheduled for 1952.[18]

On his visit to Washington in January 1952, Churchill maintained full continuity with the policy established by his predecessor, Clement Attlee. He pushed for an enhanced information exchange. He also wanted Truman to apply to the United Kingdom for consent if he should wish to use an American air base in eastern England for nuclear bombing of the USSR—an act that would make Britain a legitimate target for Soviet nuclear retaliation. Churchill did not get far on any of those issues. Vague promises were made, but there was nothing in the discussions that suggested a return to the special partnership of the war years or, indeed, any achievement to compare with Attlee's record. It was a rude awakening.[19]

In February 1952, the month after his trip to Washington, Churchill declared in Parliament that Britain would test its first atomic bomb before the end of the year—a public gesture that, for different reasons, neither Truman nor Stalin had dared to undertake. The site for the blast had already been selected—the Montebello Islands off the coast of northwestern Australia. The Americans had turned down a British request to test the bomb in the Bikini Atoll area and insisted that they would be in charge of any test conducted in Nevada. The British could not accept that. But the bomb that they were about to explode was pretty much American in design—a plutonium implosion-type device with one significant addition, a levitated pit that increased the efficiency of the explosion and permitted the use of less plutonium to achieve the same effect.

The test of the British atomic bomb, code-named "Hurricane," took place on October 3, 1952. The bomb was not mounted on a tower, as the Americans and Soviets had done, but placed in the belly of the HMS *Plym*, a river class Royal Navy frigate of 1,450 tons built in 1943 and used for anti-submarine escorts of convoys across the Atlantic. Unlike the Americans and the Soviets, who considered their atomic bombs primarily offensive weapons, the British were concerned about the damage that a foreign bomb could do to them. The danger that a ship might bring a bomb into a British port was as much on their mind in 1952 as it had been in 1940, when the British first began to think about nuclear bombs and imagined their use against a port city. The HMS *Plym* was to be sacrificed for the greater good of Britain. The 25-kiloton atomic bomb exploded in the ship's hull 2.7 meters (8.9 feet) below the water line, leaving a crater in the seabed 6 meters (20 feet) deep and 300 meters (984 feet) wide. The ship itself was reduced to what one witness described as a "gluey black substance." As far as the authorities were concerned, the test had been an unqualified success.[20]

Churchill appeared in Parliament on October 23, 1952, to declare victory and praise his Labour predecessors for their development of the British nuclear program. "All those concerned in the production of the first British atomic bomb are to be warmly congratulated on the successful outcome of an historic episode and I should no doubt pay my compliments to the Leader of the Opposition and the party opposite for initiating it," he said. Asked about the cost of the whole undertaking, Churchill was unusually gracious: "As to the cost, I have said before, as an old parliamentarian I was rather astonished that something well over £100 million could be disbursed without Parliament being made aware of it. I was a bit astonished. However," he continued, "there is the story, and we now have a result which on the whole, I think, will be beneficial to public safety."[21]

Britain, the country that had given the world Rutherford and nuclear physics, that first imagined and split the atom, was now a member of the nuclear club. It was third in line to get the bomb, but the bomb it got was its own. The empire was in shambles, but the "bloody Union Jack" on which Foreign Minister Ernest Bevin had insisted in 1946 was right there on the British atomic bomb. Who would now dare to attack Britain or not call it a great power?

Chapter 13

MANAGING FEAR

General Dwight D. Eisenhower, the sixty-two-year-old former Supreme Commander of the Allied Expeditionary Forces in Europe, became the thirty-fourth president of the United States on January 20, 1953, after a landslide victory over his Democratic opponent, Adlai Stevenson. Under the campaign slogan "Korea, Communism, Corruption," he carried thirty-nine of the forty-eight states existing at the time. Eisenhower capitalized on public frustration with the Truman administration's inability to end the Korean War, which broke out in June 1950, and at one point saw the American troops in that country all but annihilated by communist China's armed forces. He also rode the anticommunist wave raised by the explosion of the Soviet atomic bomb and the trials of the atomic spies, especially Julius and Ethel Rosenberg. Last but not least, voters were unhappy with the state of the economy and the ballooning federal budget.[1]

On the campaign trail, Eisenhower promised a new foreign policy but delayed its formulation for more than a year. When he finally presented his foreign policy vision to Congress in January 1954, it combined two ostensibly incompatible features: an increase in US military potential and deep cuts to the existing military budget. The wonder tool that Eisenhower counted on to achieve both goals was rapid development of American nuclear capabilities. Now that the technology was fully developed, building bombs was cheaper than recruiting and maintaining standing armies. A soldier through and through, whether in uniform or not, Eisenhower believed during the first years of his presidency that the bomb was there to be used, another weapon in the country's arsenal.[2]

The most urgent foreign policy problem that Eisenhower encountered upon entering the Oval Office in January 1953 was the ongoing war in Korea, which he had promised to end with an honorable peace during his presidential campaign. But how was that to be achieved? By dropping the bomb? In December 1950, when Chinese forces pushed American troops deep into South Korea, Truman declared a national emergency and went on record suggesting that his commanders might use atomic bombs against the advancing Chinese. Almost immediately, Truman retracted his statement, claiming that he and he alone had the right to order the use of the bomb. Eventually, under pressure from Britain and other European allies, he backed away from the idea of using atomic bombs. The Europeans were not only concerned about the fate of their troops in Korea but also afraid that a nuclear attack on China would provoke Soviet retaliation in Europe.

In October 1951, with the war still going badly for the United States, Truman returned to the idea of using the atomic bomb. The US Air Force began training its pilots to deliver nuclear strikes against the Chinese, but once again the idea was abandoned. The American commanders decided that atomic bombs were ineffective against dispersed armies. The changing fortunes of war were also a contributing factor. A successful counterattack by American conventional forces stopped the Chinese advance, drove them back, and stabilized the situation.

Truman, who by now had serious moral reservations against the use of atomic bombs, felt relieved. "It is far worse than gas and biological warfare because it affects the civilian population and murders them by wholesale," wrote Truman about the atomic bomb in January 1953, a few days before leaving office.[3]

General Eisenhower, the new master of the White House, returned to the idea of using the atomic bomb in Korea. He had no moral qualms about using the atomic bomb in Korea as a way of achieving American military victory without ordering general mobilization in the United States, which would have increased the armed forces and further strained the budget. Eisenhower told the National Security Council on March 31, 1953, that he was not deterred by the absence of appropriate targets for bombing. "It would be worth the cost, if through use of atomic weapons," suggested the president, "we could (1) achieve a substantial victory over the Communist forces and (2) get to a line at the waist of Korea."

But there was one concern about the use of the bomb that Eisenhower could not dismiss: like Truman, he had to consider the reaction of America's European allies. Bombing the Chinese either in Korea or in China meant risking a war with China's ally, the Soviet Union, now a nuclear power. True, the Soviets had few bombs and no way of delivering them to the United States: their long-distance aviation was almost nonexistent, and their ballistic missiles not yet developed. But they could certainly hit European targets. To a civilian adviser who said that the United States should use a couple of bombs in Korea, Eisenhower responded: "Perhaps we should, but we could not blind ourselves to the effects of such a move on our allies, which would be very serious, since they feel that they will be the battleground in an atomic war between the United States and the Soviet Union."

No matter how much Eisenhower wished, according to the March 1953 protocol of the NSC meeting, "that somehow or other the taboo which surrounds the use of atomic weapons would have to be destroyed," he decided to forgo that option. The reason came down to one simple fact—the Soviet atomic bomb. The calculations of those American scientists who had wanted Washington to share its nuclear secrets with Moscow, and of the atomic spies who had stolen American secrets in the hope that possession of the bomb by more than one power would make its use unacceptable, were finally justified. The Soviet bomb became the most important obstacle to the use of the American bomb in Korea.[4]

How to deal with the threat posed by the Soviet atomic bomb was the biggest long-term challenge to President Eisenhower. In the spring and summer of 1953, he discussed with his aides the idea of a preemptive nuclear war against the USSR before Moscow could accumulate a larger atomic arsenal. The idea was rejected. Instead, Eisenhower chose the option of constant readiness, as he wrote to Secretary of State John Foster Dulles, "to inflict greater losses upon the enemy than he could reasonably hope to inflict upon us."[5]

The question was how to do so without bankrupting the United States. An idea was proposed in October 1953 by the chairman of the Joint Chiefs of Staff, Admiral Arthur W. Radford, who suggested that the chiefs were defending the existing budget because of their concern that they would be forced to fight a coming war without relying on nuclear weapons. If they knew that the president was open to the idea of using atomic bombs, then they could plan for smaller conventional forces and thus agree to budget cuts. The National Security Council seized on this idea to reach a compromise between security and budget concerns. Nuclear deterrence thus became a keystone of the new American foreign policy.

NSC policy paper 162/2, dated October 30, 1953, read: "In the event of hostilities, the United States will consider nuclear weapons to be as available for use as other munitions." It continued: "The risk of Soviet aggression will be minimized by maintaining a strong security posture, with emphasis on adequate offensive retaliatory strength and defensive strength. This must be based on massive atomic capability, including necessary bases." Continental and local defenses were mentioned as well, but nuclear weapons came first. Eisenhower approved the paper. The doctrine, later dubbed the "New Look," prioritized the development of nuclear weapons over conventional ones.[6]

In the summer of 1953 Eisenhower transferred control over a large part of the nuclear weapons stockpile from civil authority, where it had been under Truman, to the military. He would continue the transfer in the years to come.

The new national security strategy became known as the "New Look." The success of the "New Look" relied on fear. Fear of nuclear retaliation was used as a primary foreign policy tool abroad. At home, fear of the atomic bomb had to be managed in such a way as to rally taxpayer support for the government's nuclear buildup and civil defense programs. The domestic part of the policy found its fullest expression in Operation Candor, a public relations campaign orchestrated by the White House to "overcome current public apathy to Civilian Defense" by informing the public about the threat posed by the Soviet atomic bomb.

The Candor program was a response and a correction to the public relations campaign launched during the Truman era. In December 1950, Harry Truman created the Federal Civil Defense Administration, whose primary task was not so much to protect Americans from a possible nuclear attack as to get them used to the idea that it was possible and that one could survive it. Preventing the panic that would follow the explosion of a Soviet bomb was the main purpose of the administration's public campaign. Americans, believed the initiators of the campaign, were individualistic, chaotic, and prone to panic. Something had to be done to improve their ability to withstand the psychological shock of an atomic explosion, keep up their morale, and sustain the nation's capacity to fight a nuclear war.[7]

The challenge was addressed head-on by a government-funded project called "East River." Launched in 1951, the project delivered its report in early 1953. "The very real possibility exists," read the report, "that in the event of sudden attack, mob action could readily break out. There are convenient scapegoats, and the nature of the world situation has made inevitable a long accumulation of anxiety and frustration toward Russia which could easily be translated into senseless aggression." The project warned of the possibility of "aggression toward merely unorthodox persons as symbols of communism."[8]

The solution to the panic problem, argued the authors of the report, was to neutralize nuclear terror, and utilize nuclear fear. "Civil defense education," according to a section of the report, "must make people aware that a considerable degree of fear under attack is normal and unavoidable." The task was to turn fear into a positive factor. "The fear you experience makes you more alert, stronger and more tireless for the things that you and your neighbors can do to protect yourselves." This basic message of civil defense education was to be delivered by means of massive state propaganda.[9]

Duck and Cover, a cartoon released in January 1952, is the US government's best-known civil defense training film. Its main character, Bert the Turtle, advises children to duck and cover once they see the flash of an exploding atomic bomb. Hiding under a classroom desk, as the film encouraged students to do, could help those at some distance from the epicenter of the blast but, most importantly, it trained Americans not to be afraid of an atomic bomb. An atomic explosion could smash windows as a hurricane does, or burn the skin as the sun does, but ducking behind a wall and covering one's head with one's arms could prevent serious harm.[10]

Many in the nuclear establishment believed that *Duck and Cover* did too good a job of calming American nerves. In January 1953 a panel of foreign policy and nuclear experts led by Robert Oppenheimer issued a set of recommendations to the government on the future of American nuclear policy. One of them, titled "Candor to the American Government and People," read: "The United States Government should direct public attention specifically and repeatedly to the fact that the atomic bomb works both ways. . . . Official comment on atomic energy has tended to emphasize the importance of the atomic bomb as part of the American arsenal. There is an altogether insufficient emphasis upon its importance as a Soviet weapon, and upon the fact that no matter how many bombs we may be making, the Soviet Union may fairly soon have enough to threaten the destruction of our whole society."[11]

Oppenheimer's report, which predicted a growing danger of nuclear confrontation as the United States and the USSR acquired more bombs, made a strong impression on Eisenhower and his administration, which cleared an article written by Oppenheimer for publication in *Foreign Affairs*. It compared the Soviets and the Americans to "two scorpions in a bottle, each capable of killing the other, but only at the risk of his own life." But when it came to showing the "candor" advised by Oppenheimer and his colleagues, Eisenhower was skeptical at first. Like advisors to Harry Truman, he was deeply concerned that publicizing the dangers of nuclear war too openly would cause panic. But Eisenhower's advisers on the National Security Council thought differently, believing that "candor" would "secure support of the American people for necessary governmental actions." The president eventually went along.

In the summer of 1953 the Advertising Council, created by President Roosevelt to mobilize the resources of the advertising industry in World War II, launched its Cold War Candor campaign. It included six fifteen-minute radio and television talks by Eisenhower and senior members of his administration. Fear of nuclear annihilation was to be introduced in the talks in a way that would mobilize public support for nuclear arms, not provoke panic or turn the audience against nuclear weapons altogether. The script for Eisenhower's concluding remarks to the series read: "Age of peril demands patience, determination, fortitude."[12]

By the fall of 1953, Eisenhower was prepared to bring his "candor" campaign not only to the American public but to the whole world. He wrote to Dulles: "Programs for informing the American public, as well as other populations, are indispensable if we are to do anything except to drift aimlessly, probably to our own eventual destruction." Eisenhower thought about a public campaign to explain to the world the growing danger of nuclear war as atomic bombs were about to be joined by thermonuclear (hydrogen) devices.[13]

The Americans were not the only ones who feared atomic weapons. While they and their European allies were afraid of Soviet atomic bombs, a good part of the world feared American ones. That was true even, or especially, of American allies in Europe, who were concerned about Soviet retaliation if nuclear weapons became part of the NATO arsenal. The president had to calm their fears as well.[14]

Eisenhower chose the United Nations as a venue to sell the American nuclear vision to the world. Its main parameters were outlined in an address that gained notoriety as the Atoms for Peace speech. The origins of its principal theme—the peacetime benefits to be gained from the atom—was rooted in Eisenhower's desire to change the public's perception of nuclear as a source of fear. "We don't want to scare the country to death," he told the former media executive C. D. Jackson, who brought him the first draft of a proposed speech to the United Nations General Assembly. Jackson went back to the drawing board and turned what had been a mere footnote—the peaceful uses of atomic energy—into the main focus.[15]

On December 8, 1953, Eisenhower addressed the General Assembly with what was at least the sixth draft of his speech. He began with his "educate the public" message. "I feel impelled to speak today in a language that in a sense is new, one which I, who have spent so much of my life in the military profession, would have preferred never to use," he told the assembly. He used the opportunity to assure the public at home and abroad that if an atomic attack were launched against the United States, "our reactions would be swift and resolute." A retaliatory strike would lay waste the enemy's land. That was a clear formulation of American deterrence strategy, but Eisenhower declared that he was not satisfied with such a solution. "Surely no sane member of the human race could discover victory in such desolation," continued the president. But if not nuclear deterrence, then what?

Eisenhower volunteered to lead the world to a better solution. "The United States would seek more than the mere reduction or elimination of atomic materials for military purposes," he declared, rejecting the long-standing Soviet position on banning nuclear weapons. The United States had opposed a ban when it had a monopoly on the bomb and continued to oppose it afterward, when the bomb was considered the only deterrent to keep the numerically superior Soviet forces at bay in Europe. Eisenhower now went on to say: "It is not enough to take this weapon out of the hands of the soldiers. It must be put into the hands of those who will know how to strip its military casing and adapt it to the arts of peace."

This was to be accomplished by the creation of an Atomic Energy Agency under UN auspices. The United States and the Soviet Union, suggested Eisenhower, would "begin now and continue to make joint contributions from their stockpiles of normal uranium and fissionable materials to an international atomic energy agency." How would that fissionable material be managed? Eisenhower had an answer: "Experts would be mobilized to apply atomic energy to the needs of agriculture, medicine and other peaceful activities." He added: "A special purpose would be to provide abundant electrical energy in the power-starved areas of the world."[16]

Eisenhower's speech touched a nerve. Many people were sold on the prospect of a socioeconomic revolution powered by atomic energy, although even more were terrified that nuclear weapons might destroy civilization. In the years and decades to come, the Atoms for Peace speech would acquire truly mythical status. Taken out of the context in which it was conceived and delivered, the speech would be presented as a bold new vision for the development of peaceful nuclear energy. In reality the idea was not entirely new, and the intentions behind the speech were not entirely peaceful. It was a continuation of the "educate without scaring" tactic embodied in Operation Candor at home, now offered to a foreign audience. Eisenhower's proposal harked back to the Acheson-Lilienthal and Baruch plans of the Truman era, which envisioned the creation of a UN Atomic Energy Agency or Authority, although it was adjusted to new conditions.

Key among those conditions was the loss of the American monopoly on atomic weapons, as well as the discovery of new deposits of uranium throughout the world. Consequently, the proposed new agency would not control uranium deposits—there were simply too many of them. Nor would the agency control atomic bombs, as no country, particularly the United States, was prepared to "donate" them to an international body. Instead, the new UN agency would concern itself exclusively with the peaceful development of nuclear energy. Even this proposal was not as benign as it appeared. Eisenhower and his advisers knew that the Soviet Union would strive to match the American contribution to the international agency, although that might well reduce the resources for their military. Thus, Eisenhower's proposal was a calculated move intended to arrest or slow down the development of the Soviet atomic program—a way to maintain American nuclear superiority in the absence of nuclear monopoly.

The other purpose of Eisenhower's speech was to offer hope to those concerned about an arms race that atomic energy could promote economic and social development in their countries. Over time, that was perceived as the main legacy of the speech. In the months and years to come, as Eisenhower became ever more worried about the proliferation of atomic weapons, he would turn to an old idea of the Truman era, first formulated by the authors of the Acheson-Lilienthal plan: in exchange for not acquiring nuclear weapons, countries would be offered assistance in the development of nuclear energy for peaceful purposes. That offer was not made in the Atoms for Peace speech, although it would be conflated with later mythologizing.[17]

The response to Eisenhower's speech was quite positive, not only from the British but also from the Soviet side, but it was not until July 1957 that the International Atomic Energy Agency became a reality. Not surprisingly, it was prevented from accepting donations of fissile materials. The IAEA would become a tool in the hands of the United

States, the Soviet Union, Britain, and, later, other nuclear powers seeking to discourage others from following in their footsteps and developing nuclear weapons.[18]

America's new nuclear policy was explained to the country and the world a few weeks after the Atoms for Peace speech. "American freedom is threatened so long as the world Communist conspiracy exists in its present scope, power and hostility," said Dwight Eisenhower as he began the foreign policy section of his first State of the Union address on January 7, 1954. He continued: "We shall not be aggressors, but we and our allies have and will maintain a massive capability to strike back." He meant a nuclear capability. Eisenhower regarded American nuclear weapons first and foremost as the ultimate deterrent to any potential aggression by the USSR and its allies anywhere in the world. But for nuclear weapons to work as a deterrent, the threat to use them had to be made publicly, as Eisenhower did in his address.[19]

The principles of the new nuclear policy had been laid down by the president, but it fell to his secretary of state, John Foster Dulles, to spell them out. He did so in a speech to the Council on Foreign Relations in New York on January 12, five days after the president's address. "We keep locks on our doors, but we do not have an armed guard in every home," Dulles told the audience. "We rely principally on a community security system so well equipped to punish any who break in and steal that, in fact, would-be aggressors are generally deterred." The foundation of Dulles's "community security system" was the threat of retaliation. "Local defenses must be reinforced by the further deterrent of massive retaliatory power," he argued.[20]

There was no mention of nuclear weapons in the published version of Dulles's speech, but his reference was clear. The doctrine of "massive retaliation" against communist aggression in any form, not just a Soviet nuclear strike or major military operation, was now implanted in the public mind, and Dulles became its prophet. James Reston, the Washington bureau chief of the *New York Times,* commented on Dulles's speech in the January 16 issue of the newspaper: "as clear as the Government ever says these things, in the event of another proxy or brushfire war in Korea, Indochina, Iran or anywhere else, the United States might retaliate instantly with atomic weapons against the U.S.S.R. or Red China." That was not what Dulles wanted to say, and he explained later that the response would be not automatic, but the perception was already there.[21]

For Eisenhower, the threat of massive nuclear retaliation in response to possible aggression by the USSR against its European neighbors became a useful foreign policy instrument, but it did not solve all his problems. His big concern remained the possibility of a surprise Soviet nuclear attack. Less than a year and a half into his presidency, he came to the shocking conclusion that the United States would be better off if there were no nuclear weapons in the world.

"The matter of the morality of the use of these weapons was of no significance," stated Eisenhower at a National Security Council meeting in June 1954. "The real thing was that the advantage of surprise almost seemed the decisive factor in an atomic war," recorded the notetaker. Eisenhower explained that "he was certain that with its great resources the United States would surely be able to whip the Soviet Union in any kind of war that had been fought in the past or any other kind of war than an atomic war."[22]

By mid-1954, Eisenhower had developed what looked like buyer's remorse. While his doubt about the benefits of the atomic bomb was striking, his line of thinking was hardly original. It followed the logic advocated by Leo Szilard and his fellow scientists in the early summer of 1945, when they tried to stop the Truman administration from first testing and then using the atomic bomb in the war against Japan. "Within ten years other countries may have nuclear bombs," wrote the scientists, "each of which, weighing less than a ton, could destroy an urban area of more than ten square miles (26 square kilometers). In the war to which such an armaments race is likely to lead, the United States, with its agglomeration of population and industry in comparatively few metropolitan districts, will be at disadvantage compared to the nations whose population and industry are scattered over large areas."[23]

Eisenhower had enormous power at his disposal but none to change the past. His solution would be that of building more bombs. His desire to get rid of the bombs, as they made the United States more vulnerable than it would otherwise have been, was never revealed to the public during his presidency. During his two presidential terms, Eisenhower presided over the largest nuclear buildup in world history. He inherited 841 bombs from the Truman administration. That number grew to 2,422 by 1955 and 18,638 by 1960. The Soviets, by contrast, had 50 bombs in 1952, 200 in in 1955, and 1,605 in 1960.[24]

The president's actions were informed by the passing atomic age, defined by the destructive power of the atomic bomb, while his thinking about the consequences of the coming nuclear confrontation and much of the fear he was trying to manage were informed by the new thermonuclear age, defined by the unlimited destructive power of the hydrogen bomb. Eisenhower was never able to reconcile the two.

Chapter 14

SUPER BOMB

March 1, 1954, turned out most unfortunately for the twenty-three men on board the Japanese trawler *Lucky Dragon* 5. In fact, the whole voyage—the fifth one for the boat, which had been built in 1947—was devoid of luck. The fishermen left Japan on January 22, heading for the Midway Atoll in the Hawaiian archipelago, roughly halfway between Asia and America, but lost half their nets to ocean coral. They then headed southeast toward the Marshall Islands, and early in the morning of March 1 they cast their nets. As their food supplies were low, it was supposed to be their last day of fishing.

The *Lucky Dragon* was approximately 80 miles (129 kilometers) east of the Bikini Atoll and 14 miles (23 kilometers) outside the 57,000-square-mile (148,000-square-kilometer) "danger zone" designated by the American authorities and centered on the atoll, where the United States had tested its atomic bombs since the summer of 1946. At 6:45 a.m. the men of the *Lucky Dragon* suddenly saw what they later described as "the sun rising in the West." A fireball 4.5 miles (7.2 kilometers) in diameter shot into the sky, illuminating everything on the deck of the boat and its surroundings. Eight minutes later the fishermen heard a roar, again coming from the west. They were surprised but not alarmed. They continued fishing, and indeed nothing unusual happened until about 10:00 a.m., more than three hours after the appearance of the "western sun." White particles—pulverized coral dust that looked to some like snow—soon covered the deck of the boat. One of the fishermen licked the dust but could not determine its origin. Others started to clean the deck by collecting the dust in bags. Nausea and bleeding from the gums began hours after the "snowfall." The men were clearly sick. They would call the "snow" on the deck of their vessel "death ash."[1]

The *Lucky Dragon* returned home on March 14, and the fishermen were examined at Tokyo hospitals. Japanese doctors who had treated the *hibakusha*, survivors of the bombings of Hiroshima and Nagasaki in 1945, had little doubt that the fisherman were suffering from acute radiation syndrome. One of them would die in September 1954 of complications related to treatment; others would spend months in the hospital. The Japanese turned to the US Atomic Energy Commission for explanations and help in defining the chemical composition of the fallout in order to treat the sufferers. No assistance was forthcoming. Lewis Strauss, the head of the AEC, refused to provide any data, suggesting instead that the "death ash" had nothing to do with the nuclear explosion. The fishermen, he said, had been in the danger zone spying for the Soviet Union, and now they were trying to discredit the US administration in the eyes of the international community. None of that was true.[2]

In fact, on March 1, 1954, the United States had tested a new nuclear bomb that yielded an explosion equal to a stunning 15 megatons of TNT, approximately 2.5 times the projected result. No one knew what to do with such an amazing weapon: the explosion was 1,000 times more powerful than those that had destroyed Hiroshima and Nagasaki. The bomb tested in the Bikini Atoll that day was not of the usual atomic variety but a hydrogen super bomb.[3]

The motive for the explosion of the hydrogen bomb in the Bikini Atoll lay in the panic caused in Washington by the unexpected test of the Soviet atomic bomb in September 1949, four and half years earlier. That event changed perceptions of national security among the American political and military elite. The Joint Committee on Atomic Energy met to assess the new situation on the day Truman made his announcement about the Soviet test, September 23, 1949. An increase in the production of atomic bombs was on the agenda. But was that enough? Some believed that the United States needed a super bomb one thousand times more powerful than the existing atomic bombs.[4]

It had been understood for decades that the fusion reaction produced by the heavy isotopes of hydrogen was much more powerful than the fission reaction. As early as the 1920s it had been regarded as a source of energy comparable to that of the sun and the stars. In 1934 Ernest Rutherford and two of his assistants, Mark Oliphant and Paul Harteck, working at Cambridge University, discovered the hydrogen fusion reaction. It was called a thermonuclear reaction because it required heating of the atomic nucleus. The reaction could not be achieved by bombarding atoms: the nuclei had to be brought together to initiate it, which required an enormous amount of energy and extremely high temperatures. No one knew how to produce such energy and temperatures until the idea of building an atomic bomb became a practical possibility.[5]

The first to suggest that an atomic bomb could be used to "ignite" the thermonuclear fusion reaction was Enrico Fermi. In September 1941 he shared his idea with a fellow émigré scientist, Edward Teller, a participant in Leo Szilard and Albert Einstein's "alert the government" initiative back in 1939 and in the wartime Manhattan Project. Originally skeptical of Fermi's idea, Teller eventually became all but obsessed with it. In the summer of 1944, Teller jumped at the opportunity to leave his position in the Theoretical Division at Los Alamos, where he had worked on problems involving the atomic bomb, to lead a small group of scientists working on the hydrogen bomb, which he called a "super" weapon. His place in the Theoretical Division was taken by British scholars, one of whom was Klaus Fuchs, the Soviet star agent at Los Alamos. The results of Teller's new work and the prospects of building his "super bomb" were discussed at a conference held at Los Alamos in April 1946. The issues addressed were the design of the bomb and possible thermonuclear fusion materials. Most of those who attended did not believe that production of a thermonuclear reaction was anything more than a theoretical proposition. Teller went back to research and teaching at the University of Chicago.[6]

But the idea of thermonuclear reaction was not completely forgotten. Back in 1944 Oppenheimer had proposed boosting the explosion of an atomic bomb by means of a fusion reaction. The fission atomic bomb explosion was supposed to trigger a modest thermonuclear reaction of a mixture of deuterium and tritium gas, which in turn would boost the yield of the fission explosion. In 1948 a program for the creation of a hybrid fission-fusion bomb was approved, with 1951 as the target date for its first test. Since the product would not be a hydrogen bomb per se, it did not satisfy Teller's ambition to build one, nor did it alleviate the concern of government officials that the Soviets might be ahead of the Americans in building their own "real" thermonuclear weapon.[7]

Few people were more responsive to the super bomb idea than Lewis Strauss, a member and the future chairman of the Atomic Energy Commission. On October 5, 1949, less than two weeks after Truman's statement on the Soviet atomic explosion, Strauss presented his fellow members of the commission with a memorandum: "It seems to me that the time has come for a quantum jump in our planning (to borrow a metaphor from our scientist friends)—that is to say, we should now make an intensive effort to get ahead with the Super." He believed that a new crash program of the Manhattan Project type had to be initiated if the United States were to stay ahead of the Soviets: producing more atomic bombs was not enough.

Very soon it became a key argument in favor of the Super Bomb program that not only did the Soviets have the atomic bomb, but they were about to acquire a super bomb as well, or perhaps already had one. The idea that the Soviets were working simultaneously on the A-bomb and the H-bomb was suggested by Teller and shared by some other scholars, including the father of the cyclotron, Ernest Lawrence of Berkeley. There was no evidence in the possession of the US government or scientists at the time to support that idea, but by then it did not matter. Brian McMahon, the chairman of the Congressional Joint Committee on Atomic Energy, sent a memo to the AEC and the Pentagon in which he wrote: "As you know, there is reason to fear that Soviet Russia has assigned top priority to development of a thermo-nuclear super-bomb." He suggested that the Soviets had had the bomb for two years, and "unusual and even extraordinary steps" would be required to catch up with them.[8]

In the atmosphere of panic created by the unexpected Soviet test, one could hardly risk underestimating the success of Soviet bomb development. In a way, history was repeating itself. Back in 1940 scientists had pushed for a crash program to build an American bomb, since they feared that Hitler might get a bomb first. Now, some of the same people were arguing for the production of a hydrogen bomb because the Soviets might be ahead of the United States in that project. Although the argument was compelling, the atomic scientists were no longer of one mind on the issue. Oppenheimer in particular emerged as the leader of the skeptics and those who considered the building of a new, more powerful bomb militarily useless, politically dangerous, and morally wrong.

At the deliberations of the General Advisory Committee, chaired by Oppenheimer, on October 28–30, 1949, its members decided to advise the Atomic Energy Commission against developing a hydrogen bomb. James Conant, president of Harvard University, set the tone for the discussion of the moral issues involved in producing a "super bomb" when he muttered to himself, "We built one Frankenstein." He did not finish the phrase, but it was clear that he was not eager to build another, even more dreadful weapon. In an addendum to the committee's statement, Conant wrote: "A super bomb might become a weapon of genocide."[9]

Three members of the Atomic Energy Commission were quite influenced by the arguments of the scientists, but the other two, including most notably Lewis Strauss, disagreed. Strauss appealed to President Truman directly. In a letter and memorandum submitted to the White House on November 25, he claimed that the building of a hydrogen bomb was possible and necessary. Judging by the Soviet atomic bomb test, he suggested, the USSR had the technical capability to build a thermonuclear bomb as well. And on the question of morality, he had an argument of his own: "A government of atheists is not likely to be dissuaded from producing the weapon on 'moral' grounds." In short, Strauss wanted Truman to order the commission, in which he found himself in the minority, to undertake a crash program to build the bomb. To the claim of Oppenheimer and other opponents that the new hydrogen bomb would be so powerful as to destroy anyone trying to use it, the proponents of the new bomb had a familiar answer: it was needed for deterrence in case the other side produced a super bomb of its own.[10]

Truman, who knew nothing about the hydrogen bomb until sometime in October 1949, asked Secretary of State Dean Acheson to chair a committee that would look into the issue. The Acheson Committee recommended that Truman approve the building of the bomb. He wasted no time. On the day he received the recommendation, January 31, 1950, Truman announced: "I have directed the Atomic Energy Commission to continue its work on all forms of atomic weapons, including the so-called hydrogen or super-bomb." Although he was concerned about the moral implications of the atomic bomb and his decision to use that weapon against Japan, Truman seemed quite undisturbed at taking another step on the slippery slope of nuclear arms development. "We had got to do it—make the bomb—though no one wants to use it," he remarked soon after making the announcement. "We have got to have it if only for bargaining purposes with the Russians."[11]

Edward Teller could rejoice. Together with Ernest Lawrence, he took an active part in the work of the Atomic Energy Commission. Through its dissenting members, especially Strauss, he helped to persuade President Truman to start a crash development program on the hydrogen bomb. But Teller confronted a major scientific problem in making his recommendation: among scientists, doubters about the feasibility of a super bomb significantly outnumbered its supporters. It was argued that deuterium, or hydrogen-2, one of the two stable isotopes of hydrogen, could not do the job because of projected radiation losses. Adding tritium, a radioactive isotope of hydrogen, to deal with the problem did not seem practical either, as it was an expensive element to produce, and no one knew how much of it would be needed.

The skeptics included Stanislaw Ulam, a forty-year-old mathematician and, like Teller, a refugee from Eastern Europe. Ulam came from Poland, where he had begun his studies in mathematics at the Lviv Polytechnical Institute, in today's Ukraine. He arrived in the United States in 1935 and took part in the Manhattan Project, helping to make calculations for George Kistiakowsky's explosion lenses, the design of which was partially stolen and passed to the Soviets by Greenglass. In early 1950 Ulam was invited back to Los Alamos to help build the hydrogen bomb. His calculations disproved Teller's earlier hypothesis that tritium would ignite deuterium and start a self-sustaining reaction: bad news for Teller and those in the government who backed his project.

But in January 1951 Ulam had a "eureka" moment: it occurred to him that a fusion reaction could start at lower temperatures if the fuel material were compressed by the mechanical shock produced by the explosion of an atomic bomb. That would make super-expensive tritium unnecessary and ensure a self-sustaining chain reaction with the aid of deuterium, which was much cheaper and easier to obtain. Ulam shared the idea with his wife, then with a colleague, and finally broke it to Teller. Teller immediately appreciated the importance of Ulam's proposal, accepting it and proposing improvements. The Ulam-Teller design of the hydrogen bomb was born. The rest, given the experience of the Los Alamos laboratory in building atomic bombs, was a matter of time and effort. There was no lack of the latter, but many believed that time was in short supply: the Soviets might get a hydrogen bomb of their own first, argued Teller, and many believed him.[12]

Teller's concerns were not entirely unfounded. The Soviets began their research on thermonuclear reaction in 1946, apparently in response to reports on similar work conducted at Los Alamos, and pursued it concurrently with their atomic bomb project.

In 1948, when Oppenheimer decided to start building a boosted fission weapon, in which an atomic bomb explosion would ignite fusion fuel material to boost its yield, the Soviets came up with an idea for their own fission-fusion bomb. Its author was the twenty-seven-year-old Moscow physicist Andrei Sakharov, the future human rights activist, dissident, and father of the Soviet liberal movement. Sakharov called his design a "Layer Cake" consisting of layers of tritium and deuterium on the one hand and uranium-238 on the other, placed between additional layers of plutonium and conventional explosives. Ignited by a fission explosion, a fusion reaction would in turn create a fission reaction in uranium-238, which for its part would enhance the thermonuclear reaction.

Sakharov soon found a way to produce a thermonuclear reaction without tritium, which was expensive and hard to handle. Both tritium and deuterium were to be replaced with the much more easily produced lithium deuteride. By the end of 1948, the Soviets had a workable design for a fission-fusion bomb. As they did not yet have an atomic bomb, it was too soon to produce a thermonuclear bomb dependent on the success of the atomic project. The design more closely resembled an American boosted-fission bomb than a hydrogen device, but it was a thermonuclear weapon nevertheless. Thus, in a matter of speaking, Teller and those who believed him were right: the Soviets were ahead of their American counterparts when it came to research on a thermonuclear bomb.

In August 1949, after the successful test of the Soviet atomic bomb, Igor Kurchatov, the scientific director of the Soviet nuclear project, gave his subordinates a new task before allowing them to take their first break in five years. Once they were back, he told his scientists, their next project would be the hydrogen bomb. Sakharov, the author of the

Layer Cake, was reassigned along with his colleagues from his Moscow laboratory to Sarov, the Soviet analogue of Los Alamos, in March 1950. Layer Cake was moved from the back burner to the center of the Soviet nuclear stove. As Sakharov was still working on his calculations along with scores of other mathematicians and physicists, Lavrentii Beria's secret-police empire went into overdrive, building plants for the industrial production of deuterium and lithium-6, which were needed for the production of lithium deuteride, the Layer Cake's fusion fuel. It would take a while to build those facilities, giving the Americans enough time to catch up and even get ahead.[13]

On November 1, 1952, the Americans tested their first hydrogen device on Elugelab Island in the Eniwetok Atoll of the Pacific Ocean. It was a hydrogen complex rather than a bomb. The explosive device itself was 6 meters (20 feet) in height and 2 meters (7 feet) in diameter. Weighing 82 tons, it became known as the "sausage." An entire complex of tubes and equipment, needed to keep liquid deuterium at very low temperatures and weighing close to 20 tons, was built around it. Much of the structure around the core of the "bomb" was meant to provide data about the process rather than to facilitate it. Edward Teller, who became known as the "father of the hydrogen bomb," was not present at the test. He learned about it by reading data from a seismograph in the Livermore Radiation Laboratory at the University of California, where he worked.

The "Ivy Mike," as the test of the first thermonuclear device was code-named, was a huge success. The yield of the explosion equaled more than 10 megatons of TNT, 450 times more powerful than the Nagasaki bomb. The blast left a crater on the site of Elugelab Island, where the thermonuclear installation had been constructed. A white tree-shaped cloud emerged above the obliterated island. The cloud was 50 kilometers (31 miles) high, with a crown more than 100 kilometers (62 miles) in width. The success of Ivy Mike was kept secret from the Soviets and the world at large. The explosion proved the accuracy of the concept proposed by Ulam and Teller, but a building of the size created for Ivy Mike was not a bomb that could be dropped on a potential enemy. A usable thermonuclear weapon required a fuel different from liquid deuterium, one that did not need 20 tons of equipment to make it work. The solid fuel to be used in the first American hydrogen bomb was lithium deuterium, the same fuel suggested by Soviet scientists back in 1948.[14]

In August 1953, as American scientists were still working on their first deliverable lithium deuterium bomb, the Soviets exploded their thermonuclear weapon in the Kazakh steppes. It was of Sakharov's Layer Cake design and thus more a boosted fission bomb than a true fusion bomb, which the Americans had already figured out how to make. The yield of Layer Cake was equal to 400 kilotons of TNT, a tiny fraction of Ivy Mike's yield. Only between 15 and 20 percent of the yield was generated by fusion; the rest was achieved in the old-fashioned way, by fission reaction. That feature of Layer Cake could not be significantly improved: most of the yield of boosted fission bombs had to come from fission reaction. The Soviets had yet to build their first true thermonuclear bomb.[15]

The Americans outdid the Soviets once again on March 1, 1954, when they exploded their first thermonuclear bomb. The test, conducted at the Bikini Atoll and code-named Castle Bravo, produced a blast of 15 megatons, thirty-seven times more than the Soviet Layer Cake, coating much of the world in radioactive dust. The Japanese fishermen of the *Lucky Dragon*, referred to at the beginning of this chapter, were not the only victims of the explosion. Sailors on a number of American ships in the area were exposed to high levels of radiation. Fifteen nearby islands and atolls were affected by radioactive fallout, and the inhabitants of two were evacuated. The area would see a spike in cases of thyroid cancer by the early 1960s.[16]

On April 2, 1954, as the world was digesting the news about the radiation-stricken Japanese fishermen and natives of Pacific atolls located hundreds of kilometers from the epicenter of the hydrogen bomb explosion, American television viewers were treated to an edited version of the film documenting the explosion of the first hydrogen device back in 1952. Robert Cutler, Eisenhower's special assistant for national security affairs, argued that civil defense authorities should "scare the American people out of their indifference" into treating seriously the instructions they received from the government on protecting themselves from possible nuclear attack.[17]

The film served the Eisenhower administration's political purposes, but it also did not come a moment too soon. It was in the spring of 1954 that the Soviets finally cracked the secret of building their very own hydrogen bomb. The solution to the problems of ignition and sustainable reaction discovered by Andrei Sakharov and his group of physicists and mathematicians was akin to the one proposed by Ulam and Teller, although there is no indication whatever that it was stolen from the Americans. The Soviets reached the same conclusions and solutions to their problems as their American rivals, although they did so three years later. An indication that the Soviet ideas were home-grown is that the leaders of the Soviet nuclear industry, who were in charge of the secrets stolen from Los Alamos, wanted their scientists to improve the Layer Cake model, which was beyond improvement. The top scientific director of the project, Igor Kurchatov, backed Andrei Sakharov's solution and managed to win the turf war with the bureaucracy.

The bomb was ready in the fall of 1954 and tested successfully in November of that year. The yield, equivalent to 1.6 megatons of TNT, was designed to be low for test reasons. The Soviets now had the same technology as the United States. The American atomic bomb monopoly had lasted four years, their hydrogen bomb monopoly a year and a half. By the end of 1955, ten years after the start of the atomic age and the atomic weapons race, the two superpowers had the technology to destroy each other and the rest of the world as well. They did not yet have enough bombs to do so, but producing them was only a matter of time. Some serious thinking was required on both sides to decide what to do with that power.[18]

The first Western leader to voice his concern about the hydrogen bomb publicly was Winston Churchill. He reacted to the news about the Soviet test of Sakharov's Layer Cake bomb in August 1953: "We were now as far from the age of the atomic bomb as the atomic bomb itself from the bow and arrow." The news about American possession of a deliverable hydrogen bomb, released on the eve of the Castle Bravo test, did not make Churchill any more optimistic. He later talked about the H-bomb revolutionizing "the entire foundation of human relations" and putting mankind "in a situation both measureless and laden with doom." Privately, he suggested: "Even if some of us temporarily survive in some deep cellar under mounds of flaming and contaminated rubble, there will be nothing left to do but to take a pill to end it all."[19]

The rest of the world was equally disturbed by the news emerging from the Castle Bravo explosion. The story of the irradiated crew members of the *Lucky Dragon* broke on world news agencies with the force of a Pacific tsunami. The hospitalization of the sailors and the death of one of them while under treatment proved something that Americans living near the Nevada Proving Grounds, used for nuclear tests since the late 1940s, suspected but could not prove—nuclear fallout could affect people hundreds of miles away, causing serious health problems and even death. Those were frightening thoughts. Out of them was born the Japanese 1954 blockbuster movie *Godzilla*, a monster reptile lifted from the ocean bottom by radiation to crush everything in its path. Godzilla became a stunning embodiment of the fear created by the hydrogen explosion.[20]

Atomic scientists pointed out that hydrogen bombs, unlike atomic ones, scattered fallout around the globe, endangering people in many countries. That was the gist of an article published in the *Bulletin of the Atomic Scientists* in November 1954. The AEC issued its own report, confirming the danger posed by fallout radiation and admitting that no one was in a position to determine its effects or judge what doses should be considered harmless. In July 1955, an appeal signed by the world-famous British philosopher Bertrand Russell and nine scientists, including Albert Einstein, hit the media. It was addressed to world governments, urging them to seek peaceful resolutions of conflict in an era defined by nuclear weapons.[21]

The Russell-Einstein manifesto warned the public about the danger of nuclear war in the age of the hydrogen bomb. "No one knows," wrote the scientists, "how widely such lethal radio-active particles might be diffused, but the best authorities are unanimous in saying that a war with H-bombs might possibly put an end to the human race." They seized on an old idea first expressed by H. G. Wells and then advanced by proponents of the world government and atomic scientists alike: the only way out was to eliminate war as such, which would require the limitation of national sovereignty. They wanted to abolish thermonuclear weapons, meaning hydrogen bombs, and hold a scientific conference to discuss the perils unleashed by the development of nuclear weapons.[22]

In the United States, President Eisenhower was preparing for the worst. In July 1956, assessing the results of a civil defense exercise called Operation Alert, Eisenhower warned his subordinates that it was a mistake to expect fully rational behavior of government officials, who would be as terrified as everyone else. "We are simply going to have to be prepared to operate with people who are 'nuts,'" declared the president. He wanted his aides to decide who would bury the dead. The man who had won the war in Europe for his country a decade earlier had no illusions about what nuclear war might bring and what the real situation on the home front might resemble. The world had entered the hydrogen age.[23]

Chapter 15

MISSILE GAP

In 1953, the "clock of doom" launched in 1947 by the publishers of the *Bulletin of the Atomic Scientists*, allegedly on the advice of Edward Teller, to indicate how close the world had come to nuclear annihilation was set at two minutes to midnight, the point identified as the hour of doom. The reason was simple—the arrival the previous year of Teller's thermonuclear device tested in the Ivy Mike shoot. The Castle Bravo test of an actual hydrogen bomb in 1954 ensured that the hands of the clock stayed in the same ominous position for years to come. It was just as the nuclear threat to the world reached its highest level that signals of a possible detente between East and West began to arrive from Moscow.[1]

Joseph Stalin died on March 5, 1953, raising both hopes and fears about the direction of Soviet foreign policy. The most powerful figure that emerged from Stalin's shadow in the first weeks and months after his death was the dictator's security tsar and chief of his nuclear arms project, Lavrentii Beria. But he would not last long: considered a threat to his colleagues, he was arrested in June 1953 and executed before the end of the year. Next in line to take the reins of the Soviet government was the fifty-one-year-old apparatchik Georgii Malenkov. On the day of the Kremlin coup that saw Beria arrested, Malenkov signed a decree subordinating the atomic project to himself.

The successful test of the first Soviet semi–hydrogen bomb, Andrei Sakharov's Layer Cake, in August 1953 became Malenkov's personal triumph. Before the event, he announced to the world that the American monopoly on the hydrogen bomb had already been broken and invited numerous Soviet party and state officials to witness the test. Malenkov's bravado aside, he emerged as a rather reluctant Cold Warrior. Better educated than most of his colleagues, with a genuine interest in physics and philosophy, Malenkov regarded the continuing militarization of the Soviet economy and society as unsustainable. He looked for ways to ease tensions with the United States and invest more in consumer goods and less in armaments.[2]

The new climate in the Kremlin was reflected in the Soviet reaction to Eisenhower's Atoms for Peace speech. The Soviets did not reject it as a piece of propaganda but suggested that they were prepared to discuss Eisenhower's proposal if the United States solemnly declared that it would not use atomic or hydrogen bombs. The State Department agreed to hold separate discussions on both types of weapon. The Soviets consented to such two-track nuclear diplomacy, making possible continuing deliberations about Eisenhower's idea of a joint fissile-fuel bank under UN control. In March 1954 the Americans proposed the creation of a UN agency that would administer the program. Things looked promising.[3]

Also in March 1954, Malenkov delivered a public address in which he echoed some of the themes of Eisenhower's Atoms for Peace speech, declaring that a world war, "given modern weapons, would mean the destruction of world civilization." Coupled with Malenkov's earlier pronouncement that he saw no objective reason why making war between the United States and the USSR was inevitable, this suggested an emerging modus vivendi that diverged from Soviet rhetoric of the early Cold War years. That had stressed the impossibility of peaceful coexistence with the ideologically hostile imperialist West and thus the inevitability of war, including nuclear conflict.

Malenkov was backed by Soviet nuclear scientists, including the father of the Soviet nuclear program, Igor Kurchatov, who signed a letter from Soviet academics in reaction to the Castle Bravo test. "It is clear that the use of atomic arms on a massive scale will lead to the devastation of the combatant countries," wrote the authors, adding that "humanity faces the enormous threat of termination of all life on earth." Malenkov and the scientists were prepared to assert that nuclear war could not be fought and won. Aside from its basis in fact, this was a smart propaganda move. The Soviets did not yet have a hydrogen bomb of their own and were not in a position to admit it. Calls for nuclear disarmament in the name of saving the world from nuclear Armageddon were a good way to mobilize public opinion against Washington and perhaps ward off a nuclear attack on the USSR. While the statement was welcomed abroad, it caused unexpected trouble for Malenkov at home.

In April 1954 Malenkov's rivals in the Soviet leadership, led by Nikita Khrushchev, then leader of the party but not of the government, and Viacheslav Molotov, Stalin's long-serving foreign commissar, attacked Malenkov for lack of communist zeal. He was constrained to declare that any atomic attack by the West would lead to its destruction by weapons of the same kind, mirroring the other part of Eisenhower's Atoms for Peace speech. On April 27, Molotov wrote to Dulles informing the US Secretary of State that the USSR would not participate in the Atoms for Peace program unless the United States signed a nuclear disarmament treaty. This was an old Soviet demand that the Americans had previously rejected. In practice, it meant the rejection of Eisenhower's Atoms for Peace proposal.[4]

In January 1955, Khrushchev and Molotov succeeded in removing Malenkov as the head of the government. At the Central Committee meeting that month, Molotov declared: "A Communist should speak not about 'the destruction of the world civilization' or about 'the destruction of the human race' but about the need to prepare and mobilize all forces for the destruction of the bourgeoisie." With Malenkov gone, a brief opportunity for nuclear peace between the two superpowers had disappeared. The two nuclear powers returned to confrontation, meaning that preparations for nuclear war would continue on both sides. But Malenkov's conclusion that the hydrogen bomb made war unthinkable, putting an end to the idea of world revolution by means of warfare, did not disappear without trace in the Kremlin. It was picked up and refurbished as the concept of "peaceful coexistence" between the capitalist West and the socialist East by the very man who destroyed Malenkov's political career, Nikita Khrushchev.[5]

The sixty-year-old Khrushchev had emerged as a member of the ruling group immediately after Stalin's death in March 1953. In June of that year, it was Khrushchev rather than Malenkov who organized the arrest of Beria, and now he had succeeded in removing Malenkov as well, replacing him with his loyalist Nikolai Bulganin. Khrushchev kept for himself the powerful position of first secretary of the Communist Party's Central Committee, using it to consolidate power after Malenkov's removal.[6]

Khrushchev's debut on the world stage took place in July 1955, when he attended the first post–World War II summit of Soviet, American, British, and French leaders. The purpose of the meeting, held in Geneva, was to ease international tensions. Khrushchev was eager to join in. Not yet head of government and thus not the formal head of the Soviet delegation, Khrushchev still dominated informal discussions involving the delegation, but the small talk in which he engaged—bragging in one case about Soviet achievements in crossing zebras with cows—made a less than favorable impression on the Westerners. "How can this fat vulgar man, with his pig eyes and his ceaseless flow of talk, really be the head—the aspirant Tsar—of all these millions of people and this vast country?" wrote the stupefied Harold Macmillan, the British foreign secretary and future prime minister, in his diary.[7]

Khrushchev, for his part, was pleased with his diplomatic debut. He returned to Moscow in high spirits. In Geneva he had discovered something that would guide his policy toward the West, especially the United States, for the rest of his tenure. President Eisenhower, who led the American delegation to Geneva, had said that nuclear war was now all but impossible, as the "new weapons" (meaning the hydrogen bomb) could destroy the entire Northern Hemisphere. This was a revelation to Khrushchev, who was concerned that the United States might launch a nuclear attack on his country. Now he had learned that the Americans feared nuclear war as much as the Soviets. To prevent war, he had thought it necessary to frighten the Americans with the Soviet nuclear arsenal, which he knew to be inferior to the American. Having located Washington's soft spot, Khrushchev was prepared to exploit American fear of nuclear war to the utmost.[8]

From Stalin, Khrushchev had inherited the atomic bomb. In November 1955, a few months after the Geneva summit, with Khrushchev de facto in charge, the Soviet Union successfully tested its first true hydrogen bomb. Khrushchev now had both atomic and hydrogen bombs but no way of delivering them to the United States, as Soviet long-range bombers were incapable of crossing the Atlantic, and he had no air bases close to American shores. He pinned his hopes on missile technology, and by October 1957 he had a missile capable of reaching American soil. Sputnik was launched into orbit that month by the world's first intercontinental missile, the R-7, which was capable of delivering a nuclear charge measuring up to 3 megatons of TNT, the equivalent of almost 200 Hiroshima-type bombs. "Now that we have a transcontinental missile, we hold America by the throat as well," said Khrushchev to Mao Zedong, explaining the significance of the Sputnik-heralded revolution in US-Soviet relations. "They thought America was beyond reach. But that is not true."[9]

And yet, Khrushchev was appalled when Mao made public statements declaring that he was not afraid of atomic weapons. "I'm not afraid of nuclear war. There are 2.7 billion people in the world; it doesn't matter if some are killed. China has a population of 600 million; even if half of them are killed, there are still 300 million people left." Khrushchev did not have 300 million people to begin with: the population of the USSR reached 208 million in 1959. Despite the propaganda success of Sputnik, which convinced the world that the Soviets were ahead of the Americans in missile technology, Khrushchev knew that the United States had a much larger nuclear arsenal and superior forms of delivering it. Even after the launch of Sputnik, the USSR did not have more than a couple of ballistic missiles capable of delivering atomic and hydrogen bombs to the USA.[10]

In July 1959 Khrushchev visited the missile factory in the Ukrainian city of Dnipropetrovsk, where the Soviet missile constructor Mikhail Yangel showed him what looked like an assembly line of intermediate-range R-12 missiles, known to the Americans as the SS-4 Sandal. A few years later, the Sandal would be delivered to Cuba. Khrushchev was truly impressed and declared at a meeting with workers at the plant that the Soviet Union was now producing missiles like sausages. Khrushchev's son Sergei, himself an aspiring rocket scientist, asked his father why he was bragging when he knew that the Soviet Union had only a few operational missiles. The response was: "It's not so important how many rockets we actually have: after all, we're not about to start a war. The main thing is for the Americans to believe in our power. That will reduce the probability of war."[11]

The Soviets' only hope of survival in a nuclear war with the United States was the vast territory of the Soviet Union: the assumption was that the Americans would not be able to destroy all Soviet industrial centers simultaneously. But the Americans were building more bombs, while the Soviets were only starting to develop their arsenal of intercontinental missiles. Needing more time, Khrushchev decided to bluff, pretending that the Soviet Union had parity in bombs and missiles with the United States.

The Soviet launch of Sputnik caused a panic in the United States. The American sense of relative safety was now gone, along with the doctrine of mass retaliation: the United States could no longer count on delivering a nuclear strike against the USSR in case of its invasion of Western Europe without risking a nuclear response. Many were afraid that Sputnik was just the beginning. President Eisenhower's Science Advisory Committee produced a report suggesting that by 1959 the Soviet Union would surpass the United States in quantity of ICBMs (Intercontinental Ballistic Missiles). The report, named after Horace Rowan Gaither, who chaired the commission that produced it, was leaked to the press. In August 1958 the syndicated columnist Joseph Alsop predicted that there would be 500 Soviet ICBMs by the end of 1960, and 2,000 by the end of 1963. American plans called for 123 ICBMs by mid-1959. The country's political class was confused and frightened.

The Gaither report put Eisenhower on the defensive at home. The young and ambitious Senator John F. Kennedy of Massachusetts, already aspiring to the White House, sharply criticized Eisenhower in the Senate for putting "fiscal security ahead of national security" and predicted the loss of American superiority in striking capability by 1960. The term "missile gap," which Kennedy used to describe America's alleged lag behind the Soviets in ICBM production, became the equivalent of a baseball bat with which he attacked the Eisenhower administration.

No missile gap existed, as the Gaither Report was based on mistaken intelligence estimates issued in November 1957, the month after the Soviets launched Sputnik. The actual gap hugely favored the American side: the Soviets had anywhere between 10 and 25 missile launchers against more than 100 in the United States. The CIA briefed Kennedy on the issue in the summer of 1960, but he ignored the information: the missile gap was a perfect weapon against the Republicans and their presidential candidate, Richard Nixon. He would keep using it.[12]

Kennedy promised to take money from the civil economy and invest it in the military, as Eisenhower was reluctant to do. The president remained silent, since he knew from classified sources that there was no missile gap but could not reveal the fact publicly. Secret overflights of Soviet territory by Lockheed-made U-2 spy planes, known as "Dragon Ladies," indicated that the Soviets were lagging behind the Americans. But revealing that publicly to rebut Kennedy and his supporters would have meant disclosing a top-secret intelligence operation. Eisenhower was not prepared to do so, as he did not want to jeopardize Soviet-American relations, which were finally showing signs of improvement.[13]

Khrushchev and Eisenhower found themselves being pushed toward each other by growing anti-nuclear attitudes around the world as the two superpowers competed for influence in the former European colonies, and the general public became aware of the threat posed by the hydrogen bomb. *On the Beach*, a novel by the aeronautical engineer Nevil S. Norway, who used the pen name Nevil Shute, captured the atmosphere of the time. The book, published in 1957, describes a nuclear war six years in the future that makes the Northern Hemisphere uninhabitable. The plot focuses on Australia, where the locals await annihilation by fallout. A feature film based on the book, with Gregory Peck and Fred Astaire in starring roles, was released in 1959. In both the novel and the film, the Australian government distributes suicide pills to save the population from suffering and slow death by radiation poisoning.[14]

The world wanted the two superpowers to stop testing their nuclear weapons. The first public statement calling for such a halt as a step toward nuclear disarmament was made by Prime Minister Jawaharlal Nehru of India soon after the Castle Bravo test. Nehru proposed a moratorium. Clement Attlee, the former British prime minister and then leader of the Labour opposition to the Conservative government of Winston Churchill, asked the United Nations to ban the testing of hydrogen bombs. The Soviets joined the fray, approaching the UN Disarmament Commission with a proposal to end the testing of nuclear weapons as a first step toward nuclear disarmament. The proposal did not get very far, as the Americans doubted that compliance could be adequately verified. They proposed an "open skies" policy that would allow them to overfly Soviet nuclear installations, and vice versa. The Soviets, concerned that such surveillance would reveal the American lead in long-range missiles, refused.[15]

But in April 1956 Adlai Stevenson, the Democratic presidential candidate, raised the nuclear stakes in that year's election. He proposed a ban on testing hydrogen bombs and challenged President Eisenhower to debate the issue. The Soviet government and the Federation of Atomic Scientists expressed support for Stevenson's proposal, which did not help him in the campaign. Most Americans opposed a ban on testing, which they considered vital for protecting their country. Eisenhower maintained that there should be no halt to testing without an inspection system in place to verify Soviet compliance with a future agreement.[16]

In late May 1957, the Subcommittee on Radiation of the Congressional Joint Committee on Atomic Energy began hearings about the impact of radioactive fallout on human health and the environment. On June 1, Norman Cousins, editor of the *Saturday Review* and author of the influential 1945 anti-nuclear manifesto "Modern Man Is Obsolete," helped to organize a broad liberal organization opposing nuclear tests. A committee that later became known as SANE—Committee for a Sane Nuclear Policy—took upon itself to inform the American public about the harmful impact of nuclear tests. Two days later, on June 3, newspapers published an appeal coauthored by Professor Barry Commoner of George Washington University in St. Louis and signed by 2,000 scientists and academics (eventually, the number of signatories would reach 9,000) calling on the US government to sign an international agreement banning tests.[17]

President Eisenhower felt besieged. On June 5, addressing journalists on the issue of tests, he was clearly on the defensive. He attacked the scientists who had signed the recent petition as being "out of their own field of competence." He also suggested that the whole protest had been "organized" but later qualified his comment, denying any suggestion that some "wicked organization" was behind it. The implied reference was to the Communist Party and the Soviet Union. Eisenhower repeated his tried-and-true position: no ban on testing without a verified agreement on nuclear disarmament with Moscow. He insisted on continuing tests, since any future war would involve nuclear weapons, and the United States could not fall behind the Soviets in that regard. His solution to the fallout issue was producing and testing "cleaner" bombs with reduced fallout: in his opinion, the hydrogen bomb was one of the cleanest.[18]

After rebutting scientists "out of their field of competence," Eisenhower met with indubitable experts. The group of three leading scientists included Edward Teller, the father of the American hydrogen bomb and a key proponent of continued testing. Teller came to the White House at the invitation of the AEC chairman, Lewis Strauss. The Soviets had just accepted the possibility of allowing observation posts on their territory, and Strauss, Teller, and others in their group were concerned that the president would accept an offer. Teller came to sell Eisenhower a new type of weapon, claiming that bombs under development were 90 percent clean, or free of fallout. There was a catch, however: to develop a truly clean bomb, they would need to continue testing for another six to seven years.

The so-called clean or neutron bomb was a hydrogen bomb that released most of its energy in the form of fast neutrons, reducing to zero the residual radiation that in a typical atomic bomb amounted to 10 percent of its energy and was largely responsible for the radioactive fallout. Six to seven years to develop and test the bomb turned out to be highly unrealistic but Eisenhower took the bait. Concerned about being crucified on a "cross of atoms" in public opinion for turning down what many considered a reasonable Soviet offer, he practiced his new argument on visitors. "Our tests are projected to clean up weapons and thus protect civilians in the act of war," suggested the president. He tried that approach at the next press conference without producing the desired effect. Cousins wrote to Eisenhower, urging him to receive scientists who had a different position on the ban issue, and the president even contemplated such an idea, but the AEC chairman, Strauss, advised against it.[19]

In September 1959 Khrushchev traveled to the United States at the invitation of Eisenhower and was accepted as the leader of a nuclear superpower. For him, the visit had symbolic significance: the most powerful capitalist country had recognized the first workers' and peasants' state as an equal, and he was there as their representative. Recalling his humiliation on comparing the size of his plane with those of Western leaders in Geneva in 1955, Khrushchev was eager to show off.

He flew to the United States on the Soviet TU-114 turboprop, the largest and fastest passenger airplane in the world, and the only one that could fly nonstop from Moscow to Washington. But there was a problem: in test flights the TU-114 had developed microcracks in its engine. The builders were still working on fixing the problem when Khrushchev ordered the plane into the air. As his flight might very well end in catastrophe, the Soviet navy was mobilized to form a line of ships in the Atlantic along the plane's route in case it had to make a forced sea landing.[20]

Khrushchev was prepared to take the risk: for him, prestige took precedence over safety. To his relief, the plane did not crash. To his satisfaction, President Eisenhower showed respect for the Soviet premier, calling him the leader of the communist world, an honorary title then contested by China's Mao. Khrushchev would have numerous formal and informal meetings with American political leaders, travel the country from east coast to west, and meet Americans of different walks of life and persuasions. His message was peace through fear. He declared in one of his numerous toasts: "Our two countries are much too strong, and we cannot quarrel with each other." Eisenhower seemed to be in agreement. No treaties were signed, but the two leaders agreed to continue discussions. A new world summit with British and French participation was in the works. They would soon agree on the place and date: Paris, May 16, 1960.[21]

The agreement that everyone expected from the summit was a partial ban on nuclear testing—a deal that would permit only underground tests and prohibit new explosions in the atmosphere and outer space. In the spring of 1960, the Soviets accepted as a basis for further negotiations an American proposal that the only tests not to be banned would be those registering less than 4.75 on the Richter scale. In return they asked for a temporary ban of four to five years even on such explosions. The Americans and British proposed a one-year moratorium instead, and the Soviet response was positive. Expectations were high that Eisenhower and Khrushchev, as well as the leaders of two other nuclear powers, Harold Macmillan of the United Kingdom and Charles de Gaulle of France (which tested its first nuclear bomb in February 1960), would sign an agreement at the Paris talks in May 1960.[22]

On May 1, 1960, two weeks before the opening of the summit, a Soviet surface-to-air Dvina S-75 missile shot down an American U-2 spy plane that was scanning Soviet territory for intercontinental ballistic missiles. The shooting down took place on May 1, international Labor Day, a major Soviet holiday usually marked by parades. To demonstrate his peaceful intentions, Khrushchev organized just a popular rally with no military marching through Red Square and no missiles dragged through the center of Moscow. That morning Captain Gary Powers took off from an airbase in Peshawar, Pakistan, in his Lockheed U-2-C spy plane and crossed the Soviet border on a multi-hour flight intended to locate and photograph Soviet military installations, especially ballistic missiles able to reach the United States. It was the fourth such flight approved by President Eisenhower and administered by the CIA in an attempt to uncover Soviet offensive capabilities. These were de facto high-altitude inspections, including surveillance of nuclear-bomb testing sites, because the negotiators in Geneva could not agree on admitting official representatives for on-site inspections.

The previous flight of the U-2, on April 9, had gone without a hitch. Even though Soviet radar had detected the plane, Soviet interceptor fighter planes were powerless to do anything to the U-2, which flew at an altitude of 70,000 feet, far above the fighters' range. But on May 1 the Soviets managed to shoot down the U-2 with the new surface-to-air Dvina S-75 missile as Powers was flying over the Ural Mountains not far from Kyshtym, the site of a Soviet plutonium-producing complex, where the explosion of a spent nuclear fuel tank two and a half years earlier, in September 1957, had caused the greatest nuclear catastrophe of its day. Powers, who parachuted from his plane as it rapidly lost altitude, miraculously survived the ordeal, and was captured by the Soviets.[23]

Nikita Khrushchev was furious about American overflights of Soviet territory, which threatened to expose the true status of his missile arsenal: he had almost no missiles that could reach the United States. In 1955, he had refused Eisenhower's proposal of "open skies" inspections and ordered that intruders be shot down. In the course of the 1950s, the United States lost 130 pilots in its undeclared reconnaissance war with the USSR, but the U-2 had been unreachable to the Soviets until May 1960. Khrushchev was jubilant. He wanted Eisenhower to apologize. But Eisenhower refused, issuing a statement that the airplane had been on a peaceful mission studying weather patterns and simply lost its way. He had no idea that not only parts of the plane with spy cameras but also the pilot himself had survived the crash.

Khrushchev, who produced both, wanted the apology and would have been satisfied if the president had blamed his subordinates for the misadventure, but Eisenhower refused, considering such an evasion dishonest. Khrushchev went to Paris in mid-May to take part in the opening of the summit, still hoping that Eisenhower would apologize and pledge to stop U-2 overflights. As no apology was forthcoming, and Eisenhower promised to suspend rather than stop the flights, Khrushchev left the summit in a huff, also withdrawing his earlier invitation to Eisenhower to visit the USSR—a courtesy that he extended after his visit to the United States in September 1959.[24]

The crisis had embarrassed Eisenhower, hurt the electoral chances of his vice president, Richard Nixon, who supported his boss's refusal to apologize, and created a new opportunity for John Kennedy, whose presidential campaign was gathering steam. "As a substitute for policy, Mr. Eisenhower has tried smiling at the Russians," declared Kennedy, "our State Department has tried frowning at them; and Mr. Nixon has tried both. None have succeeded. For we cannot conceal or overcome our lack of purpose and our failure of planning by 'talking tough'; nor can we compensate for our weaknesses by 'talking nice,' by assuming that the righteousness of our principles will ensure their victory."[25]

Khrushchev decided to make a huge political scandal out of Eisenhower's lies about the plane, abort the summit, and put Soviet-American relations on hold, thinking not only about the past but also about the future. A new American president would be moving into the White House. His name was John Kennedy, and Khrushchev had high hopes that he would be easier to deal with than Eisenhower.[26]

Chapter 16

BOMBE ATOMIQUE

No American president had contributed more to the proliferation of nuclear weapons than Dwight Eisenhower, who presided over the period not only of the greatest expansion of the American nuclear arsenal but also of the spread of nuclear technology and knowhow throughout the world. Even measures designed to preclude the spread of nuclear weapons, such as America's readiness to share nuclear technology for peaceful purposes within the Atoms for Peace mandate, eventually contributed to the proliferation of nuclear weapons as well, helping aspiring nuclear nations to acquire knowledge and expertise in that field.

Eisenhower made a conscious decision to engage in vertical proliferation—the increase of the American nuclear arsenal. He opposed horizontal proliferation if it meant the acquisition of nuclear weapons by other countries but favored it when it came to sharing technology with Britain and the deployment of American nuclear weapons on foreign territory. The major obstacle to the latter policy was the McMahon Act of 1946, which barred the administration from sharing nuclear technology with anyone, including allies. In 1958, however, Eisenhower managed to amend the act to allow sharing not only knowledge but also technology with the British, a practice that had already been going on despite the act.

The deployment of American nuclear weapons on the territory of NATO allies included the placement of nuclear-tipped Thor intermediate-range ballistic missiles in Britain and Jupiter medium-range ballistic missiles in Italy and Turkey.[1] Under a "dual key" arrangement, tactical nuclear weapons were dispatched to West Germany. The arrangement suggested American control over nuclear warheads and NATO allies' control over nuclear-capable artillery tubes and aircraft. With the permission of the president of the United States, NATO commanders on the ground could use nuclear warheads against the enemy. The Soviets questioned the arrangement and were concerned about the US government's ability to keep nuclear weapons under its complete control in Europe. Such doubts were shared by some members of Congress and politicians around the world.[2]

One of the reasons for America's sharing of nuclear weapons was to dissuade NATO allies from acquiring such weapons of their own. It was an uphill battle. In January 1961, when President Eisenhower turned over the keys to the White House to his successor, John F. Kennedy, he simultaneously passed on not only a greatly increased arsenal of atomic and hydrogen bombs but also concern about a world in which there was one more nuclear power than there had been when he assumed office in January 1953, and two states dangerously close to acquiring their own weapons.

That nuclear power was France, and to a degree it was France's fault that the long-awaited May 1960 Paris summit between Dwight Eisenhower and Nikita Khrushchev failed. The two leaders had agreed in principle to hold a summit meeting in Paris during the Soviet premier's visit to the United States in September 1959. It would have taken place earlier than May 1960 and thus before the fateful U-2 accident if President Charles de Gaulle, the host of the meeting, had not dragged his feet. One of de Gaulle's reasons for delaying the summit until late spring was his desire to host the leaders of the United States, the Soviet Union, and the United Kingdom, all nuclear powers, as an equal. France was preparing to conduct its first ever nuclear test and successfully exploded its first atomic bomb on February 13, 1960. The road to the summit was now open, but delay contributed to its failure.[3]

De Gaulle had long been obsessed with the idea of the French atomic bomb. In 1945, serving as prime minister of the interim French government, he already associated his country's return to the club of the world's great powers with the possession of nuclear weapons. Among those who urged de Gaulle to launch a French nuclear program was the forty-five-year-old scientist Frédéric Joliot-Curie, a Nobel Prize winner and pioneer in nuclear research. His article of April 1939 on chain reaction spurred the Germans into action. Albert Einstein mentioned his research in his letter of 1939 to President Roosevelt. Joliot-Curie was as disturbed by the possible consequences of the discovery of fission reaction in Germany as Leo Szilard was in the United States and even informed one of his Soviet colleagues, Abram Ioffe, about that discovery.[4]

Ironically, France launched its own nuclear program in response not to Joliot-Curie's research but to the German uranium project. Before Hitler's troops marched into Paris in May 1940, the French government managed to acquire uranium from Belgium and heavy water from Norway. The fall of Paris ended the French nuclear program. Joliot smuggled his research papers abroad. While some of his colleagues joined the Manhattan Project from Canada, Joliot-Curie stayed behind and joined the French resistance, taking an active part in the Paris uprising of 1944. With the war over and France liberated, Joliot wanted de Gaulle to help resume scientific work in nuclear physics.[5]

The request was strongly backed by Raoul Dautry, the prewar minister of armaments, who had worked with Joliot on the atomic bomb project in 1940. Dautry believed that to reclaim great-power status after the war, France needed an atomic bomb. De Gaulle was more than receptive. On October 18, 1945, less than two months after the bombing of Hiroshima and Nagasaki, and roughly three and a half months before de Gaulle left the office of chairman of the provisional government, he created the Commissariat à l'énergie atomique. The commissariat was charged "with the mission of developing the uses of atomic energy in various fields of science, industry, and national defense." The newly created position of High Commissioner for Atomic Energy, the head of the commissariat's scientific branch, went to Joliot.[6]

Joliot was eager to resume his research in nuclear physics but, curiously enough, he was not interested in building a bomb, for which he saw no further reason. A card-carrying member of the French Communist Party, Joliot was eager to follow the official Soviet line and supported nuclear disarmament. Besides, like many participants in the Manhattan Project, he believed that access to nuclear secrets belonged to the world rather than to individual nations.

As the Soviets stole nuclear secrets from the Americans and built their own bomb under cover of their public campaign for the elimination of nuclear weapons, Joliot limited the work of scientists associated with his commissariat to research on peaceful uses of nuclear energy. In 1950, as the Cold War intensified, the Soviets exploded their atomic bomb, and American atomic spies were tried and sentenced, some of them to death, Joliot was forced to resign as scientific head of the French nuclear project.[7]

Intentionally or not, Joliot made an enormous contribution to the development of the scientific basis for the building of the French atomic bomb. The first French nuclear reactor (a heavy-water model influenced by a Canadian analogue) went critical in 1948, and in the following years a limited amount of plutonium was produced for research purposes. With Joliot's departure, the French nuclear program progressed from the purely scientific to the industrial stage. A key role in the transformation of the program was played by the rising thirty-two-year-old French politician, Félix Gaillard, who became state secretary in charge of atomic energy in 1951. He persuaded the government to adopt a five-year nuclear program at an overall cost of twenty billion francs.[8]

Officially, France was developing nuclear energy for peaceful purposes, and no formal decision was made to acquire nuclear weapons. The military began to discuss the possibility of building a nuclear bomb in 1952. But with the army busy for most of the 1950s fighting wars in former colonies, there was no concerted effort to push the atomic bomb program forward. In 1954, however, debate about the bomb entered the public realm, the background being West German rearmament and the French defeat in Indochina. French prestige, if not security, was under attack, and building a bomb seemed the best way of regaining the country's great-power status. General de Gaulle, then in retirement, argued in favor of a French atomic bomb. In 1954, a military bureau was set up within the structure of the commissariat, and in the following year the first funds were transferred to the commissariat from the Ministry of Defense.[9]

The French nuclear establishment found a home for itself at the Marcoule nuclear site, less than 35 kilometers (22 miles) northeast of Avignon in southern France. The French opted for reactors working on natural uranium and moderated by graphite. The first of those reactors went operational in January 1956, the second in April 1959, and the third in April 1960. The last one was too large to be called a prototype, and everyone at the commissariat knew that it existed to produce plutonium. But Pierre Guillemet, whom Gaillard had appointed to administer the commissariat in the early 1950s, was not volunteering that information to the government. "Do you believe political people understand anything about nuclear energy?" he asked one of the commissariat officials.[10]

The "political people" were indeed uncertain about the direction in which they wanted to take the French nuclear program. In early 1956, Prime Minister Guy Mollet declared his opposition to the French nuclear program and support for Euratom (European Atomic Energy Community), a US-backed European initiative to develop nuclear energy and a common market for it in Western Europe. The Suez crisis that happened later that year dramatically changed Mollet's mind, and thus the French official position, on a separate nuclear project for France.

In July 1956, the Egyptian government of Gamal Abdel Nasser decided to nationalize the Suez Canal. The decolonization drive that engulfed the colonial possessions and dependencies of the two largest colonial empires, those of Britain and France, entered a new stage. In response, London and Paris came up with a plan to invade Egypt. The British and French wanted to establish control over the channel to secure their supply of oil from the Gulf. But they needed a pretext and decided to create one by persuading the Israeli government to attack Egypt and allow the two Western powers to enter the conflict as "peacekeepers."

The Israelis, who had their own reasons to enter the war, attacked Egypt as planned on October 29, 1956, and swiftly defeated the Egyptian troops on the Sinai Peninsula. Britain and France then issued ultimatums to both sides, followed by air raids against the Egyptians and the deployment of British and French ground troops. But Nasser countered by sinking forty ships in the Suez Canal, making it nonoperational for months, while the Soviet Union and the United States demanded a ceasefire and insisted on the withdrawal of the Franco-British-Israeli forces. The Americans were concerned about regional instability, worsening relations with the Arabs, and possible Soviet military involvement in the region. Washington privately threatened to destroy London financially. The threat of UN sanctions created economic panic in London and forced Britain to devalue its currency. Moscow publicly threatened the use of missiles, which was correctly understood as an allusion to nuclear weapons. The tripartite coalition forces were constrained to withdraw.

Britain and France suffered a tremendous blow to their prestige and world standing. In France, the humiliating retreat strengthened the case of those who claimed that, given the divergence of French and American interests in Africa and the postcolonial world in general, France could not rely on the American nuclear umbrella and needed nuclear weapons of its own to resist Soviet political and military pressure.[11]

In Paris, Prime Minister Guy Mollet went virtually overnight from opposing the French nuclear arms program to supporting it. The protocol signed in the aftermath of the Suez crisis between the Commissariat à l'énergie atomique and the Ministry of Defense set clear goals for the country's military nuclear weapons program. Within five years, France was supposed to have a bomb and test it too. In conjunction with the crisis, a secret Committee on Military Applications of Atomic Energy was formed and a key decision made to build a nuclear reactor for France's Middle Eastern ally and collaborator in nuclear research, Israel. France decided not only to acquire its own nuclear bomb but also to become a proliferator of nuclear technology even before obtaining the weapon.[12]

Mollet was soon out of office, but that made no difference. In April 1958, the new prime minister and former state secretary in charge of nuclear energy, Félix Gaillard, signed an order authorizing the building of the bomb, with a test date scheduled for the spring of 1960. Gaillard survived in office for slightly more than half a year but left a strong mark on the history of his country. The French nuclear die was cast.[13]

General Charles de Gaulle returned to the pinnacle of French political power in May 1958. The comeback took place in the midst of a coup led by the military and backed by French settlers in Algeria, who wanted to preserve the country's great-power status and empire. De Gaulle disappointed the plotters by ending the French presence in Algeria but met their expectations when it came to great-power status. For him, France's return to the world stage was synonymous with nuclear power. As he replaced the parliamentary republic with a semi-presidential one and launched the Fifth French Republic, de Gaulle did everything in his power to ensure the realization of the Fourth Republic's last geopolitical wish, the acquisition of an atomic bomb.[14]

The new president's task was not just to build and test the bomb but also to "sell" the idea of nuclear-armed France to the world, which was preoccupied after Castle Bravo with concern about nuclear fallout and thermonuclear apocalypse. When it came to convincing the world that France needed the bomb, there was no more important constituency for de Gaulle than the government of the United States, France's liberator in World War II, provider of Marshall Plan funds in the postwar years, senior partner in the NATO alliance, and nuclear superpower keeping France under its own atomic umbrella. Given the cost of building the bomb, de Gaulle hoped to obtain American technological assistance—a tall order.

The nonproliferation of atomic bomb data had been the core of American nuclear policy since the McMahon Act of 1946, which prohibited sharing nuclear secrets with any country, including America's wartime ally, Britain. But in the wake of the startling Soviet launch of Sputnik, Eisenhower was eager to find a way of sharing American nuclear technology and weapons with NATO allies. Amendments to the McMahon Act made it possible to share technology with the British. There was also sharing with the Canadians, who had taken part in the Manhattan Project during the war. France, with the exception of a few scientists, had not been party to that undertaking; more importantly, Congress was very reluctant to share anything with Paris, citing the volatility of the political situation in France and security concerns, given the communist and pro-communist views of many French scientists working on the nuclear program.[15]

De Gaulle's chance to make his argument to the Americans came on July 5, 1958, less than two months into his new tenure, when Secretary of State John Foster Dulles paid a visit to Paris. Dulles came to sell de Gaulle on the idea of sharing control over American nuclear weapons with NATO and thus to dissuade him from developing an atomic bomb of his own. He proposed the development of a scheme whereby, "in the event of a major attack on French or United States forces in Europe, nuclear weapons available to NATO would be used immediately without having to depend on a United States political decision, concerning which the French might have some doubts." He was "prepared to see French forces fully trained in the use of such weapons and French equipment adapted to deliver them." Dulles was careful to suggest that it was up to France to decide whether it wanted to develop its own nuclear weapons, but in his opinion "if one after another NATO state were to embark on an independent nuclear program, it would indeed be wasteful and would seriously dissipate our total resources."

De Gaulle wanted none of that, telling Dulles that France was "on the way to becoming a nuclear power." He was bullish on the time frame for that development: "France would have an atomic explosion within some months." Would it not be a waste of resources? De Gaulle had a ready answer: "If France were given nuclear weapons or produced them thanks to United States assistance, this would be an economy and thus a reinforcement of the alliance." Later in the meeting, de Gaulle returned to the idea of dual control over American weapons: "Any nuclear arms made available under NATO planning on French soil must be under the direct responsibility of France, with the United States participating in this control." De Gaulle's condition was one that Dulles was not prepared to accept: Congress would never approve such a deal.[16]

In the next few weeks, de Gaulle met with key figures in the French nuclear program and asserted his support for what was to be an exclusively French undertaking. He annulled plans for a joint Franco-German project contemplated by his predecessors and officially confirmed on July 22, 1958, the government's decision made earlier that year to proceed with building the bomb. "The nuclear force is above all an instrument of policy, a means to an end, which is not so much security as independence, a diplomatic advantage that reinforces the status of this country and expands the role it may play," declared de Gaulle on that occasion. He wanted the bomb in his diplomatic arsenal first and in his military second.[17]

De Gaulle's decision to go his own way with regard to nuclear weapons was part of his broader agenda of renegotiating France's relationship with the United States and NATO. He wanted a larger say for France in deciding NATO political and military strategy, as well as greater NATO involvement in the Mediterranean and Africa, areas of strategic importance to France, which was still struggling to maintain its colonies and influence in the region.

In September 1959, when Eisenhower flew to Paris to reduce tensions between the two countries, de Gaulle proposed that the president turn NATO into a predominantly political organization run by a new incarnation of the Big Three: the United States, the United Kingdom, and France. Security issues, on the other hand, would be the primary responsibility of individual member nations. For France, that meant not only the right but also the need to acquire nuclear weapons. There was no progress on a nuclear weapons–sharing arrangement with the United States. Eisenhower returned home to meet Nikita Khrushchev, who was paying his first and last visit to the United States. De Gaulle would soon announce plans to build a *Force de frappe*—a strike force consisting of air-, sea-, and land-based nuclear weapons.[18]

In early 1960, Eisenhower made a public statement declaring his desire to further amend the McMahon Act of 1946. The plan was to allow sharing with allies the kind of weapons already in the possession of the Soviet Union. "I have always been of the belief that we should not deny to our allies what the enemies have," said Eisenhower. Congress would not budge. There would be no sharing of American nuclear weapons either with France or with other NATO allies. The Soviets knew that the French were moving ahead with their atomic bomb project. In late 1959, when Deputy Foreign Minister Valerian Zorin entered the discussion about France's nuclear status, he said that the Soviets did not object but followed up with a question: "Do you think it is safe in such a small country?" It was clearly a threat and a signal that the Soviets were not happy but, like the Americans, could not do much about French nuclear ambitions.[19]

On February 13, 1960, as planned, France successfully tested its first atomic bomb in the Algerian part of the Sahara Desert. The yield was equivalent to more than 60 kilotons of TNT. Between February 13, 1960, and April 25, 1961, there would be four atmospheric tests in that location. France had joined the nuclear club on its own terms, creating the *Force de frappe* announced earlier. "France cannot be France without greatness," wrote de Gaulle. In his view, as France lost its colonies, nuclear weapons would ensure its continuing status as a great power in Europe and around the globe.[20]

"If France must have allies, she has no need of a protector," declared the French president. He clearly had in mind the United States, although that was not his only concern. In his opinion, France had regained status equal to that of Britain. Like the United Kingdom, the Fifth French Republic had built a bomb as much to deter the Soviets as to achieve independence from the United States and assure its great power status.[21]

Chapter 17

CHINA SYNDROME

Unlike the Americans in relation to the British, the Soviets had no international debts to pay for the creation of their atomic bomb. True, they had stolen many secrets from the Americans but had no intention of paying them back. And they had no equivalents of Britain or Canada demanding their part of the atomic pie that those countries had helped to create. Also, unlike the American NATO allies, the East European clients of the USSR, organized since 1955 in the Warsaw Pact, had no great-power ambitions and lacked intellectual, financial, and technological capabilities to build a bomb.

Thus, the Soviets kept their nuclear weapons to themselves, not only not sharing the technology with their allies but also refusing to move atomic bombs beyond their own borders. They had overwhelming superiority in conventional forces in Central Europe, and there was no incentive to move their intermediate-range missiles to East Germany or Czechoslovakia, as that would not have put them within range of the United States. The Soviets practiced nonproliferation of both technology and nuclear weapons at a time when the Americans were only declaring such goals. But as Eisenhower began to develop his Atoms for Peace program, sharing nuclear technology for energy production with the rest of the world, the Soviets felt that they had to respond in kind.

In January 1955, Moscow made official offers to a number of socialist countries to cooperate in the field of nuclear energy research. The countries receiving such offers included the East European satellites Czechoslovakia, Romania, and East Germany, as well as a brand-new member of the socialist camp, the People's Republic of China. No communist leader was more encouraged by the invitation than Mao Zedong, who for his own reasons decided that month to build a bomb of his own, a Chinese atomic bomb.[1]

Communist China was fourteen months old when on November 30, 1950, it was threatened with an atomic bomb by no less a figure than Harry Truman, the president of the United States himself. A few days earlier, Chinese forces had attacked American and South Korean divisions preparing to cross the North Korean border in pursuit of Korean communist forces that had invaded South Korea in June of that year. With the Chinese on the offensive and the Americans in retreat, Truman was desperate. At a press conference held that day, Truman declared: "We will take whatever steps are necessary to meet the military situation, just as we always have." When asked by a reporter, "Will that include the atomic bomb?," the president answered: "That includes every weapon we have.... The military commander in the field will have charge of the use of the weapons, as he always has."

The suggestion that the Americans were prepared to use the bomb and that it would be up to General Douglas MacArthur, the commander of American and allied troops on the peninsula, to decide when and how to do so, sent shivers down politicians' spines all over the world. The Truman administration had to distance itself from the president's statement, assuring the American public and the world at large that there were no plans to use the bomb and that it was up to the president, not the commander in the field, to make any such decision. MacArthur, who had proposed to invade Chinese territory, was fired in April 1951, but with American and allied forces still at a disadvantage, Truman ordered in the same month that atomic-capable B-29 bombers be deployed to the American base on Okinawa, accompanied by nine unassembled atomic bombs. Military strategists were busily making plans to bomb Chinese cities, there being no North Korean targets deemed sufficiently important to bomb with atomic weapons.[2]

Truman left the Oval Office in January 1953 without having ordered an atomic attack on China or North Korea, but with the war dragging on. His successor, President Eisenhower, did not share Truman's view that the atomic bomb was a "terrible weapon." He considered it part of his arsenal and was open to suggestions from military commanders in Korea concerning the use of tactical nuclear weapons, small atomic bombs, or nuclear shells on the peninsula. Nor did he rule out the possibility of using nuclear weapons against China. The Chinese either knew about or suspected Eisenhower of readiness to use the atomic bomb. Soon after his inauguration, the Chinese news agency denounced his plans to use nuclear weapons in Korea and beyond. Eisenhower himself believed that it was his threat to use nuclear weapons that helped end the Korean War in July 1953. Whether that was the case or not, there is little doubt that communist China's foreign policy was born in the shadow of American atomic brinkmanship.[3]

Like the leaders of other new states born or reborn after the carnage of World War II, Mao became interested in nuclear weapons before the official birth of the People's Republic of China. The new communist state was proclaimed on October 1, 1949, a few months after a Chinese delegation visited the Soviet Union and its leader asked to see the Soviet nuclear facilities. The response was negative, as the Soviets were about to test their first atomic bomb and were in no hurry to share their secrets with their new ally. In December 1949, Mao Zedong traveled to Moscow to attend Stalin's seventieth birthday celebrations and negotiate what became known as the Sino-Soviet Treaty of Friendship, Alliance, and Mutual Assistance. The atomic bomb was one of the things on his mind.

Signed in February 1950, less than ten months before China's entry into the Korean War, the treaty implied but did not explicitly guarantee the extension of the Soviet nuclear umbrella to Beijing. At China's insistence, the Soviets amended the treaty text to state that in case of attack by a third party, the signatories would assist each other "with all means at their disposal." But that was as far as they were prepared to go. The Chinese probably did not realize that the protection they sought did not amount to much, as the Soviets then had hardly more than five atomic bombs (they had tested their first one in the fall of 1949) against America's 369.[4]

The Soviet nuclear arsenal, no matter how small it was, gave Mao Zedong more confidence to enter the Korean War, which pitted China against the world's first and most powerful nuclear state, than he would otherwise have had. He stated publicly that the Soviet Union's possession of nuclear weapons would deter the Americans from using their atomic bombs. He was right: the Soviet bomb frightened America's European allies sufficiently to make Washington's use of the atomic bomb too expensive politically, no matter what military advantages it might have brought. Privately, Mao believed that the United States had too few atomic bombs to prevail in a war with China. He was probably wrong on that account.[5]

The claim that wars could not be decided by atomic weapons was the key element of the Chinese foreign policy credo and state propaganda during the Korean War. In that regard Mao followed the line established by Stalin after the bombing of Hiroshima and Nagasaki: a country without nuclear weapons had to convince its adversaries and induce its people to believe that they could prevail against a nuclear-armed nemesis. Meanwhile, the Chinese authorities did all they could to fuel the anti-nuclear movement abroad. More people signed the 1950 Stockholm Peace Appeal in China than in any other country. Of the 450 million signatures collected in seventy-five countries, 204 million came from China. That was not due to the propaganda effort alone: ordinary Chinese citizens were more afraid of the atomic bomb than their counterparts anywhere else, and for good reason—they fought a war against a nuclear superpower.[6]

American threats to use nuclear weapons against China did not go away with the end of the Korean War and resumed in 1955 in the midst of the First Taiwan Strait Crisis, when communist Chinese forces seized an island claimed by the Taiwanese nationalist opponents of the People's Republic. The shelling of the Taiwan-controlled islands of Quemoy (Jinmen) and Matsu (Matzu) in the Taiwan Strait and public calls from Beijing to "liberate" Taiwan prompted Eisenhower to send a nuclear-capable flotilla to the region. The American military proposed atomic bombing of mainland China, and Secretary of State John Foster Dulles made a number of statements in that regard. Winston Churchill called on the United States not to use nuclear weapons. He was supported by other European NATO members concerned about global nuclear war and possible Soviet attack on their countries.[7]

The crisis was resolved in May 1955 as the shelling of the Taiwan-controlled islands stopped—perhaps, as some believe, owing to the American public threats to use nuclear weapons. What is known for certain is that the crisis, if not the threats per se, jumpstarted Mao's program to acquire nuclear weapons. The key decision was made in January 1955 in the midst of the crisis. That month the Soviet government made public its offer to assist China with research on peaceful uses of the atom.[8]

The Chinese prime minister, Zhou Enlai, speaking to members of the State Council in late January 1955, suggested that the Soviet decision was the result of extensive consultations with Beijing. He called for the mobilization of scientific resources for research on nuclear energy and a mass campaign propagating its importance to the general public. Ostensibly, it was all about peaceful uses of nuclear energy as a means of avoiding nuclear war, but the true driving force of both initiatives—the decision to build the bomb—was obvious to anyone who heard or read the speech. "We are now in the atomic age," declared Zhou Enlai.

"We have to understand atomic energy, whether used for peace or war. We have to master atomic energy. We are far behind in this area, but, with Soviet help, we have the confidence and determination that we can catch up." He later continued with a direct reference to nuclear weapons: "We have Soviet help, and if only we apply ourselves seriously, we will be able to master atomic weapons."

Proposing a Chinese nuclear arms program, Zhou Enlai had to deal with the same challenge that the American administration encountered earlier in the decade: how to use fear in order to mobilize people in support of their own nuclear program without sowing panic over possible annihilation by a hostile nuclear force. "What after all is the power of atomic weapons?" Zhou asked his audience. "Many people are not clear. As a consequence, this has given rise to two types of attitudes in the world: one is ignorance and the other is terror. Our Chinese people believe there is nothing special about the atomic bomb and ignore it, looking down at it with derision. [But] it is incorrect to ignore it, and most of the world's people are terrified by it." According to him, the Americans were terrified of the atomic bomb and wanted to frighten the Chinese as well.⁹

What should be the Chinese response? Zhou praised Mao's approach to the problem, telling the audience about a conversation that Mao had had with the Finnish ambassador to Beijing, Carl-Johan (Cay) Sundstrom, a few days earlier. According to the protocol of the meeting, Mao told his visitor: "The United States cannot annihilate the Chinese nation with its small stack of atom bombs." He continued: "Even if the U.S. atom bombs were so powerful that, when dropped on China, they would make a hole right through the earth, or even blow it up, that would hardly mean anything to the universe as a whole, though it might be a major event for the solar system."¹⁰

Mao was clearly reacting to the advent of the hydrogen bomb, even if his assurance that the universe would remain intact was cold comfort indeed. The implication of Mao's words was obvious to the Chinese and foreigners alike: China was not afraid of nuclear weapons. There were not enough atomic bombs in existence to destroy China, and use of the hydrogen bomb was unthinkable, as it would destroy all life on the planet.

In Zhou Enlai's version, Mao's words sounded somewhat different. "Mao said that the very greatest harm would merely be that a hole would be blown clean through the earth, and, if there was a hole blown through the earth, and one entered from China, the other side would be none other than the United States." A version of the "China Syndrome" was born. In the American nuclear industry, reference to that syndrome was an inside joke about fuel from a nuclear reactor burning its way through the earth and popping up in China. It gave its name to the Hollywood 1979 blockbuster about an imagined nuclear accident at a US plant. Mao's version of "China Syndrome" suggested something different: a nuclear destruction of China would mean the destruction of the world.[11]

Like Eisenhower before him, Mao was looking for money to finance his nuclear program and found it in the same place the American president did—in the budget for conventional military forces. In April 1956, speaking to a session of the Politburo, the country's ruling body, Mao called for a drastic reduction of conventional military spending to free resources for building the industrial infrastructure needed to develop the bomb. "Do you really want an atomic bomb, are you totally committed to wanting it, or are you only partially committed to wanting it, and not totally committed to wanting it?" Mao asked his audience. He then gave the answer: "If you really want it, if you are totally committed to wanting it, you'll cut the proportion of military expenditures and concentrate more on economic development."[12]

The success of the program relied not only on finding the money but also on acquiring the technology, and there the great hope was the Soviets, who Mao believed would share their knowledge with the Chinese. Beijing went out of its way to present itself as a reliable and useful partner of Moscow in the nuclear project. Mao reversed his original decision to keep all uranium extracted in the country for Chinese needs and allowed the export of uranium to the USSR. Another good-will gesture inspired by nuclear hopes was the approval of Soviet communications and infrastructure projects in northeastern China. In return, Chinese scientists went to work at the newly founded Soviet institute for nuclear research in Dubna near Moscow, and Chinese students were invited to study nuclear physics at Soviet universities. Altogether six agreements were signed between the two countries to bolster bilateral cooperation in the nuclear field.[13]

In October 1957, shortly after the successful launch of Sputnik, Soviet and Chinese officials agreed on the provisions of a new Defense Technical Accord between the two countries. To great jubilation in Beijing, the Soviet Union agreed to provide China not just with the technology required to build the bomb but even with a prototype of the bomb itself. With such assistance, the Chinese road to nuclear power promised to be the shortest available. But just as the two communist militaries and economies were drawing closer, relations between the two communist parties and their leaders, Mao Zedong and Nikita Khrushchev, were becoming more complicated. Mao regarded Khrushchev's de-Stalinization campaign, which the Soviet leader launched in February 1956, as a threat to his own power, and the Soviet invasion of Hungary later that year as a threat to Chinese independence. Nor did he subscribe to Khrushchev's contribution to Marxism-Leninism—his postulate that in the nuclear age war was no longer inevitable. That looked to Mao like appeasement of the imperialist West.[14]

In November 1957, the month after Khrushchev promised him a prototype of the bomb, Mao arrived in Moscow to take part in a meeting of leaders of the world's communist parties. Addressing the gathering, he challenged Khrushchev's views on peaceful coexistence with the West. Nuclear war was no disaster for socialism, asserted Mao: "Can one estimate how many people would be lost in a future war? Possibly, it would be one-third of the whole world's population of 2,700 million, or just 900 million people. I think this is even too few if the atomic bombs are really dropped. Of course, this is very scary. But it would not be so bad even if it were half." He then asked, "Why?" and answered his own question by repeating what he had said to Premier Jawaharlal Nehru of India earlier in the year: "If half of humankind is destroyed, the other half will still remain, but imperialism will be destroyed completely, and there will just be socialism in the entire world, and in half a century or a whole century the population will grow again, even by more than half."[15]

The Chinese alliance on which Khrushchev had counted, and for the sake of which he was prepared to do almost anything, including sharing the bomb with Beijing, was not just in crisis but was actually falling apart. The Chinese did not appreciate Soviet nuclear generosity, and if they maintained that attitude, then the master of the Kremlin would face not just a political but also a security problem.

Khrushchev recalled Mao telling him in the course of the visit that he did not expect the Soviet Union to retaliate or become involved in a conflict if China were attacked by the United States. The alliance was clearly in trouble, and the Soviets rethought the whole idea of giving the Chinese a bomb. They made that decision by March 1958 and officially informed the Chinese in June 1959 that no bomb would be supplied. The official reason given was that if the West learned about the transfer of nuclear bomb technology, test-ban talks in Geneva could be derailed. The Soviets were careful to present their refusal as a temporary measure. "It will still take China at least two years to produce fissionable material, since it is necessary to complete a great deal of work to mine uranium ore and establish an atomic industry," stated the letter from the Central Committee of the Soviet Communist Party to its Chinese counterpart. "Only at that time will a whole tranche of nuclear weapons technical data be necessary." The Soviets promised continuing help with the building of the nuclear industrial complex and the production of fissionable materials.[16]

Khrushchev tried further mitigating the situation by declaring that China could count on the Soviet nuclear umbrella, but the Chinese were already moving ahead with their own nuclear bomb. The Second Taiwan Strait Crisis, which occurred in 1958, demonstrated the American commitment to the defense of Taiwan and contributed to the Chinese decision to switch from the research stage of their bomb project to the industrial one. That year the Chinese government established the

Tongxian Uranium Mining and Hydrometallurgy Institute, concerned with the extraction of uranium, as well as the Beijing Nuclear Weapons Research Institute, China's equivalent of the Los Alamos laboratory. Work began on the construction of the Baotou Nuclear Fuel Component Plant and the Lanzhou Gaseous Diffusion Plant, whose task was the production of enriched uranium. The Soviets were supposed to provide the technology required to build a plutonium bomb. When they demurred, the Chinese began to build a uranium bomb.[17]

In July 1960, speaking to delegates participating in a Central Committee conference, Mao acknowledged that the Soviet-Chinese nuclear alliance was no longer in effect. "Over the past decade the Soviet people have helped us in [our] development, and we cannot forget this," declared Mao. He continued: "We must resolve to work on [pursuing] advanced technology. Khrushchev won't give us advanced technology. Fine! If he had given it to us, it would have been a difficult debt to repay." Construction began that year of a nuclear reactor at the Jiuquan Atomic Energy Complex and a nuclear test site at Lop Nur in western China.[18]

The Chinese road to the bomb was intertwined with the Soviet road in more ways than one. Both journeys began with fear of an American nuclear attack and efforts to counter Washington's use of nuclear weapons as a tool of diplomacy. Mao, like Stalin before him, was dismissive of nuclear weapons before he acquired the bomb but, also like his Soviet counterpart, treated them as deterrents to war—even means of ensuring world peace—once they were acquired. But the political and ideological differences between the two communist states after the start of de-Stalinization in the USSR, compounded by their competition for leadership in the communist and anticolonial movements of the 1950s, led to the end of their nuclear cooperation and prompted China to undertake industrial production of the bomb on its own.

In April 1960 China signaled its opposition to the signing of the test ban treaty that some hoped to see approved at the Paris summit in May 1960. When the agreement was finally negotiated and signed by Khrushchev and Kennedy in 1963, Prime Minister Zhou Enlai reaffirmed his country's opposition to the treaty. He stated: "All the countries that signed on to the treaty are restrained by the treaty, that means that all peace-loving countries that signed have no right to possess nuclear weapons, even while they still have to face nuclear blackmail and the danger of annihilation by nuclear weapons."[19]

The emergence of China as a potential nuclear power changed calculations in Washington. In the fall of 1960, still on the presidential campaign trail, Senator John Kennedy expressed his concern about a nuclear-armed China and proliferation in general. "There are indications because of new inventions that 10, 15, or 20 nations will have a nuclear capacity, including Red China, by the end of the presidential office in 1964. This is extremely serious," he asserted in his all-important presidential debate with Vice President Richard Nixon. That concern followed Kennedy into the White House. He expected China's acquisition of the bomb to be "historically the most significant and worst event of the 1960s."[20]

Kennedy's worries were based on the assumption that the Chinese were building a plutonium bomb and the belief that once they acquired it, they would be able to shift the balance of power in East Asia in their favor, as well as provoke the Indians to build their own bomb, thereby contributing to the further proliferation of nuclear weapons. While his worries about a plutonium bomb and a change in the balance of power turned out to be unjustified, the Chinese bomb did indeed convince the Indians to produce one of their own, an acquisition that in turn would make Pakistan launch its nuclear program and acquire a bomb.[21]

When in September 1961 Kennedy addressed the United Nations General Assembly, putting forward a nuclear disarmament program, its first item was "signing the test-ban treaty by all nations." In his mind, the treaty was supposed to be a comprehensive agreement involving all nuclear powers, especially those countries working on their atomic programs. That was a tall order. The British were on board, but others were not. De Gaulle not only would not sign the Partial Test Ban Treaty but would eventually withdraw from NATO. As he said by way of explanation, "The world situation in which two super-States would alone have the weapons capable of annihilating every other country . . . could only paralyze and sterilize the rest of the world by placing it either under the blow of crushing competition, or under the yoke of a double hegemony that would be agreed upon between the two rivals."[22]

Those words were spoken in 1966. By that time, China had already successfully tested its first uranium bomb. For de Gaulle, France's nuclear weapons meant avoiding superpower hegemony. The same was largely true of the Chinese. The two countries embarked on their production of nuclear weapons and continued their tests against the wishes of the nuclear superpowers. Each of them perceived a military threat to itself from one of the superpowers and sought to establish its independence from the other. Fear of possible attack, as well as national pride and desire for self-reliance in defense, drew two otherwise very different countries and leaders, Charles de Gaulle and Mao Zedong, toward the same goal of acquiring nuclear weapons. Both men regarded them first and foremost as tools of diplomacy rather than warfare and means of deterrence rather than attack.

Chapter 18

CUBAN GAMBLE

John F. Kennedy assumed the office of president of the United States in the midst of the hydrogen age, and his first pronouncements on nuclear weapons were very different from the early statements of his predecessor, General Eisenhower. If the latter began by considering the atomic bomb as one more weapon in the American diplomatic and military arsenal, Kennedy regarded nuclear bombs, now including thermonuclear ones, as devices that could end human civilization.

While there was a striking difference between the early Kennedy and the early Eisenhower in their thinking about nuclear weapons, there was little between the late Eisenhower and the early Kennedy. The hydrogen bomb changed the thinking of the American political elite, Republican or Democratic, young or old. Like Eisenhower after Castle Bravo, Kennedy realized that nuclear war could have apocalyptic consequences. He was especially concerned about an accidental outbreak. "Every man, woman and child lives under a nuclear sword of Damocles, hanging by the slenderest of threads, capable of being cut at any moment by accident or miscalculation or by madness," declared Kennedy in September 1961.[1]

Nikita Khrushchev knew how concerned Kennedy was about the possibility of nuclear war. His response was to blackmail the young president. "If the US wants to start a war over Germany, let it be so," Khrushchev told Kennedy when they met in Vienna in June 1961 for their first and only summit. Upon his return to Washington, Kennedy asked his advisers to tell him the probable consequences of a Soviet nuclear attack on the United States. He was told that overall losses were estimated at 70 million people, more than one-third of the American population at the time. When the president was told that just one missile hitting an American city would produce a death toll of 600,000, he remarked that the figure equaled the total losses of the Civil War. "And we have not gotten over that in a hundred years," reflected Kennedy.[2]

In Vienna, Khrushchev failed to bully Kennedy into withdrawal from Berlin. Back in the United States, Kennedy responded to his threats over Berlin with a bellicose speech of his own, announcing a dramatic increase in the size of the American armed forces. He rebutted Khrushchev by citing the ancient philosopher Epicurus: "A man who causes fear cannot be free from fear." Khrushchev decided to resolve the crisis by building the Berlin Wall, preventing free movement in and out of East Berlin, but nuclear blackmail did not disappear from his political toolbox. The geographic focus of its implementation changed quite dramatically, however, from Berlin to Cuba.[3]

By the spring of 1962, Cuba had become a thorn in the political flesh of both leaders. Kennedy, having inherited from the Eisenhower administration a CIA plan to invade the island with the help of Cuban exiles, was the first to suffer.

The goal of the CIA plan was to overthrow the revolutionary administration established two and a half years earlier by Fidel Castro and his leftist allies. That revolution, which led to the nationalization of

American assets on the island, was more anticolonial than communist, but when it came to Latin America the Washington establishment saw little difference between those motives. Kennedy, who came into office vowing to fight communism and improve relations with Latin America, was caught between his inaugural promises. He authorized the invasion but was determined to keep the American role in it secret, which prevented him from using American forces to back the venture. The Bay of Pigs invasion, as the US-backed operation of anti-Castro exiles came to be known, ended in complete defeat three days after it began, on April 20, 1961.[4]

April 1961 became one of the worst months of Kennedy's presidency and one of the best in Khrushchev's tenure. On April 12 the Soviets successfully launched the first man into space, cementing the public perception of their lead in the missile field, and their ability to deliver nuclear bombs to American soil. In fact, there was no missile gap or, rather, the existing missile gap favored the Americans, not the Soviets. Khrushchev wanted to bridge it by developing a new type of intercontinental missile not powered by the difficult-to-handle liquid fuel used by the missile that had put Sputnik into orbit. But as 1961 gave way to 1962, he realized that his scientists and engineers were still years behind the Americans, who were about to start deploying the solid-fuel Minuteman missiles, which could be fired on very short notice, while Soviet missiles needed hours for fueling.

In April 1962, one year after the successful voyage of the first man in space, Khrushchev fired his commander of missile forces, and the country's leadership adopted a new program of missile building. With regard to Cuba, however, Khrushchev's April 1962 was nothing like April of the previous year. The American economic blockade brought the Cuban economy to the brink of disaster, and Khrushchev's reluctance to deliver on his promise of significant economic and, more importantly, military assistance created a rift between the offended Castro and the leadership of Cuba's Communist Party. Mao Zedong emerged on the Cuban horizon as an alternative to the "indecisive" Khrushchev. The Soviet leader, expecting another American invasion, was concerned that he might lose Cuba either to Kennedy or to Mao.[5]

A month later, in May 1962, Khrushchev came up with a solution to both problems: he would install Soviet medium- and intermediate-range missiles, which he had in abundance, on Cuba, dubbed by the Soviets the "Island of Freedom." That would be the equivalent of hitting three birds with one stone: his missiles could easily reach the United States, protect the island from a new American invasion, and make the Castro regime less inclined to flirt with Mao, whose relations with Khrushchev were going from bad to worse.

Khrushchev's plan was to install the missiles in secret and, once they were battle-ready, to announce their deployment to Kennedy and the world. He did not intend to attack the United States but was counting on the fearsome power of nuclear weapons as a deterrent: Washington would now think twice not only about invading Cuba but also about attacking the Soviet Union. As Khrushchev saw it, he was not doing anything that the Americans had not already done. In the last years of the Eisenhower administration, the United States had placed medium-range missiles not only in Britain and Italy but also in Turkey, adjacent to the southern border of the Soviet Union. Khrushchev was now paying the Americans back with interest: the distance between Cuba and Florida was only 90 miles (145 kilometers).

Castro gave his approval for the deployment of Soviet nuclear-armed missiles, citing his readiness to help the socialist camp and, more specifically, Moscow. In his mind, that would put the Soviets ahead of their American nuclear rivals. Khrushchev insisted that his motive was first and foremost to save Cuba from the Americans. The first Soviet R-12 medium-range missiles began to arrive in Cuban ports in late August and early September 1962.[6]

The United States did not detect the Soviet ballistic missiles in Cuba until October 14, 1962. Kennedy was alerted to their presence two days later. He could not believe his ears: Khrushchev, who had publicly declared that he would not put "offensive weapons" on the island and privately assured Kennedy that he would do nothing to undermine him before the congressional elections of November 1962, had deceived him.

Given this betrayal of trust, direct negotiations were considered to be ineffective, and it took Kennedy and his advisers, eventually organized in the ExCom—the Executive Committee of the National Security Council—a whole week to come up with a response to Khrushchev's move. Kennedy advocated an airstrike on the Soviet missile installations; his brother Robert, the federal attorney general, favored an outright invasion—the option backed by the military commanders. Secretary of Defense Robert McNamara, at odds with the Chiefs of Staff, proposed a naval blockade of the island. He was concerned—rightly, as it turned out—that it was too late for an airstrike or an invasion, as the missiles might be battle-ready.[7]

At 7:00 p.m. on October 22 a somber Kennedy, having received intelligence that the Soviet missiles were battle-ready, addressed the nation. In Moscow it was 3:00 a.m. on October 23. Learning of the forthcoming address, Khrushchev summoned the Presidium members to the Kremlin. "The point is that we do not want to unleash a war. We want to intimidate and restrain the USA vis-à-vis Cuba," Khrushchev told his nervous colleagues as they awaited Kennedy's speech, not knowing what to expect.

Khrushchev then asked what the Presidium should do in case of attack. One possibility was to declare that the USSR had a defense treaty with Cuba, making an American assault on Cuba tantamount to an attack on the Soviet Union. Another was to declare that the missiles were Cuban, and that Havana might respond to an invasion with nuclear weapons. In that case Washington would have to negotiate with Havana, not Moscow, and in the worst-case scenario there might be a nuclear war between the USA and Cuba, with the USSR remaining on the sidelines. Khrushchev liked the idea, but one of his close allies, Anastas Mikoyan, disagreed: the Americans would panic and hit the island with all their might, including nuclear weapons. It would be difficult for the USSR to do nothing.

As they were discussing a possible response to the attack, the Soviet defense minister, Marshal Rodion Malinovsky, prepared a telegram to the commander of Soviet troops in Cuba, General Issa Pliev: he was to resist the attack by all means possible. It then dawned on Khrushchev that he was actually giving the commander on the ground permission to attack the American mainland with nuclear weapons. "If 'all' means without reservations, that means missiles too, that is, the outbreak of thermonuclear war!" exclaimed Khrushchev. They revised the text of the telegram. Now General Pliev was prohibited from using ballistic missiles without orders from Moscow, but the telegram gave him de facto clearance to use tactical nuclear weapons if attacked. Pliev now had the power to start a local nuclear war. What would stop it from going global was never discussed. They decided to postpone sending the telegram until they heard Kennedy's speech.[8]

To Khrushchev's relief, Kennedy did not announce an imminent attack, opting instead for a naval blockade. In a private message to Khrushchev, Kennedy wrote: "I have not assumed that you or any other sane man would, in this nuclear age, deliberately plunge the world into war which it is crystal clear no country could win and which could only result in catastrophic consequences to the whole world, including the aggressor." Khrushchev's response combined reassurance with threat. "We confirm that armaments now on Cuba, regardless of classification to which they belong, are destined exclusively for defensive purposes, in order to secure Cuban Republic from attack of aggressor," wrote Khrushchev after a sleepless and nerve-wracking night spent in the Kremlin.[9]

That night Khrushchev ordered the captains of ships carrying nuclear warheads and missiles near the Cuban shores to increase their speed and asked those farther away, who would not make it to the island before the imposition of the naval blockade, to turn back. But he refused to turn back ships delivering non-military cargos, presenting Kennedy with the dilemma of what to do with those cargo ships as they approached the quarantine line.[10]

On the evening of October 23, John Kennedy sent Robert to meet with the Soviet ambassador to Washington, Anatolii Dobrynin. The Kennedy brothers wanted to know what orders the captains of the Soviet ships had received. Robert Kennedy told Dobrynin that they would have to intercept the ships. "But that would be an act of war," responded Dobrynin. The next day Dobrynin and his staff, their nerves on edge, were glued to television screens as they watched live American reports about a Soviet tanker approaching the quarantine line. The announcer counted down the miles remaining. Dobrynin expected American destroyers to open fire; instead, they let the tanker through, as no one expected it to be carrying anything but oil. "There was a general sigh of relief," recalled Dobrynin decades later. He served as Soviet ambassador in Washington for twenty-four years and considered that day the most memorable of his long tenure in office.[11]

The same day, October 24, was no less dramatic in the White House. After some hesitation, John Kennedy was persuaded by his advisers to authorize the interception of two Soviet cargo ships suspected of carrying ballistic missiles. As US Navy commanders believed that the cargo ships were protected by Soviet nuclear-armed submarines, the president agreed that the submarines should be attacked first if the ships refused to stop. The decision did not come easily. His brother Robert later described the moment in his memoir: "He opened and closed his fist. His face seemed drawn, his eyes pained."[12]

The immediate crisis was resolved when news unexpectedly reached the White House that the ships Kennedy had just ordered to be intercepted had turned back much earlier. Relieved, Kennedy aborted the order. "We are eyeball to eyeball, and I think the other fellow just blinked," commented Secretary of State Dean Rusk. They did not know that Khrushchev had blinked long before, on the night of Kennedy's television speech, when he ordered most of the ships bringing missiles to Cuba to turn back. The lack of timely information almost caused the outbreak of a shooting war. Unknowingly, Kennedy had in fact ordered the interception of ships moving away from the quarantine line, not toward it.[13]

Meanwhile, the Strategic Air Command (SAC) raised its Defense Readiness Condition (DEFCON) to level 2, just short of level 1, which meant open warfare. Nuclear-armed B-52s were now in the air, flying to Europe and the eastern Mediterranean. They were prepared to strike targets in the Soviet Union. The DEFCON level was raised on October 23 and implemented the next day, at 10:00 a.m. EST. Soviet intelligence reported the news to Moscow, where it was already late evening, October 24. According to a senior Soviet diplomat, Khrushchev "shat his pants." He was in a panic, veering from fury to depression. "The confusion that reigned there was only covered up by Khrushchev's blustering public statements," recalled the same diplomat.[14]

On October 25, the day after the Soviets intercepted the DEFCON 2 order, Khrushchev summoned his colleagues once again. He argued that they should remove the missiles from Cuba. "It's not capitulation on our part because, if we shoot, they'll shoot back," argued Khrushchev. He then proposed a plan of action. "Kennedy says to us: take your missiles out of Cuba. We answer: give firm guarantees and promises that the Americans will not attack Cuba. That's not bad. We could take out our R-12s and leave other missiles there." Khrushchev was suggesting that tactical nuclear missiles not detected by the Americans should stay on the island.[15]

Khrushchev's aides, frightened that his adventurism would make them a target of American nuclear bombers, unanimously supported the proposal. They wanted the nightmare to end. "I will never forget," recalled Khrushchev's successor, Leonid Brezhnev, "how Nikita [Khrushchev], in a panic, would send a telegram to Kennedy, then 'en route' order it to be stopped and recalled. And why? Because he wanted to screw over the Americans. I remember he was shouting at the [Central Committee] Presidium: 'We can hit a fly in Washington with our missiles!' That fool Frol Kozlov [under Khrushchev in practice the second secretary of the Central Committee] echoed him: 'We are holding a gun to the Americans' head!' And what happened? A shame! We nearly plunged into nuclear war."[16]

But Khrushchev would not have been Khrushchev if he had not

tried to bargain. "We must take into account that the US did not attack Cuba," he said, addressing his long-suffering Presidium colleagues on October 27. "We will not eliminate the conflict if we do not give satisfaction to the Americans and do not tell them that our R-12 missiles are there," continued Khrushchev, before adding: "And if we receive in return the elimination of the [American] bases in Turkey and Pakistan, then we will end up victorious." There were no American missiles in Pakistan. Khrushchev apparently confused missile installations with the airbase from which the Americans used to send their U-2 flights over the USSR. By the time Khrushchev's aides translated his rambling into a letter, Pakistan was gone, but Turkey remained. The idea had come to Khrushchev from the American media, where it was discussed without protest from the White House: Kennedy had contemplated the exchange from the very start of the crisis.[17]

Just as Khrushchev finally made a proposal to which Kennedy was receptive, things began to spin out of control for the two leaders. On the morning of October 27, exhausted after a sleepless night, General Issa Pliev, the Soviet military commander in Cuba, took a nap. Just then, his subordinates fired a surface-to-air missile at an American U-2 airplane surveying the island, shooting it down. They had been expecting an American air attack at any minute, and Castro had already ordered his air defenses to shoot at low-flying American airplanes. The two Soviet generals who gave the order without consulting Pliev or Moscow believed that with the U-2 registering the locations of their missiles, an American assault was imminent. The officers who executed the shot braced themselves for immediate retaliation.[18]

With the U-2 down, the Cuban crisis had degenerated into a shooting war. It almost turned into a nuclear one when, late at night on October 27, the captain of a Soviet submarine in the Sargasso Sea ordered a strike against American ships with a nuclear torpedo. Only the interference of a senior officer on the submarine stopped the launch. In Moscow, Khrushchev woke up to the news about the downing of the U-2. Terrified of what might come next, he was prepared to withdraw his demand about the Turkish missiles and take instead Kennedy's pledge not to invade Cuba as the price of his withdrawal. At that moment news arrived that the no less desperate Kennedy had offered a secret deal involving a swap of the Soviet missiles in Cuba for the American ones in Turkey.

Khrushchev jumped at the offer. Soviet acceptance was broadcast on Radio Moscow to save time and prevent any new accident: a note passed through diplomatic channels might take up to twenty-four hours to reach the White House.[19] "We instructed our officers—these weapons, as I had already informed you earlier, are in the hands of Soviet officers—to take appropriate measures to discontinue construction of the aforementioned facilities, to dismantle them, and to return them to the Soviet Union," read the key sentence of Khrushchev's typically lengthy response. It was met with jubilation in Washington and, as Castro stated later, with "indignation" in Havana.[20]

Fidel Castro, whose revolution was anti-imperial first and communist second, felt betrayed—his fate and that of his country had been decided by the superpowers without consulting him. He refused to allow the inspection of missile sites on Cuban territory—one of the conditions of the Khrushchev-Kennedy deal. He also opposed the withdrawal of Soviet bombers, which Kennedy considered part of the "offensive weapons" formula that he used in correspondence with Khrushchev. Finally, Castro wanted the Soviet tactical nuclear missiles, of which Kennedy was unaware, to stay on the island.

Khrushchev had a problem on his hands—a rebel client who had first accepted nuclear weapons and was now resisting their removal. He sent a special representative—his top aide, Anastas Mikoyan—to convince Castro to go along with the deal struck by Moscow and Washington. Castro was forced to allow the withdrawal of the missiles, bombers, and tactical nuclear missiles—they were Soviet property, after all—but still refused to allow the inspection of the Cuban missile sites. The Soviets had to show the missiles loaded on their ships to the Americans, and American inspections of ships headed back to the USSR humiliated the Soviet military, turning its commanders against Khrushchev. In October 1964 Marshal Rodion Malinovsky, the Soviet defense minister, backed the palace coup that removed Khrushchev from power.[21]

Khrushchev's nuclear gamble backfired. His attempt to blackmail Kennedy with the threat of nuclear war failed, and Khrushchev himself became so frightened that he was prepared to remove the missiles in exchange for nothing but a promise from Kennedy not to invade the island. Khrushchev had confidently expected the missiles to prevent American aggression not only against Cuba but also against the Soviet Union. But once the missiles were discovered, he abandoned his scheme. Kennedy, on the other hand, abandoned his plan for an airstrike on the missile sites once he learned that the weapons were battle-ready. No matter what mistakes they made, the two leaders were frightened of launching a nuclear war by accident and did all they could under the circumstances to avoid it.

The conditions under which the missiles were removed from Cuba pointed to ways in which the nuclear threat could be addressed in the future: the two superpowers might simultaneously remove their missiles from certain areas and reduce their nuclear presence in certain parts of the world. The missile swap of 1962 showed that nuclear-armed states could become political hostages of the countries that agreed to accept their weapons. Not only did Khrushchev have trouble with Castro, but Kennedy was concerned about the reaction in Turkey and NATO generally to his unilateral decision to remove missiles from Turkey. It was not only impossible to deploy nuclear weapons on the territory of an ally without the ally's consent but also difficult to withdraw them unilaterally without damaging political relations. Khrushchev learned that lesson the hard way: Cuba would become the first and last country in which the Soviets deployed their nuclear arsenal.

More than anything else, the crisis pointed to the extreme danger of nuclear brinkmanship. Khrushchev, for one, lost his appetite for nuclear bluffing. The two leaders were sufficiently frightened of each other and of their nuclear arsenals to take a new look at their relations and the arms race of which they had become hostages.

Chapter 19

BANNING THE BOMB

The US-Soviet discussion concerning a nuclear test ban resumed on the same day that the most acute stage of the Cuban missile crisis came to an end—October 28, 1962. Writing to Khrushchev upon receiving his radio address pledging the removal of the missiles, Kennedy welcomed it as "an important contribution to peace." "Perhaps now," he continued, "as we step back from danger, we can together make real progress in this vital field. I think we should give priority to questions relating to the proliferation of nuclear weapons, on earth and in outer space, and to the great effort for a nuclear test ban."[1]

Kennedy wanted both a test ban and a nonproliferation agreement. With the test-ban negotiations going on since 1958, he hoped that both goals could be achieved with one treaty. When he moved into the Oval Office in January 1961, the test-ban talks were deadlocked and all but derailed by the failed Paris summit of 1960, although the two countries observed the moratorium on all nuclear tests. Kennedy and Khrushchev resumed the dialogue by correspondence in March 1961. For Khrushchev, the test ban was never about horizontal nonproliferation. Besides,

Kennedy wanted a comprehensive ban on tests, whether above or below ground or in the world's oceans. Compliance could easily be verified in all spheres except underground. To detect low-yield underground tests, inspectors would have to be admitted to both countries.[2]

Kennedy, like Eisenhower before him, was prepared to do so, while Khrushchev was unwilling to allow more than three inspections per year. Kennedy, knowing that he would have to convince the skeptical Congress to ratify the deal, pushed for twenty. Khrushchev would not budge, and his representatives suggested that the Americans would use inspections for espionage. The Soviet Union was lagging behind in the development and production of nuclear arms, and Khrushchev recalled in retirement: "We could not allow US and its allies to send their inspectors crisscrossing around the Soviet Union. They would have discovered that we were in a relatively weak position, and that realization might have encouraged them to attack us."[3]

With the Cuban missile crisis coming to an end, the two leaders decided to give the test ban a new try. "Mr. President, we have now conditions ripe for finalizing the agreement on signing a treaty on cessation of tests of thermonuclear weapons," wrote Khrushchev on October 30, 1962, only two days after accepting the American conditions for ending the crisis. "It would be very useful to agree on ending tests after such strain when people lived through great anxiety," he continued. "It would be a great reward for the nervous strain suffered by the peoples of all countries. I think that your people felt as much anxiety as all other peoples expecting that thermonuclear war would break out any moment. And we were very close to such war indeed. That is why it would be good to give satisfaction to the public opinion."[4]

Khrushchev was prepared to go much further than banning tests of hydrogen bombs. He proposed complete nuclear disarmament—an old Soviet diplomatic strategy—and a nonaggression agreement between NATO and the Warsaw Pact, leading to the dissolution of the military blocs. That was a new overture. The dissolution of NATO would have left Europe at the mercy of the superior Soviet forces, and it is unlikely that Khrushchev counted on the acceptance of that proposal. But after the Cuban debacle he needed something truly spectacular to restore his standing at home and abroad. He was under attack not only from Fidel Castro but also from Mao Zedong for caving in to the Americans. His promise to Kennedy to keep secret the swap of missiles in Cuba for those in Turkey left him with no propaganda tools to fight back.

Khrushchev was desperate to show that he had obtained something substantial from Kennedy in exchange for the removal of Soviet missiles from Cuba. While a nonaggression pact and the dissolution of the military blocs were long shots, signing a test-ban agreement seemed more feasible. There was one caveat, however: Khrushchev was not prepared to compromise on the issue of inspections. He reminded Kennedy that they had already agreed on banning tests in the atmosphere, in outer space, and underwater. They disagreed only on underground tests, and Khrushchev asked Kennedy to encourage his subordinates to find a compromise that would not involve on-site inspections. "We shall not accept inspection, this I say to you unequivocally and frankly," asserted Khrushchev.[5]

In the correspondence that followed, Khrushchev repeated his previous offer of three inspections per year. But there was some softening of his position as well. He proposed to install so-called black boxes, automatic monitoring stations—an idea, as he wrote to Kennedy, suggested by British scientists and apparently supported by their American colleagues participating in the Pugwash Conferences on Science and World Affairs. Khrushchev suggested three stations in the United States and three in the Soviet Union, proposing seismically active regions in the USSR as locations. "We believe that now the road to agreement is straight and clear," wrote Khrushchev.[6]

Kennedy reduced his demand for inspections from twenty to eight and agreed to start formal negotiations. They began on January 14, 1963, first in Washington and then in New York; besides the Americans and the Soviets, they included the British. Both Kennedy and Khrushchev sent signals of good will to encourage the negotiating process. On January 22, Kennedy met with his foreign policy and nuclear teams and advised Glenn Seaborg, chairman of the Atomic Energy Commission, to postpone already planned nuclear tests until the conclusion of the talks. Seaborg was anything but happy but had to oblige.[7]

Khrushchev was also on his best behavior. In December he received Norman Cousins, a leader of the American "ban the bomb" anti-test movement, in the Kremlin. Cousins was on a mission from both Kennedy and Pope John XXIII, trying to improve Soviet-American relations and the position of Catholics in the Soviet Union. At the pope's request, passed on by Cousins, Khrushchev released from the Gulag the leader of the Ukrainian Catholics, Archbishop Yosyf Slipy. Before the end of January, Slipy was already in Rome, finally free after more than seventeen years of imprisonment.[8]

But the negotiations that both Kennedy and Khrushchev had counted on to make a breakthrough to agreement ended in impasse. The question of the number of inspections turned out to be impossible to resolve, with neither party prepared to compromise. The Soviet representatives, bound by instructions from Moscow, could not offer more than three inspections per year, while the Americans demanded no less than eight. Kennedy knew that he would not be able to convince Congress to ratify the treaty with fewer than eight inspections per year, as there was concern that the Soviets would cheat and continue their tests.[9]

News of the impasse on testing was met with disappointment by the growing "ban the bomb" movement. By then, the damage to human health caused by atmospheric nuclear testing was an open secret. Proof came with the results of the Baby Teeth Survey, released in November 1961. The idea for the survey came from an article by the Johns Hopkins University biochemist Herman M. Kalckar, who suggested that children's teeth could be analyzed to measure the absorption of reactivity by the human body. The survey was launched in December 1958 by Barry Commoner, professor of plant physiology at Washington University in St. Louis, and by the Greater St. Louis Citizens' Committee for Nuclear Information.

The committee launched a campaign to collect as many primary teeth as possible. They were after strontium-90, which the human body, especially the fast-growing bodies of children, mistook for calcium and stored in bones and teeth. The goal was set at 50,000 teeth per year, and the survey was supposed to continue for ten years. The campaign was highly creative in encouraging children to donate their teeth by creating membership in Operation Tooth Club and awarding donors with a button inscribed "I Gave My Tooth to Science." The preliminary study showed a steady rise in the amount of strontium-90 in the teeth of children born in the 1950s—the later a child was born, the greater the amount of strontium. Comprehensive results showing a dramatic increase of strontium in the teeth of American babies made a strong impression not only on the public but allegedly on President Kennedy as well.[10]

Kennedy contemplated the idea of sending to Moscow his ultimate fixer—his brother Robert—to meet with Khrushchev and try to reach a compromise on the test ban. He settled eventually on Norman Cousins, who had established good rapport with Khrushchev back in December and proved his ability to show results: the release of the Ukrainian archbishop was a clear indication of that. In April 1963 Cousins visited Khrushchev in Pitsunda, near Sochi on the Soviet Black Sea coast. Khrushchev refused to yield, but the visit did not go unappreciated. Khrushchev apparently decided that Kennedy was not yielding either. He had to look for a way out of the deadlock.[11]

On April 24, 1963, after returning from his Black Sea vacation, Khrushchev proposed to the Presidium of the Central Committee, the country's ruling body, that efforts to achieve a comprehensive agreement on testing be abandoned. A partial ban should be sought instead, excluding underground tests and covering only those carried out in the air, outer space, and underwater. A similar idea had been closely associated with hopes for the signing of a treaty by Khrushchev and Eisenhower in May 1960. It continued to be discussed in the West and was also proposed by some Soviet scientists. Khrushchev appeared to have lost hope of negotiating a better deal with Kennedy than he might have reached with Eisenhower.[12]

A month later, the same idea was put forward in Washington. On May 27 two US senators, the Republican Thomas E. Dodd and the

Democrat Hubert H. Humphrey, introduced a resolution supported by thirty-four of their colleagues in support of a test-ban agreement excluding underground explosions. When on May 30 Kennedy and Prime Minister Harold Macmillan of Britain proposed to send a high-level delegation to Moscow to continue talks, Khrushchev agreed. With underground tests off the table, so was the question of inspections.[13]

On June 10 Kennedy delivered one of his most celebrated speeches, a commencement address at American University in Washington. Partly written by Norman Cousins, the speech struck a conciliatory note when it came to ideological conflict between the two countries. "No government or social system is so evil that its people must be considered as lacking in virtue," declared Kennedy. He also echoed Khrushchev's repeated claims that nuclear war could destroy both superpowers, suggesting the equal status of their nuclear arsenals. The speech sent a signal that Kennedy was warming up to the idea of banning only atmospheric tests. "I now declare that the United States does not propose to conduct nuclear tests in the atmosphere so long as other states do not do so," said Kennedy. "We will not be the first to resume." He did not commit to stopping underground tests, limiting his pledge to those that could be verified.[14]

Khrushchev responded to Kennedy in a speech given in East Berlin on July 2. He declared that because the position taken by the Western powers precluded the conclusion of a comprehensive deal, the Soviet Union was prepared to sign a limited agreement banning tests in the atmosphere, outer space, and underwater. The statement was met with applause, as was the accompanying proposal to sign a nonaggression pact between East and West. If Kennedy regarded the test ban treaty as a prelude to future nonproliferation agreements, Khrushchev conditioned it on the signing of a nonaggression pact.[15]

The task of decoupling the two proposals fell to the high-level British-American delegation headed on the American side by Averell Harriman, the World War II–era ambassador to Moscow, whom Khrushchev treated with respect as someone who had represented Franklin Roosevelt and reminded him of the good old days in US-

Soviet relations. The negotiations began in Moscow in Khrushchev's presence on July 13, 1963. Khrushchev wanted a nonaggression pact, while Harriman proposed a nonproliferation agreement. The two proposals canceled each other out, and a partial test ban was the result. The document was initialed on July 25 and signed on August 5 in Moscow. Khrushchev finally had something to show his opponents in exchange for his nuclear retreat from Cuba.[16]

In the United States, Kennedy had yet to convince Congress to ratify the agreement. He found a ready partner in Norman Cousins and his National Committee for a Sane Nuclear Policy. Its activists became salesmen of the partial test-ban deal with Moscow. On August 2, 1963, SANE ran a full-page advertisement in the *New York Times*, one of many actions initiated or supported by the committee. Titled "Now it's up to the Senate . . . and You!," the ad read: "A test ban treaty would put an end to widespread radioactive fallout from nuclear testing. Present and future generations would be spared additional reproductive damage and bone cancer. Little can be done about what has already occurred." The campaign was a huge success. Congress ratified the agreement by a vote of 80 to 14.[17]

The signatories of the partial test-ban treaty included three nuclear powers—the United States, the Soviet Union, and the United Kingdom. Already nuclear France and potentially nuclear China refused to join: they felt that they were behind the "big three" and needed testing to catch up with the established nuclear powers. The signing of the treaty did not lead to the nonproliferation that John Kennedy had hoped for. His top concern was China.

On December 14, 1962, when Kennedy wrote to Khrushchev signaling his readiness to return to the discussion of the test ban after the Cuban debacle, he focused specifically on China. "I hope that in your message on this subject you will tell me what you think about the position of the people in Peking on this question," wrote Kennedy. "It seems to me very important for both of us that in our efforts to secure an end to nuclear testing we should not overlook this area of the world." Khrushchev ignored Kennedy's China question. The long letter on the test ban treaty that Khrushchev sent to Kennedy on December 19, five days after Kennedy had asked his question about Beijing, contained no mention of China at all.[18]

In January 1963 Kennedy told his advisers that he was concerned about the way in which the Chinese might wield the power of their atomic bomb once it was built. He still hoped that the test-ban agreement would include Beijing and assumed that the Soviets shared his concerns about China. There was little doubt that Khrushchev was unhappy with Mao, but there is no indication that he saw the treaty as a way to slow down or stop China's progress toward acquisition of the bomb. In the summer, Kennedy asked Averell Harriman, his representative at the negotiations, to find out whether the USSR would agree to join the United States in an effort to stop China from acquiring nuclear weapons or allow the Americans to pursue that goal on their own. Khrushchev showed no interest in the informal offer. There was little doubt that Khrushchev was unhappy with Mao, but there is no indication that he saw the treaty as a tool to impede China's progress. Khrushchev was losing hope that Soviet-Chinese relations could be dramatically improved, but he did not want to see them worsen. If he were seen to be allying himself with the United States against communist China, that would be a major blow to his standing in the communist world.[19]

Kennedy's aides, especially national security adviser McGeorge Bundy, believed that the United States had only one option—a military attack on China's nuclear industry infrastructure to stall the country's progress toward the bomb. In September 1963 Bundy met with a guest from Taiwan, General Chiang Ching-kuo, a son of Chiang Kai-shek and head of his security services. Ching-kuo visited the CIA, where he discussed American transportation and logistical support for a raid by Taiwanese commandos on China's fuel-production plant in Baotou in Inner Mongolia. Bundy, who took part in the discussions, felt that the general should meet the president to discuss the matter further.

The meeting took place on September 11. Kennedy was cautious. He asked his guest "whether it would be possible to send 300 to 500 men by air to such distant . . . atomic installations as that at Baotou, and whether it was not likely that the planes involved would be shot down." Chiang responded that the CIA considered the operation feasible. That probably did not sound very reassuring to Kennedy, still scarred by the Bay of Pigs disaster, in which the CIA had promised him quick results from another commando-type operation. Chiang returned to Taiwan without Kennedy's approval for the operation.[20]

On October 7 Kennedy signed the partial test-ban treaty, which contained nonproliferation clauses. On November 22 he was gone, the treaty becoming his last major contribution to international politics. The assassination of John Kennedy did not change American policy toward China or nonproliferation in general. McGeorge Bundy, who retained his position under President Lyndon Johnson, continued working on the option of a strike against Chinese nuclear facilities. In the fall of 1964, with the arrival of intelligence reports suggesting that the Chinese were about to test their first atomic device, the urge to act became stronger, but so did Johnson's desire for caution. In the late summer and early fall, in the midst of his first and last presidential race, he was not taking any risks. There would be no Taiwanese commandos and no American-backed attempt to interfere with the Chinese nuclear program.[21]

The Chinese successfully tested their first atomic bomb on October 16, 1964. China's status as a nuclear power and the threat that it presented was no longer Khrushchev's concern—he had been removed from office two days earlier by his underlings, who cited the Cuban nuclear adventure among his many sins. "May it be our farewell salute to Khrushchev!" a smiling Zhou Enlai told Mao Zedong. To the relief of the Americans, the Chinese weapon was a uranium and not a plutonium bomb. But what Kennedy had feared as the worst development of the 1960s had come to pass: China had acquired a nuclear weapon. The official statement issued by Beijing that day held to Mao's definition of the bomb as a paper tiger.[22]

"This was our view in the past, and this is still our view at present," went the statement. "China is developing nuclear weapons not because we believe in the omnipotence of nuclear weapons and that China plans to use nuclear weapons. The truth is exactly to the contrary. In developing nuclear weapons, China's aim is to break the nuclear monopoly of the nuclear powers and to eliminate nuclear weapons."[23]

Development of the atomic bomb as a step toward the elimination of nuclear weapons seemed to be a unique Chinese response to the nuclear scientists' and nuclear powers' doubletalk on the subject. Some American scientists had wanted to drop the bomb on humans to warn the world about the dangers of nuclear war; other scientists, concerned that one country with the bomb was worse than two, had passed atomic secrets to the Soviets in an effort to break the American monopoly; Eisenhower had built up the American nuclear arsenal while promoting Atoms for Peace; and now China was claiming to seek the elimination of nuclear weapons by joining the nuclear club.

There was no indication that China would now sign the test ban treaty. The statement attacked it as "a big fraud to fool the people of the world, that it tried to consolidate the nuclear monopoly held by the three nuclear powers and tie up the hands and feet of all peace-loving countries." The statement mentioned three, not four, nuclear powers, omitting France, which had not signed the treaty. Beijing and Paris were now in the same boat or, more precisely, outside the boat built by the original Big Three of the nuclear age.[24]

Rhetoric aside, the Chinese built their own atomic bomb for reasons similar to those of their predecessors: security, international standing, and national pride. After so many explicit and implicit American threats to use the bomb against China, the Chinese felt that they not only needed one of their own but had a right to it. After the successful test of their atomic bomb, they would keep working on improving its design while doing their best to acquire a hydrogen bomb as well. They successfully tested one in June 1967.

The Soviet leaders would soon regret Khrushchev's refusal to join forces with Kennedy to stop the Chinese from acquiring their own atomic bomb. In March 1969 the Soviet Union and the People's Republic of China, two communist powers now armed with nuclear weapons, clashed over control of a few tiny islands in the Ussuri River on the Soviet-Chinese border. The Soviets were outnumbered but possessed superior armaments that allowed them to repel the Chinese attack. With war in the air, Soviet commanders considered the use of nuclear weapons and a strike against Chinese nuclear installations. The Soviets in the Far East were in a position comparable to that of the Americans in Central Europe: inferior in ground forces, they pinned their hopes on nuclear weapons as a deterrent.

In August 1969, when the Chinese once again crossed their border with the USSR into the republic of Kazakhstan, Moscow decided to engage in indirect signaling that it was determined to use nuclear weapons if such border provocations continued. The Soviets informed their East European allies and leaked information to the Chinese that unless Beijing desisted from such behavior, they were prepared to deliver a nuclear strike against Chinese nuclear facilities, including the Lop Nur testing range. The Chinese got the message two days before the expiration of the informal Soviet ultimatum in September. Subsequent negotiations resulted in an agreement not to attack each other with conventional or nuclear weapons, but little trust remained between the two former allies.

The Soviets decided not to include nuclear weapons in their planning, as they did not believe that nuclear bombs could stop millions of Chinese troops from flooding across their border. Even so, the signaling worked. In October 1969 Mao and his closest lieutenants left Beijing in anticipation of a possible Soviet nuclear attack on the capital. In close succession, the Chinese conducted tests of an atomic and a hydrogen bomb to showcase their nuclear potential. Despite this show of force, it took months before the Soviet-Chinese negotiations produced a ceasefire and made it possible for Mao to return to his capital.[25]

American and, to a lesser degree, Soviet efforts at nonproliferation had failed spectacularly, to the detriment of both the Americans and the Soviets. But new hope emerged with an initiative from countries that did not have nuclear weapons and did not want to acquire them. The non-nuclear countries were concerned about vertical proliferation—the buildup of nuclear arsenals and thus the threat of nuclear war posed by the superpowers. Those countries were prepared to put limits on horizontal proliferation, or their own right to acquire the bomb, in exchange for reining in the superpower race.

The initiative began in 1958, when the Irish foreign minister, Frank Aiken, submitted a draft resolution titled "Further Dissemination of Nuclear Weapons" to the First Committee of the UN General Assembly. The resolution claimed that proliferation was making nuclear disarmament more difficult to achieve and called on the parties to the UN-sponsored disarmament talks, including the United States and the Soviet Union, to stop proliferation and ask other states not to acquire nuclear weapons. The resolution was put to a vote but not adopted. The United States was in the process of deploying nuclear weapons to Europe under the auspices of the "dual key" program and was not interested.

In 1959 Aiken submitted a new version of his resolution that took account of American interests, but this time it met with rejection from the Soviet Union: the new version allowed the United States to continue its transfer of weapons to other countries as long as they remained under American control. In 1960, Aiken revised the resolution once again: this time it was supported by the Americans but not by the Soviets. When Sweden resubmitted the resolution in the following year, calling on non-nuclear countries to refuse nuclear weapons, it was the turn of the United States to oppose it: such a resolution would discourage the European members of NATO from accepting and hosting

American nuclear weapons. That would have been detrimental to the success of the American-sponsored Multilateral Force (MLF), a fleet of ballistic-missile submarines manned by NATO crews and armed with US Polaris missiles.[26]

In June 1965 India and Sweden, two countries then considering the acquisition of nuclear weapons, came up with a new proposal. They put forward a draft nonproliferation treaty bringing together two elements: the refusal of non-nuclear states to acquire nuclear weapons would be combined with the agreement of nuclear states to establish some form of control over the size of their nuclear arsenals. The Cuban missile crisis had shown the world how uncontrolled and dangerous great-power nuclear competition could become.

The timing of the new resolution turned out to be perfect. With the Multilateral Force project defeated by disagreements between the United States and its NATO allies on the funding and basing of submarines, the Americans were prepared to change their position on the nonproliferation initiative of the non-nuclear nations. The Soviet Union was also interested in the treaty: now not only Washington but also Moscow was frightened by the Chinese atomic bomb and Beijing's rapid progress toward a hydrogen weapon.[27]

With the two superpowers favoring the treaty, the non-nuclear countries joined in as well. The turning point was the agreement of nuclear powers to recognize the non-nuclear states' right to develop nuclear arms programs. In return for giving up that right, the non-nuclear states were promised nuclear technology for non-military purposes and a continuing effort on the part of the nuclear states to work toward nuclear disarmament. The two superpowers did their best to promote the treaty to their allies and clients. The American nuclear umbrella over Europe and Japan eased the concerns of NATO members, while the Soviets, for their part, had little difficulty in persuading the East European countries to join. Many countries believed that they were more secure without nuclear weapons, now that their neighbors and rivals were prohibited from acquiring the bomb by an international agreement policed by the superpowers.[28]

President Lyndon Johnson signed the treaty on July 1, 1968. In Moscow two other veterans of the Cuban missile crisis, the US ambassador to the Soviet Union, Llewellyn Thompson, a participant in the ExCom meetings in the fall of 1962, and the Soviet foreign minister, Andrei Gromyko, put their signatures to the treaty. The Non-Proliferation Treaty went into effect on March 5, 1970, with forty-three countries submitting their depositions of acceptance and ratification. The UN International Atomic Energy Agency, charged with the dual task of promoting the use of nuclear energy for peaceful purposes and curbing the spread of nuclear technology for warfare, now acquired the legal teeth it needed to accomplish the latter mission. It was tasked with implementing the nonproliferation regime though inspections of nuclear facilities in the non-nuclear states.[29]

The Non-Proliferation Treaty turned out to be effective in slowing down the acquisition of nuclear weapons. It could not undo the proliferation process of the past or influence the behavior of countries that were not party to it. At the time of signing, there were three countries that fit both categories—France, China, and, last but not least, Israel, which had never admitted ownership of the bomb.

Chapter 20

STAR OF DAVID

Israel's road to the atomic bomb began in 1948, the year the modern state of Israel declared its independence. David Ben-Gurion, the founding father of Israel, also established the country's nuclear program. The atomic bomb was his best hope for the survival of Israel and the guiding star of his foreign policy vision.

The driving force behind Ben-Gurion's project was first and foremost fear that the Jewish state and nation might be annihilated or, in other terms, it was the security of the state rather than pride or international standing that fueled the Israeli drive toward the bomb. That fear was informed by the recent trauma of the Holocaust and the extreme animosity between Israel and its Arab neighbors that followed the declaration of the country's independence from Britain in May 1948. The war that accompanied the birth of Israel as an independent state lasted until February 1949 and involved five Arab states. It ended with victory for Israel, which extended its territory beyond the borders allocated to it by the UN mandate, but made permanent enemies out of its neighbors.[1]

Ben-Gurion was convinced that Israel had to hold the annexed territories in order to survive and that the outbreak of a new conflict was only a matter of time, since the Arabs did not accept the outcome of the war. His worst nightmare was a simultaneous attack of all Arab armies on Israel. Anxiety and fear led Ben-Gurion, like many leaders before and after him, to work for the acquisition of nuclear weapons. He needed the atomic bomb not to deal with threats posed by a nuclear power but to cope with the prospect of conventional war against a numerically much more powerful enemy.[2]

The Israeli nuclear program began only a few years later, in unison with the French one. Ben-Gurion began looking for experts in weapons of mass destruction as early as April 1948, and in November of that year he wrote a pamphlet informing his troops, among other things, of the "tremendous power" concealed in the atom. About the same time, he became interested in the Palestine-born Jewish scientist Moshe Sordin, who was working on the French nuclear project, and 1949 became the year of action. Israeli scientists were invited to France to work at a nuclear facility; Israeli students were sent to the world's best universities to study nuclear physics, with Enrico Fermi as one of their teachers; and geologists began their survey of the Negev Desert in search of uranium.[3]

The Israeli project not only began with the help of the French nuclear program but in many ways shared its trajectory. Both Charles de Gaulle and David Ben-Gurion would establish the research components of their countries' nuclear projects, lose political power, and then come back to continue the projects and bring them to fruition—the building of a bomb. If de Gaulle was out of power by 1946 and returned in 1958, Ben-Gurion, who began his tenure as prime minister in 1948 and temporarily lost power in 1954–55, returned to serve as prime minister until 1963. Ben-Gurion's first major crisis after his return to office was that of the Suez Canal. Israel shared the crisis with France, and for both countries it was a key moment in their atomic bomb programs.

The French and British, as discussed earlier, wanted Israel to start a war with Egypt that would provide the two European powers with a pretext to regain control of the canal after its nationalization by the government of Gamal Abdel Nasser. Israel had multiple reasons to go to war, from trying to defeat Nasser, who was regarded as the only Arab leader capable of uniting Israel's neighbors in an attack on the new state, to securing the passage of Israeli shipping through the Straits of Tiran, Israel's access route to the Indian Ocean, which had been blocked by Egypt in 1950. But when the French defense minister spoke with an Israeli official on the day after Nasser's announcement of the nationalization, asking whether Israel would take part in a tripartite invasion of Egypt, the official responded that his country would do so only under "certain circumstances." Those circumstances were of nuclear nature.

The Israeli official in question was Shimon Peres, a thirty-three-year-old rising star of Israeli politics and future prime minister. At the time he was director-general of the Israeli Ministry of Defense. His job was supplying arms for the Israeli military, and his mission to acquire the atomic bomb. He had no authority to make promises about war and peace, and it is not clear what circumstances he had in mind, but one of his biographers suggests that he saw an opportunity to acquire a nuclear reactor—his long-term goal in dealing with the French—and jumped on it.

Israeli participation in the war was formally negotiated at a conference in Sèvres on October 22–24, 1956, less than a week before the start of the invasion. Before the signing of the agreement, Peres finalized his deal with the French—the building of a reactor in Israel and supplying it with uranium. The war that began with Israel's invasion of the Sinai Peninsula on October 29 did not bring the results for which the governments in London, Paris, and Tel Aviv had hoped. The Soviet threat of a missile attack, along with American protests and threats of financial ruin, forced the allies to withdraw, dealing a major blow to the prestige of Britain and France, but not Israel.[4]

Unlike France and Britain, Israel emerged from the Suez crisis with a significant gain. In the midst of the crisis, Shimon Peres managed to persuade French officials to agree to build a much bigger reactor than had been promised before the start of the conflict. Threatened with Soviet nuclear missiles, the Israelis wanted to know whether the French were prepared to provide security guarantees in case Israeli troops withdrew from Sinai. The French were reluctant. Peres asked for a bigger reactor as a form of guarantee—the Israeli nuclear program was supposed to enhance the country's security after its withdrawal from Sinai. The French, who wanted to keep Israel on their side but did not want to commit to its defense, agreed.

The agreement was signed in October 1957 after lengthy negotiations and a pledge from Peres that the new enlarged reactor was intended for peaceful purposes only. The French, knowing very well what the Israelis wanted to do with the reactor, shielded themselves from a possible international scandal behind Peres's pledge. They agreed to build a 40 MWt reactor capable of producing up to 15 kilograms (33 pounds) of plutonium per year, as well as an underground chemical facility for the extraction of plutonium. The official documents, however, referred to a 24 MWt reactor, while the plutonium-processing plant was not mentioned in the agreement at all.

The British chipped in as well or, rather, they could not resist making one million pounds from the sale of heavy water for which they had no use. But they had to be careful to avoid being accused of proliferation. The heavy water needed to run the reactor was acquired by the Israelis from Britain through the intermediacy of Norway, again in exchange for a pledge to use the reactor for peaceful purposes only. The construction of the brand-new reactor began in early 1958 in the Negev Desert near the city of Dimona. While the facility became known as the Dimona reactor, it was officially called the Negev Nuclear Research Center. In August 2018 it was named after Shimon Peres in recognition of his role in the Israeli nuclear project and the construction of the Dimona center.[5]

The Israeli nuclear project faced a great deal of domestic opposition, both from government officials and from scientists who either believed that a state born out of the ashes of the Holocaust had no business developing weapons of mass destruction or felt that they had been excluded from making key decisions imposed on them afterward. Some doubted whether Israel could afford such a project and expenditure. But Ben-Gurion ran the nuclear project as President Roosevelt and Prime Minister Attlee had managed theirs before him—only a select group of ministers were informed about what was going on, and money was allocated in secret from key figures in government and parliament. The amounts were enormous by Israeli standards.

Building the reactor was a costly undertaking: Peres estimated the cost at $80 million. The real price tag is still unknown, as a good part of the cost was covered by donations from the Jewish diaspora that never made it into the official budget. In Peres's estimate, the diaspora's share was $40 million, half the official cost. During the construction of the reactor and the plutonium facility, the official research and development budget of the defense ministry increased more than fourteen times.[6]

Israeli reliance on the French became an issue a few months after the start of work in the Negev Desert. In May 1958 Charles de Gaulle came to power in Paris, determined not only to push forward the French nuclear project but also to make it purely French. The country's ties with Euratom—the European Atomic Energy Community—were cut, negotiations with the West Germans on a joint project stopped, and cooperation with Israel slated for termination.

Although de Gaulle demanded an end to the relationship with Israel as well, Jacques Soustelle, the minister of nuclear energy, did his best to shield the Dimona project from cancellation. Moreover, the French and Israelis exchanged information on the construction of the bomb: the French blueprint was shared with Israel in 1959. After Soustelle broke with the government in 1960 because of his opposition to de Gaulle's decision to grant independence to Algeria, the official French attitude to the Israeli project prevailed, cutting existing ties between two nuclear establishments.

Another, even more powerful factor was also at play. France's successful test of an atomic bomb in February 1960 made the country a nuclear power and all but ended its role as a proliferator of nuclear data and technology. The French now had nothing to gain and everything to lose from cooperation with Israel, as they would face criticism for increasing membership in the exclusive nuclear club and accusations from the Americans, Arabs, and others that they had made the Israeli bomb possible. In May 1960, the French informed the Israelis that they were ceasing cooperation and wanted to make sure that the reactor they were still in the process of building would not be used for the production of plutonium. They asked the Israelis to declare that the reactor was intended for peaceful purposes only and open it to international inspection.

In June 1960, an upset Ben-Gurion went to Paris to see de Gaulle in person and plead for the continuation of the Dimona project, which by now included not only the reactor but also the underground plutonium-producing facility. He assured de Gaulle that Israel would not build a bomb, but that was not enough for the French president. The French were prepared to pay compensation for breaching the agreement, but they wanted to end their cooperation and prevail on Israel to make the Dimona project public. Ben-Gurion did not take the money. Instead, Peres negotiated with French officials to reach a compromise.

While the government withdrew from cooperation, French companies were allowed to complete their contracts. Israel promised to announce the existence of the Dimona center to the world. De Gaulle washed his hands, and Ben-Gurion saved his atomic bomb project. Peres continued to build the reactor and the plutonium facility. The uranium oxide or so-called yellowcake used as fuel for the reactor, which had been supplied by France, would be now supplied, again secretly, by Argentina.[7]

Ben-Gurion started looking for a way to reveal Dimona to the world. The plan was to announce the creation of a new university in Beersheba with a research complex including a brand-new "experimental" reactor.

The Israelis wanted John McCone, the head of the US Atomic

Energy Commission, to attend, but the invitation served as a wakeup call for the Americans. Until then, the key figures in the US government had been under the impression that Israeli efforts were limited to nuclear research. In 1957 a US-Israeli agreement was signed to arrange for American contractors to build a 5 MWt pool-type reactor in Israel. But in the months preceding the invitation, US intelligence began to receive information about suspicious developments in the Negev Desert. It turned out that the Israelis were engaged in two-track nuclear diplomacy, with the American track indeed limited to research, the French one leading toward the bomb.[8]

The CIA director, Allen Dulles, briefed the National Security Council about the Israeli bomb problem on December 8, 1960. "This complex," he said, "probably included a reactor capable of producing weapons grade plutonium." He did not believe that either the Soviets or the Arabs would accept the Israelis' "peaceful purposes" explanation. "The Arab reaction to the Israeli facility will be particularly severe," continued Dulles. Vice President Richard Nixon was concerned about the global proliferation of nuclear weapons. "People will ask if Israel can do it, why not Cuba?" remarked Nixon.

Undersecretary of the Treasury Fred Scribner noted a possible domestic problem as well. "Israel might have been able to build this expensive nuclear facility because of funds which reach that country from Jewish charitable organizations in the U.S.," said Scribner. "These contributions are deductible from U.S. income taxes and the Treasury has experienced difficulties in the past because some of the charitable funds are diverted to government operations in Israel." Indeed, the Ben-Gurion–Peres funding scheme for the bomb included American donations.[9]

Ten days later, on December 19, Secretary of State Christian Herter, accompanied by a number of senior officials, went to see President Eisenhower to report on the Israeli reactor and share his suspicion that it was funded "through diversion from private and public aid to Israel." Eisenhower estimated the cost of the project at anywhere between $100 million and $200 million: because Israel was receiving American assistance, he wanted to know where the money was coming from. He supported the proposal to put the Israeli plant under International Atomic Energy Agency control, as Israel was a signatory to the Vienna convention on peaceful uses of nuclear energy. Herter hoped that it would still be possible to stop the Israelis from building the bomb.[10]

Meanwhile, the still secret Dimona complex "exploded" in a major media scandal. On December 16, the London *Daily Express* broke the news: the Israelis were building a nuclear weapons facility. A few days later the *New York Times* reported growing concern in American government circles about Israel's development of the capacity, with French assistance, to build an atomic bomb. On December 21, alarmed Ben-Gurion addressed the Knesset. He revealed the existence of the Negev center and the construction of the 24 MWt reactor, which, he assured the deputies, was "designed exclusively for peaceful purposes." Among its tasks, continued Ben-Gurion, was the training of Israeli cadres for a nuclear power station to be constructed in ten to fifteen years. The Americans were not convinced.[11]

With the change of administration in the White House, the Israeli bomb was passed on as a problem from Eisenhower to John F. Kennedy and his advisers. Christian Herter warned Kennedy that the countries most determined to obtain nuclear weapons were Israel and India. A Special National Intelligence Estimate prepared by the outgoing administration discussed the negative consequences of an Israeli bomb. They included very hostile reaction in the Arab world, with blame directed against the United States and France, as well as the strengthening of the military and political alliance between the USSR and the United Arab Republic, a state that then comprised Egypt and Syria—all to the detriment of American interests in the region. Beyond that, other countries might be encouraged by the Israeli example and try to acquire bombs of their own. The estimate suggested that the Israelis might have their first weapons-grade uranium by 1962.[12]

Kennedy wanted American experts to inspect the Dimona site. A similar request had been made by the Eisenhower administration, but Kennedy ensured that the Israelis would comply with the American demand. He conditioned his first meeting with Ben-Gurion on the results of the inspection. If it showed that there was a military program under way, there would be no meeting. The Israelis, who depended on the United States in economic, military, and political terms, had little choice but to comply. Two US Atomic Energy Commission representatives visited the Dimona reactor on May 20, 1961.

The Israelis made sure that they saw little and understood even less. After a tour of the facility, they filed a report stating that the nuclear research center at Dimona was "conceived as a means for gaining experience in construction of a nuclear facility which would prepare them [the Israelis] for nuclear power in the long run." They never visited the underground plutonium-processing facility, which was still under construction, and missed the crucial point that, with modifications, the power of the reactor could be doubled.

A meeting between Kennedy and Ben-Gurion could now be arranged, and it took place on May 30, 1961, at the Waldorf Astoria Hotel in New York. Ben-Gurion assured the president that the Israeli nuclear project was meant to serve peaceful purposes. Because southern Israel needed fresh water, energy from the reactor would be used to desalinate seawater. But there was a caveat. According to the Israeli memorandum of the conversation, Ben-Gurion added: "But we will see what will happen in the Middle East. It does not depend on us." He did not expect the Soviets to supply nuclear arms to the Egyptians but believed that Egypt might acquire nuclear weapons on its own in ten to fifteen years.

Kennedy was not pleased. He was concerned that the Israeli nuclear program would raise tensions in the region, while the United Arab Republic "would not permit Israel to go ahead in this field without getting into it itself." He wanted to reassure the Arabs and the world in general that Israel was not building a bomb. Ben-Gurion did not object to publishing the results of the American inspection of Dimona but was not open to the idea of inspections by representatives of "neutral" countries. They moved on to other subjects.[13]

Dissatisfied with the results of the meeting, Kennedy would continue American inspections of Dimona. The US intelligence estimate issued in 1961 predicted that Israel would be able to "produce sufficient weapons grade plutonium for one or two crude weapons a year by 1965–66, provided separation facilities with a capacity larger than that of the pilot plant now under construction are available." The State Department asked the Atomic Energy Commission to be on the alert for visiting Israeli scholars working on nuclear issues and suggested "discreet surveillance" of Dr. Israel Dostrovsky, a prominent Israeli chemist and key figure in the country's atomic project.[14]

In September 1962 AEC inspectors paid another visit to the Dimona complex. They were completing their visit to the US research reactor provided to Israel in the late 1950s as part of the Atoms for Peace program when they were invited for a tour of Dimona. The tour lasted 45 minutes, and according to a State Department memo the experts "found no evidence Israelis preparing produce weapons." The Israelis counted the impromptu visit as one of the inspections that the Kennedy administration had demanded of them. The State Department officials were only too ready to fall for the deception, but the US intelligence services were more cautious. They found those conclusions insufficient to prove "whether in fact the reactor might give Israel a nuclear weapons capability."[15]

While the AEC visitors predicted that the Dimona reactor might become operational in two or three years, US intelligence assumed that it could go critical before the end of 1963. In April of that year, Kennedy insisted on a regular biannual inspection of the Dimona site by American experts. Ben-Gurion wrote to Kennedy, describing Israel's difficult security situation in light of a recent pronouncement by Arab countries on fighting for the "liberation of Palestine," and the possibility of "another Holocaust." He was seeking security guarantees. Kennedy was not responsive. He insisted on the inspection, wanted access for American experts to all parts of the Dimona complex, and warned Ben-Gurion that the American "commitment to and support of Israel" could be "seriously jeopardized" if his demands were not met.[16]

In June 1963 Ben-Gurion stepped down as prime minister, passing on the Dimona project and Israel's troubled relations with Kennedy to his successor, Levi Eshkol. Kennedy resumed his pressure, insisting that visits of American inspectors to Dimona were "as nearly as possible in accord with international standards, thereby resolving all doubts as to the peaceful intent of the Dimona project." Eshkol asked for time to study the problem. Kennedy did not have much time left. He was assassinated in November; the Dimona reactor was launched in December 1963, and no American visit to the site took place that year.[17]

American inspectors finally made their way to Dimona in January 1964 on the basis of an agreement reached in the last months of Kennedy's presidency. They wrote in their report that the reactor had gone critical a month earlier. Israel now had the main component of its nuclear bomb project, a reactor that could produce plutonium. In 1968 the CIA reported to President Lyndon Johnson that Israel had built its atomic bomb. In the following year Israel made American inspections of the Dimona complex all but impossible. Israel would become the first and only country to adopt a policy of "strategic ambiguity," neither admitting nor denying its possession of nuclear weapons.[18]

By the start of the 1967 Six-Day War between Israel and the allied forces of neighboring Arab states, Tel Aviv most probably had a couple of primitive atomic bombs, although there was never a need to consider using them to win the conflict. Warfare was initiated by Israel, caught the Arabs by surprise, and led to a resounding victory. After the war, in possession of a growing arsenal of atomic bombs, Israel stuck to its policy of strategic ambiguity. Officially it denied possession of nuclear weapons, as happened in 1970, when unnamed Israeli officials responded to a report in the New York Times claiming that Israel possessed nuclear weapons, stating: "Israel is not a nuclear state and will not be the first country to introduce nuclear weapons into the Middle East." Some within the Israeli military establishment questioned that policy, arguing in favor of an admission of Israeli nuclear status to deter a possible Arab attack.[19]

The attack came on October 6, 1973, Yom Kippur or the Day of Atonement, the holiest day of the year in Judaism, which gave its name to the war. It would be the last Israeli war witnessed by Ben-Gurion, who died a few weeks after the end of the conflict. As in the Six-Day War, the side that attacked first had the advantage of surprise, and by October 8 the Israelis found themselves in a desperate position. They were fighting on two fronts: in Sinai, where the Egyptian army had crossed the Suez Canal, and in the Golan Heights, where the Israelis faced Syrian mechanized divisions. On that day the Israeli defense minister, Moshe Dayan, suggested that the existence of the "Third Temple" or the state of Israel was in jeopardy.

According to some accounts, Dayan urged Prime Minister Golda Meir to permit the arming of missiles and fighter jets with nuclear warheads to be used as weapons of last resort. She allegedly gave permission to activate atomic weapons. Indeed, the Israelis took steps to increase the readiness of their nuclear capabilities, but experts are divided on whether that was done with or without the knowledge of the government and top military officials. Nor is there agreement on whether the increase in readiness was intended as a signal to Arab countries, the Soviet Union, or the United States—in particular, whether it was an attempt to blackmail the United States into increasing assistance to Israel and resupplying its army with American weapons.[20]

While there are doubts that so-called nuclear blackmail of Washington was indeed the goal, the nuclear preparations in Israel did not go unnoticed there. It is not clear what role that knowledge played in the decision to resupply the Israelis, but American weapons were promptly shipped to Israel, changing the course of the war. The Israelis were soon on the counteroffensive, defeating their adversaries both in the Sinai Peninsula and in the Golan Heights. The Soviets responded by resupplying their Arab allies. They also threatened to deploy their own airborne troops in the region, but never did so. There are reasons to believe that Soviet nuclear-armed missiles arrived in the Mediterranean region. American troops worldwide went on the DEFCON 3 level of readiness, with no. 5 being normal peacetime conditions and no. 1

being open warfare. Washington was now signaling to Moscow that further escalation of the crisis could directly involve the two superpowers.[21]

The Soviets did not get involved. By October 25 the war was over, with another resounding Israeli victory. Nuclear weapons were not used, but there is little doubt that they, either Israeli, American, or Soviet, became a factor in the conflict. The war also revealed a major problem with the undeclared Israeli bomb: it turned out to be powerless to deter a combined Arab attack. The policy of strategic ambiguity, closely associated with the way in which Israel acquired nuclear weapons, may have been an important factor in the failure of Israel to deter its enemies, first in 1967 and then in 1973.[22]

In 1974 President Ephraim Katzir made a statement that came closest to an official admission of the existence of the Israeli bomb and the government's willingness to use it as a deterrent. Israel had long wanted to develop "a nuclear potential," stated the president, and it now possessed one. He explained that his country would not be the first to introduce nuclear weapons into the region but had the potential to develop them quickly if required.[23]

By the time Katzir pronounced those words the threat posed by the nuclear weapons became a pillar of the new global equilibrium, defined by the doctrine of "Mutually Assured Destruction," or MAD. In a manner of speaking, the world went mad to stay alive.

Chapter 21

MAD MEN

"But this is absolute madness," US president Merkin Muffley, a character in Stanley Kubrick's timeless black comedy, *Dr. Strangelove Or: How I Learned to Stop Worrying and Love the Bomb,* tells the Soviet ambassador. "Why on earth would you build such a thing?"

The ambassador, Alexi De Sadesky (a reference to Marquis de Sade and his famous phrase, "It is always by the way of pain that one arrives at pleasure"), responds that the nuclear attack on the USSR will automatically trigger a response from the Soviet "Doomsday Machine," a nuclear weapons system that cannot be reprogrammed. "When it is detonated," explains the ambassador, "it will produce enough lethal radioactive fallout so that within ten months, the surface of the Earth will be as dead as the moon!" The two countries end up as hostages of their nuclear strategies, and the film ends with mass nuclear explosions accompanied by Vera Lynn's song of 1939, "We'll Meet Again," with its not particularly optimistic line, "Don't know where, don't know when."[1]

Dr. Strangelove was released in January 1964 to an enthusiastic public reception and was nominated for four Academy Awards. The filmmaker caught the mood, fears, and tragic ironies of the time. The idea of a doomsday machine came directly from the writings of Herman Kahn, a military strategist and futurist who served as one of the inspirations for the Dr. Strangelove character. It was he who came up with the idea of a "doomsday machine"—a potentially effective but also destabilizing nuclear deterrent that might destroy all life on earth. Donald Brennan, one of Kahn's subordinates at the Hudson Institute, came up with the concept of "mutually assured destruction" in 1962. He posited that the two superpowers had enough nuclear power to annihilate each other, making war with atomic weapons all but suicidal.[2]

Before long, mutually assured destruction became known as MAD, an acronym coined by the Hungarian-born mathematician John von Neumann. Along with his countrymen Leo Szilard and Edward Teller, von Neumann made a major contribution to the Manhattan Project and then helped Teller build the hydrogen bomb. His other lasting addition to the nuclear vocabulary was the term "kiloton" to measure the explosive power of nuclear weapons. Von Neumann developed an equilibrium strategy that led to the formation of MAD as a concept. It became the foundation of American nuclear strategy before the end of the 1960s.[3]

"It's not mad," argued Robert McNamara, the defense secretary under Kennedy and Johnson. "Mutually assured destruction is the foundation of deterrence." The key Republican foreign policy figures were of the same opinion. Henry Kissinger, Richard Nixon's national security adviser and co-architect of American foreign policy, was a great believer in MAD. He wrote: "With no advantage to be gained by striking first and no disadvantage to be suffered by striking second, there will be no motive for surprise or preemptive attack. Mutual vulnerability means mutual deterrence. It is the most stable position from the point of view of potential all-out war."[4]

The new theory was largely responsible for the new disposition of nuclear forces in the world. The 1960s witnessed a dramatic change in

Washington's relative nuclear power. In 1960 there were 22,000 nuclear warheads, and the United States owned 93 percent of them, the rest being divided among the Soviet Union, the United Kingdom, and France. By 1970 there were more than 38,000 nuclear weapons, with the United States accounting for 68 percent. While the number of American nuclear warheads reached its peak at more than 30,000 in the mid-1960s and then began to decline, the Soviets kept building up their stockpile. By 1969 they had reduced the gap, with about 10,000 warheads compared to approximately 26,000 in the United States. The Soviet buildup also extended to missiles and bombers capable of delivering nuclear bombs. If in 1964 the United States had 1,880 "delivery vehicles" versus 389 Soviet ones, by 1970 the Soviet Union had more ICBMs than the Americans.[5]

The Soviets were allowed to change the numerical balance of ICBMs for one simple reason: as the 1960s progressed, fewer and fewer people in Washington believed that more nuclear bombs were needed to win the Cold War. Robert McNamara stated after his terms in the US government that the US and the USSR had nuclear parity during the Cuban missile crisis even though the United States had approximately seventeen times as many nuclear weapons as the Soviets—roughly 5,000 American weapons to 300 Soviet ones. The notion of parity was based on the assumption that some of the 300 Soviet warheads would survive a first strike and, once launched at American targets, would kill millions of Americans. It was too high a price to pay for entering into a nuclear war.

Nuclear-armed submarines, a significant number of which would always be on duty concealed beneath the surface of the ocean, guaranteed that both superpowers would manage to deliver a second strike. The Americans developed that capability first, but the Soviets caught up in the course of the 1960s. Now neither power could attack the other without opening itself to counterattack and sacrificing millions of its citizens.[6]

In 1967 President Lyndon Johnson, concerned by the continuing Soviet nuclear buildup, proposed talks on limiting strategic arms (long-range intercontinental missiles and bombers) to the Soviet government, which agreed in the following year. Nixon and his administration picked up exactly where their Democratic predecessors had left off. Nixon mentioned the talks in his first press conference as president in January 1969, and before the end of the year American and Soviet negotiators met in Helsinki to begin the Strategic Arms Limitation Talks (SALT).

The two sides had an incentive to talk that went beyond nuclear strategy: their economic, social, and political problems were mounting. With the Vietnam War going on and plans for Johnson's Great Society requiring ever more funds, he could hardly afford to get involved in an expensive new arms race with the Soviet Union. The same was true of his successor, Richard Nixon. Although Johnson's Great Society plans were scrapped, the war continued, requiring new expenditures.

The Soviet industry was doing relatively well, but the chronically underperforming agricultural sector, which Stalin had collectivized by means of man-made famine and oppression, required more and more money, and even then the Soviet Union had to buy grain abroad. Despite Khrushchev's boast to Nixon back in 1959 that the Soviets would surpass the United States in scientific and technological progress, it was becoming ever more obvious in Moscow that even a modest program would require Western technology and, if possible, Western loans.

The vulnerabilities of both regimes brought about the politics of détente. In the words of John Lewis Gaddis, détente "had been meant to lower the risk of nuclear war, to engage a more predictable relationship among the Cold War rivals, and to help them recover from the domestic disorders that had beset them during the 1960s."[7]

The task of the negotiators was not to achieve nuclear disarmament or even the reduction of nuclear weapons—the promise they had put into the Non-Proliferation Treaty—but to cap the nuclear weapons buildup by limiting not warheads but the missiles that delivered them. The negotiators targeted both land-based ICBMs, instruments of the first strike, and Submarine-Launched Ballistic Missiles (SLBMs), instruments of the second strike and the backbone of the MAD strategy.

In 1969 the United States had 1,054 ICBMs, 656 SLBMs, and 576 bombers capable of delivering nuclear bombs. The Soviets had more ICBMs—approximately 1,600 according to American estimates made in the autumn of 1971—but fewer SLBMs and significantly fewer bombers. The basic idea was to freeze the number of launchers at existing levels and halt the costly arms race in that sector: nuclear weapons were now costing the two countries a total of $50 million per day.[8]

The limitation of strategic weapons capable of delivering nuclear warheads was not the only issue discussed by the participants in the SALT negotiations. No less important and much more complicated was agreement on anti-ballistic missile (ABM) systems. If the continuing rivalry in strategic weapons was seen as a waste of resources that did not change the strategic balance between the superpowers, ABMs were considered not only wasteful but also exceptionally dangerous weapons systems that threatened the very foundations of the MAD principle by undermining the strategic balance between the United States and the Soviet Union.

The ABM systems were developed with the idea of protecting urban centers and nuclear and missile installations from a first strike. Being defensive in nature, they were not supposed to cause concern on the opponent's side. In fact, however, they created fear that if one side could protect itself from missile attack, then it could launch a first strike, hoping that its ABM system would help it survive the retaliatory second strike. The balance of nuclear terror was under threat. The ABMs undermined MAD and thus the idea of limiting strategic weapons. The negotiators would spend days, months, and eventually years trying to bring the ABM systems under control in order to save MAD.[9]

Richard Nixon expected the SALT talks to lead to a summit meeting with the Soviet leaders that could help him solve the numerous foreign and domestic problems besetting his administration. His key concern was not the Soviet nuclear buildup that his predecessor had sought to arrest but the seemingly unending Vietnam War, which was causing huge problems at home and abroad. With the first US-Soviet summit since the 1961 Vienna meeting between Kennedy and Khrushchev on the horizon, Nixon wanted a "package deal," hoping to link the signing of SALT with the settlement in Vietnam that the Soviets were supposed to help him achieve.

The new Soviet leader, Leonid Brezhnev, also wanted a summit for reasons that included but transcended nuclear arms control. He was slowly but surely emerging as the dominant member of the political triumvirate formed after the ouster of Khrushchev, which included premier Aleksei Kosygin and the chairman of the Soviet parliament, Nikolai Podgorny. Kosygin was all for the summit, but Podgorny, a strong supporter of communist Vietnam, opposed Soviet-American rapprochement. Also opposed was the defense minister, Andrei Grechko. Brezhnev needed a productive summit to cement his position as the top Soviet leader.

On the American side, the SALT talks were conducted by a team headed by Gerard D. Smith, the director of the Arms Control and Disarmament Agency. But preparations for Nixon's summit were handled by his national security adviser, Henry Kissinger, who wanted the summit to consolidate his reputation as a diplomatic wizard. He had gained it, inter alia, by arranging Nixon's sensational visit to China the previous year. Despite Nixon's instructions that there should be no summit without the Soviets agreeing to pressure North Vietnam to stop its ongoing invasion and destabilization of the South, Kissinger pushed ahead with preparations for a meeting that, unlike the botched one in Vienna, was supposed to hold no surprises. The key issues were to be decided beforehand and agreements announced at the summit.[10]

In the course of talks conducted in Moscow in April 1972, unbeknownst to the American SALT negotiators, Kissinger and Brezhnev agreed that the number of ABM systems would be limited to two per country: one to protect the capital, the other to protect one of the missile bases. They also reached a basic agreement on the limitation of strategic weapons. Kissinger had signaled American readiness to accept a number of Soviet ICBMs significantly higher than the American, but it was only in Moscow that he managed to include SLBMs in the deal. They were being negotiated in the SALT talks, but Kissinger took over the issue in his back-channel negotiations with the Soviets.

The US Navy was far ahead of its Soviet counterpart in the number of submarine-based missiles, but the Soviets had launched an ambitious program to catch up, and the Pentagon wanted to stop that surge. Kissinger offered the Soviets a deal that limited Soviet submarine-based missiles to 950: since the Americans had more than 650 SLBMs at the time, that figure gave the Soviets a significant theoretical advantage, but it greatly exceeded the number that the Soviets thought they would be able to build in the near future. Kissinger "sold" the agreement to the Pentagon by claiming that the Soviets were planning to build many more than 950, so that he was in fact "arresting" their SLBM buildup.

Brezhnev accepted the figures on SLBM launchers suggested by Kissinger. When Kissinger told Brezhnev that he wanted to hear what the Soviet leader thought about SLBMs, Brezhnev responded: "Nothing." He then added jokingly: "What can I say about them? They travel under water, we can't see them, they're silent." An interpreter then read a Soviet note on SLBMs accepting Kissinger's offer. "The figures are agreed. There is no problem about figures," responded the excited Kissinger. He then added in the jovial spirit of the meeting: "I will show you what a bad diplomat I am. Gromyko wouldn't do this, but I think the submarine matter is acceptable in principle." Not to be outdone in diplomatic nicety, Brezhnev answered: "This shows what a strong diplomat you are. I agree our Foreign Ministry would never do that, but that's an example of how bad it is."[11]

With Brezhnev agreeing to the SLBM numbers suggested by Kissinger and the two ironing out the deal on ABMs, the road to the summit was open. Everything was now ready for the signing of a historic deal that would limit the arms race for the first time in the history of the nuclear age.

It was just then that the geopolitical rivalry between the two superpowers interfered, threatening to topple the summit and kill the SALT treaty. Concerned about the emerging rapprochement between Washington and Moscow, Hanoi ordered a major offensive against South Vietnam, driving Nixon to order new bombing of North Vietnam and mining of North Vietnamese ports to cut off supplies of Soviet and Eastern Bloc weapons and ammunition. Soviet and East European ships docked in Vietnamese ports were also hit by bombs, killing Soviet and allied sailors and putting the summit in jeopardy.

Nixon considered calling off the summit before Moscow could do so. The fate of the Paris summit of 1960, boycotted by Khrushchev to the great embarrassment of Eisenhower, Nixon's boss at the time, was fresh in Nixon's memory. But Brezhnev prevailed over his hardline colleagues in the Politburo and insisted on holding the summit despite the growing crisis over Vietnam. The decision was made at a special meeting of the Central Committee to which hundreds of Soviet officials were summoned to listen to Brezhnev's arguments in favor of the summit. Chief among them was the signing of the SALT treaty.

With the Central Committee supporting his position, Brezhnev removed from power Podgorny's principal ally and opponent of the summit, the party boss of Ukraine, Petro Shelest. He would be accused of Ukrainian nationalism, his removal helping Moscow to launch a campaign of repression against Ukrainian dissidents in that key Soviet republic. On his trip to the Soviet Union, Nixon would not only visit Moscow but also make a stopover in Kyiv, the capital of Ukraine, there to be welcomed by Shelest's successor, Volodymyr Shcherbytsky, a Brezhnev loyalist.[12]

When Nixon arrived in Moscow on May 22, 1972, Brezhnev complained to him how difficult it was to continue with the meeting in the midst of the new Vietnamese crisis. But the two leaders took an instant liking to each other, and negotiations on the points already settled between Brezhnev and Kissinger proceeded with little disagreement and some humor, demonstrating good will on both sides. The difference between this summit and the one between Kennedy and Khrushchev in 1961 could not have been more profound.

"As regards the ABM question, this now appears to be cleared up," remarked Brezhnev. "Twelve hundred is OK with us." Nixon intervened, reminding his Soviet counterpart about deploying his ICBMs away from the western Soviet borders. "Fifteen hundred kilometers [932 miles]," responded Brezhnev, pretending a grudge. "You mean we should put it in China?" Nixon signaled his readiness to compromise: "Well, as the General Secretary will find out, I never nitpick." But Brezhnev, playing the gracious host, agreed with the original proposal. "Fifteen hundred kilometers is all right. The most important point is not the mileage. You wanted us to move eastward, and so now we agree. It would be easier for us to accept twelve hundred, but fifteen hundred is all right too, and we won't speak of it anymore."[13]

For all Brezhnev's bonhomie, he readily agreed only to items already arranged or cleared with his jealous colleagues. Otherwise, he relied on the assistance of memos and experts present in the room. On the American side, decisions were made unilaterally by Nixon. Like Brezhnev, he was eager to take credit for the achievement of arms limitation, but unlike his communist counterpart, he had a free hand in conducting foreign policy in Moscow: Secretary of State William P. Rogers had been sidelined by Kissinger and informed about some parts of the agreement but not about others. Gerard Smith, the chief arms negotiator, was summoned to Moscow only to take part in the signing ceremony of the agreement concluded without him.

Signed on May 26, 1972, the nuclear arms limitation agreement became a sign of new times shaped by the concept of mutually assured destruction. MAD made it easier to agree on nuclear arms than on other issues traditionally dividing the Cold War adversaries. The two "MAD men" tried to use the arms deal to advance other items on their international agendas, admittedly to no avail. Nixon wanted a settlement in Vietnam but did not get one. Brezhnev wanted an agreement on the inadmissibility of nuclear war. Like Khrushchev's proposed "addendum" to the test ban treaty in the form of a "nonaggression pact," the "no nuclear war" initiative was shot down by the Americans before the summit. They were by no means prepared to denounce the use of nuclear weapons in Europe, given the preponderance of Soviet conventional forces there.[14]

SALT ended up being a turning point in Soviet-American relations on a number of levels. First, as Nixon later claimed, it heralded the end of the era that had begun in 1945. The reference was to the uncontrolled arms race initiated by the Trinity test and the bombings of Hiroshima and Nagasaki. Second, the treaty used the "mutually assured destruction" principle as the foundation of a new policy that recognized both sides as equal, meaning equally capable of destroying each other in case of nuclear conflict. Also, as discussed above, the talks uncoupled nuclear diplomacy from geostrategic competition in the Third World.

The Moscow summit of 1972 opened the door to new agreements on arms limitation and even future reduction of strategic arsenals. It not only set a precedent but also introduced satellite surveillance as a mutually acceptable way of verifying fulfillment of the agreement. Scandals concerning U-2 overflights and endless wrangling about the number of on-site inspections were avoided thanks to the development of better cameras and the availability of ballistic missiles to put spy satellites into orbit. Missiles in space were now helping to police missiles on earth.[15]

The SALT agreement was regarded from the very beginning as a steppingstone to a more comprehensive agreement on the limitation of strategic weapons. Nixon had high hopes that he would be able to sign a new treaty during Brezhnev's visit to the United States in the following year. But even though further visits—Brezhnev's transatlantic journey in the summer of 1973 and Nixon's return to Moscow in the summer of 1974—resulted in a number of important treaties, including amendments to the ABM treaty, reducing the number of anti-ballistic defense systems in each country from two to one, and limiting the power of underground tests to 150 kilotons. Nixon never signed a comprehensive agreement with Brezhnev.[16]

The key reason for that was the opposition to such an agreement in Congress, where many questioned the complex math of the asymmetrical provision in the SALT I treaty that allowed the Soviet Union to have 2,358 ICBM and SLBM launchers as opposed to America's 1,710. The Senate passed a resolution proposed by Henry Jackson, an influential Democratic senator from Washington State, otherwise known as the senator from Boeing, which prohibited any future agreements that allowed Soviet numerical superiority in any weapons class. The resolution manifested not only congressional mistrust of the Soviets but also its rejection of Nixon-Kissinger secret diplomacy and the horse trading between different classes of weapons that it involved. Negotiations of a new treaty became excessively complicated.[17]

The impeachment of Richard Nixon and his resulting resignation in August 1974 did not derail the continuing Soviet-American nuclear rapprochement based on the concept of MAD. The "MAD man's" torch was picked up on the American side by Nixon's successor, Gerald Ford. Brezhnev met with Ford in November 1974 in the Soviet Far East near the city of Vladivostok. They agreed to limit each other's strategic arsenals to 2,400 ballistic missiles and 1,320 Multiple Independently Targeted Reentry Vehicles (MIRVs). According to the Soviet ambassador to Washington, Anatolii Dobrynin, who played a key role in the preparation of both the Moscow 1972 and the Vladivostok 1974 summits, the meeting marked the climax of détente. From there it was mostly downhill, a process punctuated by Brezhnev's rapidly deteriorating health: he suffered two seizures, one before the start of the talks.[18]

When the comprehensive agreement, dubbed SALT II, was finally ready for signature in June 1979, Brezhnev faced not Ford but Ford's successor, Jimmy Carter, on the other side of the table. The chemistry that had developed between Brezhnev and Nixon, two pragmatic if not cynical politicians, would not be repeated in relations between Brezhnev and the new American president. After suffering a number of strokes, Brezhnev lost the "animal magnetism" that had so surprised Nixon in 1972 and could barely speak without a note or stand without the support of an aide. Carter believed not only in the need for control, reduction, and eventual destruction of nuclear arsenals but also in human rights, and as far as Brezhnev was concerned, that created a problem.

If in 1972 the Soviets had lectured the Americans on their war in Vietnam, after 1976 it was the Americans who raised Soviet violations of international norms. The Helsinki Agreements signed by Brezhnev and Ford in the summer of 1975 provided international justification, if not a legal basis, for the activities of human rights watch groups, and Carter and his national security adviser, Zbigniew Brzezinski, were not ready to look the other way at Soviet domestic behavior. Besides, there was growing pressure from the Senate to resume the nuclear arms race, which many senators believed America was losing. The CIA was accused of underestimating the threat posed by Soviet rearmament, and a group of independent experts led by the Harvard historian of Russia and the USSR, Richard Pipes, confirmed those accusations—wrongly, as it later turned out. The conclusions of the Pipes group were leaked to the press soon after Carter's victory in the presidential elections and strongly promoted the rise of critical attitudes toward détente in Washington.[19]

Leonid Brezhnev was concerned about the arms race and infuriated by the American use of human rights in foreign policy. "And now even after Helsinki, Ford and Kissinger and various senators are demanding to arm America even more, they want it to be the strongest. They keep pressuring us because of our Navy, or because of Angola, or they come up with something else," complained Brezhnev to his foreign policy advisers. There was domestic pressure on Brezhnev to continue the military buildup. "Then [Minister of Defense Andrei] Grechko comes to me and says they increased this, they are threatening to 'raise' that. Give me more money, he says, not 140 billion but 156 billion," confided Brezhnev to his aides. "What am I supposed to tell him? I am the chairman of our country's military council, I am responsible for its security.

The minister of defense tells me that he is not responsible if I don't approve the funds. So I approve [the increases] again, and again, and again. And the money goes flying. . . ."[20]

But despite all obstacles, the new treaty was signed by Brezhnev and Carter at the Vienna summit in the summer of 1979. It was based on the agreement reached by Brezhnev and Ford in Vladivostok back in 1976. The treaty banned the development of new ICBMs and set a limit of 2,400 on the overall number of strategic nuclear weapons. Each side set the goal of reducing that number to 2,240 by January 1, 1981. The number of MIRV land, sea, and air launchers was limited to 1,320 on each side, and a verification schedule was introduced—a major concession on the part of the Soviet Union. At the time the treaty was signed, the USSR had 2,504 missile launchers versus 2,058 in the United States. To the satisfaction of American critics of SALT I, SALT II envisioned a reduction of Soviet launchers and allowed for an increase of American ones.[21]

Détente in relations between the Soviet Union and its allies and the United States and American allies in Europe was rapidly approaching its end. The negotiations and the signing of the agreement took place against the background of the announced deployment of a new class of Soviet SS-20 "Saber" intermediate-range missiles in 1977 and the deployment in Europe two years later of American Pershing II intermediate-range ballistic missiles (IRBMs). Carter postponed congressional ratification of the agreement on the basis of what turned out to be false information about the posting of a new Soviet brigade in Cuba. By the time the new Cuban mini-crisis was resolved—the brigade had been dispatched there on conditions negotiated at the end of the 1962 crisis—the Soviets had invaded Afghanistan, making the passage of any agreement with them a virtual impossibility.[22]

SALT II was never submitted to Congress and never ratified by it. Nevertheless, both countries stuck to the agreement and observed it for the next seven years, showing once again that nuclear diplomacy was relatively autonomous if not independent of the geopolitical rivalries of the two superpowers. That era of two-track diplomacy would come to an end in the 1980s, the decade that would also challenge the foundations of MAD and the geostrategic equilibrium on which the concept had been based. The MAD years turned out to be among the sanest in the history of the nuclear age.

Chapter 22

SMILING BUDDHA

In September 1971 Richard Nixon visited the birthplace of the American plutonium project, the Hanford Works in Washington State. He came to Richland, a quintessential atomic city created by the Manhattan Project and sustained by the nuclear arms race, to announce government funding for the development of the Hanford Fast Flux Test Facility, a plutonium-fueled fast breeder reactor that would produce more fissionable fuel than it consumed. It was a nuclear revolution that America was eager to lead.

Nixon thanked the Hanford Works designers and engineers for their contribution to the cause of peace: by making nuclear weapons they were deterring war, went the MAD logic of the time. But he was also eager to speak about nuclear energy in different terms, ones that resonated with the Atoms for Peace logic of his former boss, Dwight Eisenhower. Nixon told a personal story: he used to live in San Clemente, California, on the Pacific coast, where many of his neighbors were afraid of the nearby nuclear power plant. But Nixon told the crowd that in the nearby plant he saw the energy of the future, which would help to preserve a "clean and beautiful environment." He assured the plutonium makers of Hanford that they had a bright future in the new American economy based on the peaceful uses of nuclear energy. "Do not be afraid of it, build it for peace," declared the president. The crowd applauded.[1]

In Nixon the American nuclear energy sector found its most enthusiastic supporter since President Eisenhower. By the early 1970s, the results of the Atoms for Peace program were there in plain sight to be admired and appreciated: the cooling towers of nuclear plants producing electricity had changed the skyline of dozens of American cities and towns. "At present," declared Nixon in April 1973, "there are 30 nuclear power plants in operation in the United States; of the new electrical generator capacity contracted for during 1972, 70 percent will be nuclear powered." Nixon saw an even brighter future ahead. "By 1980," he declared in the same speech, "the amount of electricity generated by nuclear reactors will be equivalent to 1.25 billion barrels of oil, or 8 trillion cubic feet of gas. It is estimated that nuclear power will provide more than one-quarter of this country's electrical production by 1985, and over half by the year 2000."[2]

In April 1973, as Nixon was delivering his speech, American Atomic Commission engineers were getting ready to use nuclear power to release the energy hidden in the depths of the earth. They placed three atomic bombs, each with a 30-kiloton yield, or twice the yield of the Hiroshima bomb, in shafts up to 6,560 feet (2,000 meters) deep located 50 miles (80 kilometers) north of Grand Junction, Colorado. The explosions were supposed to release "tight gas" trapped in rocks of low permeability. Plans were ready to proceed with at least 140 underground nuclear explosions in order to release up to 300 trillion cubic feet of natural gas stored beneath the Rockies. That gas would satisfy American energy needs for generations.[3]

The Plowshares program, which focused on the use of nuclear explosions for digging channels and releasing underground natural gas, promised not just good news but great news for American consumers. American demand for oil was on the rise, while its production was in decline, making the country more and more dependent on foreign oil. By the fall of 1973, the United States found itself in an oil crisis caused by an embargo imposed on Washington by the Arab oil-producing countries in retaliation for American support of Israel in the Yom Kippur War. Within a few short months, world prices for a barrel of oil almost quadrupled from $2.90 in October 1973 to $11.65 in January 1974. Drivers wanting to fill their gas tanks had to wait in lines extending for miles, making the politicians' promises of energy self-sufficiency through the development of nuclear energy more attractive than ever before. Nixon called for 1,000 nuclear plants before the year 2000.[4]

None of that would ever materialize. Nixon and the world were in for a huge surprise. The decade that had begun with optimistic promises of Atoms for Peace ended with the industry and the concept itself in steep decline. The cost of nuclear reactors, which had difficulty competing with oil and gas under normal conditions without government backing, was one reason for the dramatic fall in applications for new permits to build reactors in the late 1970s. Another was the growing safety concerns that paralyzed the industry after the partial meltdown of the reactor at Three Mile Island in March 1979. The Hanford Fast Flux Test Facility, which Nixon visited in 1971, was supposed to give birth to a new generation of plutonium breeder reactors but the project was terminated before the end of the decade because of cost overruns and its potential for proliferating nuclear arms technology.[5]

Hopes of using nuclear explosions to unleash the energy of the planet also failed to materialize. The Colorado detonations did release natural gas, but it was too contaminated with radioactive particles from the nuclear explosions to be delivered to people's kitchens. The Plowshares program, which depended on the expected bonanza of natural gas from the Rockies, was terminated, and no further tests were conducted. Such great hopes for the program as the digging of a new deeper and wider Panama Canal and creating a new harbor in Alaska expired even earlier. Those projects were called off when the test explosions produced amounts of nuclear fallout rivaling or exceeding those released by military tests.

The Soviets abandoned their own projects for channel digging in 1971, after three nuclear explosions produced a radiation cloud that crossed Soviet borders. The canal, 700 meters (2,297 feet) long, 380 meters (1.247 feet) wide, and between 10 and 15 meters (33 and 49 feet) deep, became a lake that received the name Nuclear. While Soviet efforts to find peaceful uses for nuclear explosions continued until the end of the Cold War, with precious little to show for the money spent, the American program was fully terminated in the late 1970s after thirty-one tests and $770 million expended.[6]

Just as it seemed that nothing could save global hopes of putting nuclear explosions to practical use, ostensibly good news arrived from Delhi, India. On May 18, 1974, the country's scientists and engineers conducted a successful test of India's first atomic device, heralded by the Indian government as a nuclear explosion for peaceful purposes.

As far as the nuclear powers that initiated the Non-Proliferation Treaty were concerned, the news was anything but good. Not only had another country acquired a nuclear arms capacity, but it had done so with the help of the Atoms for Peace initiatives, which were supposed to prevent non-nuclear nations from acquiring atomic weapons. It was never explained what channel the Indians were about to dig or what gas they intended to release. But it was an established fact that the Indians were beneficiaries of the US-backed Atoms for Peace programs, which had helped them gain access to nuclear technology and train their first generation of experts in the field, who would later build and explode a bomb.

News of the explosion came as a complete surprise to the American authorities. The Indians managed to conceal their intentions, if not their capabilities. Intelligence reporting on the Indian nuclear program had diminished in the months leading up to the test, as the White House was preoccupied not only with handling Nixon's relations with China and the Soviet Union but first and foremost with Vietnam. The US embassy in Delhi reported to Washington a few months before the explosion that the probability of the Indians having an "early test" were at a "lower level than previous years."[7]

Like the state of Israel, India as an independent country was born in the wake of World War II, non-nuclear but with strong nuclear ambitions on the part of its founding father. "As long as the world is constituted as it is, every country will have to devise and use the latest devices for its protection," declared India's future prime minister, Jawaharlal Nehru, in 1946, a year before the declaration of independence. He continued: "I have no doubt India will develop her scientific researches and I hope Indian scientists will use the atomic force for constructive purposes. But if India is threatened, she will inevitably try to defend herself by all means at her disposal." Nehru's pronouncement contradicted the statements of his mentor, Mahatma Gandhi, who called the atomic bomb "the most diabolical use of science" and suggested that it could not stand against the power of truth and non-violence. Nehru apparently had his doubts.[8]

By the time Nehru spoke in public about the development of India's nuclear potential, the country in the making had already established a nuclear research center of its own. The Tata Institute for Fundamental Research came into existence in December 1945 at the initiative and under the leadership of the thirty-six-year-old physicist Homi Jehangir Bhabha. A graduate of Cambridge University, he received his doctorate in nuclear physics in 1935. He cut his teeth as a scientist at the Cavendish Laboratory before returning to India on the eve of World War II. Bhabha managed to convince the leadership of the Indian National Congress, India's largest pro-independence organization, to use trust funds in its possession to finance nuclear research. In 1948, the second year of the country's independence, Bhabha established India's Atomic Energy Commission with Nehru's help. He became its first chairman.

From the days of its birth, India embarked on its own nuclear project. It was driven largely by considerations of prestige, as well as expectations of cheap electrical energy and economic development, but also with an eye to military potential: the government act that established the commission provided for the secrecy of its operations. In 1954 Nehru helped Bhabha launch the construction of the Atomic Research Center, an Indian analogue of Los Alamos that would be concerned with designing the bomb. Nehru also created the government Department of Atomic Energy, appointing Bhabha as its secretary and subordinating the department directly to the prime minister's office. In taking personal responsibility for the atomic project, Nehru was following in the footsteps of Roosevelt, Stalin, Attlee, and Ben-Gurion.

The Indian project resembled the French in two ways: the decision to build the bomb was made over a considerable period and, while the military component was always part of the calculation, for a long time it was hardly the dominant one. In the international arena and at home the Indian government positioned itself as an opponent of nuclear weapons. The reasons for that were mainly domestic—Indian politics and political tradition.[9]

In 1954, in the wake of the Castle Bravo test of the first hydrogen bomb, Jawaharlal Nehru became the first world leader to call for a moratorium on nuclear testing. A year later his nuclear point man, Bhabha, was elected president of the International Conference on the Peaceful Uses of Nuclear Energy, which took place in Geneva. At Geneva and beyond, India emerged as the leader of the nonaligned movement of countries that refused to join either the American or the Soviet camp during the Cold War. The Soviet economic model, as well as the USSR's way of dealing with its nationality question, generated considerable interest in India, but Moscow's close alliance with communist China, which extended to the nuclear field, could not but disturb Delhi.

India's relations with the United States were beset by problems of a different kind. At the first glance, India was becoming everything that President Eisenhower wanted to see in his Atoms for Peace partners: interested in nuclear research and dedicated to peaceful uses of nuclear energy. This was only partially true. While becoming one of the world's most powerful spokesmen for nuclear disarmament, India's first prime minister believed that if others were armed with atomic bombs, India also had the right to acquire them, along with the unqualified right to develop nuclear energy for peaceful purposes.

Nehru's response to Eisenhower's Atoms for Peace initiative was negative from the start, consistent with India's rejection of the Baruch Plan back in 1946. As far as Nehru was concerned, the proposed international agency controlling access to nuclear technology and a fissile fuel bank bore the hallmarks of a new colonialism. Nor did Nehru see how the initiative was supposed to keep countries not taking part in the program from acquiring nuclear weapons. Weighing on his mind was a potentially nuclear China. American-Indian relations also suffered from US policy toward India's other regional rival, Pakistan, to which Washington began providing military assistance in 1954. Although the United States attempted to remedy the situation by offering military assistance to India as well, Nehru refused the offer.[10]

In 1955, India signed a deal with Canada for the construction of a reactor based on the National Research Experimental Reactor (NRX) at Chalk River, originally designed for the production of plutonium. The United States agreed to supply heavy water, believing that the reactor would be used exclusively for the peaceful development of nuclear energy. This arrangement would serve as a model for the French-Israeli deal on building the Dimona reactor, signed in 1957. The Indian reactor was launched in 1960. The Americans, as agreed, provided the Indians with 21 tons of heavy water for the reactor, which turned out to be capable of producing enough weapons-grade plutonium to build two atomic bombs per year.[11]

By 1961, Bhabha was busy constructing a plutonium-producing plant using American technology made available through the Atoms for Peace program and an American contractor. The International Atomic Energy Agency statute had been modified at Bhabha's insistence back in 1956, loosening safeguards on the use of nuclear technology. The argument that won the day was anticolonial in nature—an international agency had no right to interfere with the economic development of individual countries. As Nehru continued to declare in the international arena that India was dedicated exclusively to peaceful uses of nuclear energy, the construction of a costly plutonium-processing plant whose product was intended exclusively for an atomic bomb went on uninterrupted.[12]

The Sino-Indian war over the old imperial border in the Himalayas that broke out in October 1962 provided new incentives for the development of the Indian "atoms for war" project. The Soviet Union, on which India counted as a mediator, originally took a pro-Chinese stand, being concerned about Beijing's policies in the midst of the Cuban missile crisis. Nehru turned for assistance to the United States but Washington was busy dealing with the missile crisis, and the Indian army was ill equipped to withstand the continuing Chinese assault. Nehru was forced to end the war on China's terms. In the aftermath of the war India doubled its defense budget, and opposition politicians called on Nehru to build a nuclear weapon. Nehru, who had done all he could to develop the military component of India's nuclear project, found himself in a tough spot. He could not admit Indian atomic ambitions without endangering cooperation with the nuclear powers. Instead, he declared that India had the knowledge and ability to build an atomic bomb but would not do so, as it was advocating an end to nuclear testing.[13]

The year 1964 became a turning point in the Indian nuclear program. In May of that year Nehru died; in June the completed plutonium-processing plant received its first unseparated plutonium from the Canadian-designed reactor; and in October the Chinese successfully tested their first atomic bomb. Before the end of the month the new prime minister, Lal Bahadur Shastri, following intensive debate within his own party, the parliament, and society, decided to develop "peaceful nuclear explosives." This was no more than window dressing for the West that allowed India to continue cooperating with the nuclear powers on its atomic project. It also saved money by extending the building of the Indian bomb over a longer period of time. Shastri's policy amounted to a compromise between Gandhi's aversion to nuclear weapons, Nehru's public commitment to their peaceful use, and India's growing desire to acquire a bomb as a deterrent vis-à-vis nuclear-armed China.[14]

Indira Gandhi, the daughter of Jawaharlal Nehru and a close assistant during his many years at the helm of the Indian state, took office as prime minister in January 1966 deeply committed to the nuclear disarmament policy and sincerely doubting whether the acquisition of the atomic bomb would indeed help India deter China. But her attitude changed fairly soon after she took power, and she followed her father's strategy, maintaining publicly that India should not build an atomic bomb but tacitly supporting the project. In May 1968, on Indira Gandhi's instructions, her government pulled out of nuclear Non-Proliferation Treaty talks, sending a clear signal that Delhi at a minimum did not want to be bound by restrictions on the acquisition of the bomb that would come with the treaty. India refused to sign the treaty once it was ready, leaving the country's atomic options open.[15]

But the Indian bomb project encountered political as well as technical problems. Bhabha, the founding father of India's nuclear program, died in an aircraft accident in January 1966, the month in which Indira Gandhi became prime minister, leaving the bomb option without a strong if divisive advocate. The plutonium-producing plant he had built was turning out much less plutonium than expected. In 1970 the plant developed a serious leak and had to be shut down temporarily. But the idea of building the bomb was already there, and the Indian scientists pushed on. Once the reactor was relaunched, its fuel was used to produce plutonium. By 1972 not only did India have enough plutonium to build a bomb, but its physicists and engineers had finally figured out how to do it.[16]

India's progress did not go unnoticed. In January 1972 the American embassy in Vienna reported to Washington that India had the capacity to produce a "nuclear explosion." The diplomats pointed out that there were problems with the safety of underground tests but assumed that a test could eventually take place. It would benefit the Indian government politically, as the government could keep the country's "nuclear hawks" at bay by demonstrating that India had acquired the requisite technology to build the bomb and "could, if necessary, match Chinese and other putative enemies any time it chose." The embassy officials believed that it was now up to the Indian government whether to conduct a test. "We see nothing US or international community can presently do to influence GOI [government of India] policy directions in atomic field," read the report.[17]

Indira Gandhi, still maintaining her public stand against nuclear weapons, gave a secret go-ahead for the construction of an explosive nuclear device in September 1972. The explanation given later pointed to the United States rather than China as the possible trigger. In the midst of the Indian-Pakistani war of 1971, the United States attempted to intimidate India by sending a carrier battle group led by the nuclear-armed battleship *Enterprise* into the Bay of Bengal. The Soviet Union responded by sending nuclear-armed submarines to intimidate the Americans. It seems likely that the nuclear showdown in the Bay of Bengal prompted India's decision to proceed with building the bomb. The key factor was that by the end of 1971, when the decision was apparently made, India had all the prerequisites—political, financial, scientific, and technical—to produce a nuclear weapon.[18]

The test of the explosive nuclear device, subsequently called "Smiling Buddha," took place on May 18, 1974, at a military base in Rajasthan, India's territorially largest state. The estimated yield of 8 kilotons was lower than projected, but that did not matter. By all accounts, India's "peaceful" device had made it a nuclear-weapons state. The American embassy in Vienna explained the timing of the test by internal Indian politics—the need to improve the government's status during an economic downturn and considerations of national pride, expressed as a desire for India "to be taken seriously" abroad.[19]

The international outcry produced by the test convinced Indira Gandhi not to proceed with additional tests or build a nuclear arsenal, but everyone knew that her country now had that capacity. In the late 1980s Delhi would change its mind and proceed with the weaponization of its nuclear technology. This time the threat to be deterred came not from China or the United States but from Pakistan. The test of 1974 took place in a shaft and thus did not violate the Partial Test Ban Treaty, to which India remained a signatory. Because India did not sign the nuclear Non-Proliferation Treaty, the test did not violate its commitments under other treaties. Nevertheless, the "peaceful" explosion was a signal that countries safeguarding the nonproliferation regime and subscribing to the treaty had to tighten their rules and strengthen their safeguards. The Indian test came about as a result of excessive sharing of scientific knowledge and technology with a country that had repeatedly declared its commitment to exclusively peaceful uses of nuclear energy.[20]

The American intelligence services produced a report that sought to explain why they had failed to warn Washington about the coming test. One of the conclusions was that the Indian nuclear program was not of much interest to policy makers in Washington, an attitude that affected the priorities of the spy agencies. That view had to be changed, suggested the authors of the report. "The proliferation issue is a significant one for the US government. The nth country intelligence problem should be given much more priority than it seems to have enjoyed to date."[21] The "nth country problem" was intelligence jargon for the diffusion of nuclear weapons. Thus, the suggestion was to increase intelligence gathering on the possible acquisition of nuclear weapons by other states.

The Ford administration conducted its own review of American nonproliferation policies and committed itself to strengthening the nonproliferation regime by preventing or delaying the acquisition of "peaceful nuclear explosive" devices by non-nuclear states. In November 1975, at the initiative of the United States, London hosted a "private" meeting of representatives of the key nuclear powers, parties to the NTP agreement, the United States, the Soviet Union, and Great Britain. Also in attendance were representatives of France, a nuclear power that was not party to any of the nuclear treaties, and Canada, Germany, and Japan, which did not have nuclear weapons. On the agenda was the formulation of a policy to prevent the transfer of nuclear technology that could be used to develop nuclear weapons and thus a repetition of the Indian situation, for which Canada was regarded as the main culprit.

The meeting resulted in the formation of a Nuclear Suppliers Group, which presented the director general of the International

Atomic Energy Agency in Vienna with "Guidelines for Nuclear Transfers" in January 1978. The restrictions that the guidelines imposed on countries exporting nuclear technology and knowhow went far beyond the safeguards enacted by the IAEA. The anticolonial arguments that India had used so successfully in the IAEA to develop its nuclear program did not work in the new group of the nuclear "haves."[22]

In March 1978 Jimmy Carter signed into law the Non-Proliferation Act adopted by Congress, which tightened the safeguards to be associated with future deals involving the US government and mandated the renegotiation of previous deals that violated those safeguards. Clients of American companies such as Westinghouse were denied equipment or fuel unless they complied with the new regime. The appeals of Yugoslavia, where Westinghouse was building a nuclear power plant, against renegotiating the existing agreement were in vain. The International Atomic Energy Agency, which served as the formal supplier of the plant, was powerless to change the American position now enacted in law.[23]

In November 1978 attempts began to rein in Pakistan's nuclear ambitions and stop the proliferation triggered by the "Smiling Buddha" explosion. The Indian "peaceful nuclear explosive" device clearly made a strong impression on Islamabad, as did India's refusal to join Pakistan in declaring the region a nuclear-free zone. The Pakistani nuclear program began after the country's dismal performance in its 1965 war with India. The call to go nuclear came not from the always powerful military, which did not want to upset its relations with the United States, but from politicians trying to capitalize on the rise of patriotic and anti-Indian sentiment among their electorate.

The first politician to demand the bomb was the future prime minister of Pakistan, Zulfikar Ali Bhutto. "If India builds the bomb, we will eat grass or leaves, even go hungry, but we will get one of our own. We have no other choice," declared Bhutto in 1965. Three years later Pakistan refused to sign the Non-Proliferation Treaty, leaving itself a free hand to acquire nuclear weapons. The 1971 war with India resulted in the loss of eastern Pakistan, which became the independent state of

Bangladesh—a national humiliation that strengthened Pakistani proponents of the bomb. The Indian "peaceful explosion" of 1974 gave a further boost to Pakistani nuclear ambitions, with the country's government now led by Bhutto, an early proponent of the bomb, as did India's refusal to join Pakistan in declaring the region a nuclear-free zone.[24]

Measures to stop the proliferation chain reaction triggered by the "Smiling Buddha" explosion were applied to Pakistan in November 1978. Bhutto sought international assistance to develop his nuclear arms program. The United States and the United Kingdom approached the other members of the Nuclear Suppliers Group, including France, with a request to prevent or delay Pakistan's acquisition of technology that could be used to produce nuclear weapons. The French canceled their deal with Pakistan and refused to supply the promised reprocessing plant. But they soon informed the Americans that the Pakistanis were looking for another country that could deliver equipment required for the plant and had approached Italy, Japan, and Spain.

In April 1979 the United States terminated its economic and military aid to Pakistan. Pakistani officials protested, pointing out that India had never been subjected to such harsh treatment for the development of its nuclear program. The decision to withhold aid resulted in rising anti-American sentiment in Pakistan, setting the stage for the burning down of the American embassy in Islamabad by a group of radical students in November 1979. Although two American Marines died in the assault, President Carter lifted the sanctions before the end of the year. The reason had nothing to do with the Pakistani nuclear program or American nonproliferation policies. Washington's calculations were suddenly changed by the Soviet invasion of Afghanistan, which took place on Christmas Day 1979.[25]

The United States needed Pakistan in its camp. In January 1980 Jimmy Carter recalled the SALT II treaty from the Senate. Instead of sanctions on Pakistan, he imposed sanctions on the Soviet Union. On the horizon was a new nuclear arms race between the two superpowers.

Chapter 23

STAR WARS

Soon after 10:00 a.m. on March 8, 1983, a helicopter took off from the White House lawn and headed for Andrews Air Force Base in Prince George's County, Maryland. The fortieth president of the United States, Ronald Reagan, was leaving Washington, DC, just as it was rocked by protesters. Close to five thousand of them were demonstrating at the Capitol.

When the Democratic congressman from Massachusetts, Edward Markey, asked the crowd: "Do you want to freeze the arms race?," the crowd responded with a powerful "Yeah." When Markey continued the quiz: "Do you think President Reagan is going to freeze the arms race?," the demonstrators responded with an even more powerful "No!" Leaving the capital all but in the hands of the protesters, Reagan flew to Orlando, Florida. He was going to present his view of the arms race and his opposition to "freezing" it to the National Association of Evangelicals, hoping for a more receptive audience down south.[1]

At the core of the controversy was the deployment of new American nuclear-armed Pershing II missiles at bases in western and central Europe. Pershings were intermediate-range missiles not covered by the SALT I and SALT II agreements, which dealt only with long-range ballistic missiles. In 1977 the Soviet Union decided to replace its R-12 medium-range and R-14 intermediate-range missiles of the types that had been sent to Cuba (the R-14s never reached the island, as they were turned back after the implementation of the US naval blockade) with a newer model. The new intermediate-range ballistic missile was known as the RSD-10 Pioneer or, in the NATO specification, the SS-20 Saber. Sabers were armed with three 150-kiloton MIRVs (Multiple Independently Targeted Reentry Vehicles) warheads. They were capable of delivering surgical strikes against NATO nuclear forces in Europe at a distance of more than 5,000 kilometers (3,107 miles).

In the opinion of American and West European commanders, the Sabers could change the strategic balance on the continent. The SALT I agreement created strategic parity between the two superpowers, leaving the superiority of Soviet ground forces in Europe unaddressed. The Americans no longer had the degree of nuclear deterrence that they had enjoyed in the first decades of the Cold War and were now about to lose it altogether. Moscow hoped that after leaving Vietnam, the United States would be forced to leave Europe as well. The deployment began in 1976, but with the shock of Vietnam fading into the past and taking the spirit of détente along with it, American readiness to keep withdrawing had diminished significantly by the end of the 1970s.[2]

The call to react to growing Soviet nuclear power in Europe came from Bonn, the capital of the Federal Republic of Germany (West Germany). In October 1977, Chancellor Helmut Schmidt of West Germany raised in public the question of "disparities between East and West in nuclear and tactical weapons." The Germans now fully accepted the post–World War II borders in Europe and welcomed Soviet natural gas in their kitchens but were concerned about the new generation of Soviet missiles deployed in the western lands of the USSR. Schmidt wanted the Americans to stop relying on their intercontinental missiles to deter possible Soviet aggression in Europe and bring in their own intermediate-range missiles. Prime Minister James Callaghan in London and President Valéry Giscard d'Estaing in Paris shared Schmidt's concerns.

In January 1979, the three European leaders met with Jimmy Carter on Guadeloupe Island in the Caribbean. The Soviet SS-20 missiles were on the agenda. The Europeans wanted the United States to update its nuclear capabilities in Europe, and Carter took notice. In December 1979, a few months after the signing of the SALT II agreement with Brezhnev and a few weeks before the Soviet invasion of Afghanistan, the NATO commanders decided to deploy in Europe 108 Pershing II intermediate-range ballistic missiles, each with a nuclear warhead blast yield of up to 80 kilotons. They were to be accompanied by 464 Gryphon ground-launched cruise missiles. But there was a caveat: NATO proposed to the Warsaw Pact that if the Soviets agreed to end the deployment of their SS-20 missiles and removed those already deployed, no American Pershings or cruise missiles would be delivered to their European NATO partners.[3]

Helmut Schmidt, who had been the first to sound the alarm about the deployment of the Soviet Sabers, hoped that the Soviets would take the deal: while his constituency was concerned about the Soviet missiles, many did not want American missiles in their backyard either. The Soviets responded with their own proposal to freeze the current nuclear arsenals in Europe, which meant limiting the number of SS-20 missiles to the 120 already installed. NATO was prepared to reduce proportionally the number of missiles on each side but was not interested in the "freeze" option, considering it a propaganda ploy. But the offer served the USSR well in its propaganda war with the West. Mass demonstrations rocked West European cities. West Germany, where all the Pershing II missiles were supposed to be installed, took the lead, with the deployment of the Pershings contested in court and 300,000 people descending on Bonn in October 1981 to protest the future placement of American missiles in their country.[4]

The deployment of the Pershings and cruise missiles was scheduled for November 1983, and anti-missile mobilization intensified as the deadline approached. Brezhnev, speaking in October 1979, had accused the West of wanting to "upset the balance of forces that has taken shape in Europe and to try to ensure military superiority for the NATO bloc." Reagan, speaking in November 1981, claimed that "the Soviet Union has developed an increasing, overwhelming advantage. They now enjoy a superiority on the order of 6 to 1." The ratio was bogus, but Reagan apparently believed in it.

Reagan was determined to close the new "missile gap" that existed in his imagination and needed time to do so. For the time being, he made the Soviets an offer that everyone knew they would not accept. The so called zero-zero option proposed to cancel the deployment of the American missiles if the Soviets removed not only their SS-20s but also the outdated R-12 and R-14 missiles that they wanted to replace with the new ones. The Soviets, as expected, rejected the proposal. Their major concern was that a deal on Soviet and American missiles was insufficient, as it did not cover either American bombers capable of delivering nuclear weapons or British and French missiles and nuclear warheads. Given the Soviet rejection, Reagan received his hoped-for moral justification for the deployment of Pershings and cruise missiles.[5]

In the US, Reagan's policy was met with resistance from activists of the American "freeze" movement, whose agenda was much broader than the European one. The American protesters, whose ranks included veterans of the SANE campaigns of the 1950s and 1960s, demanded a freeze on the production, testing, and deployment of nuclear weapons in both the United States and the Soviet Union. The Soviets did their best to exploit the US "freeze" movement to avoid resuming the nuclear arms race, which they could ill afford.

In December 1982, a few days before the Christmas break and less than a month after assuming office, the new Soviet leader, the former KGB chief Yurii Andropov, inaugurated what would become known as a Soviet "peace offensive." He proposed an immediate freeze on nuclear arsenals while the two countries looked for ways to reduce their strategic weapons. He wanted cuts as deep as 25 percent and put forward what he called an "honest 'zero option' ": in exchange for non-deployment of the Pershings and cruise missiles, the Soviets proposed to remove their old intermediate-range missiles and reduce the number of their SS-20s to the level of missiles deployed by Britain and France. In that case there would be no American missiles in Europe, but there would be parity between the Warsaw Pact and NATO nuclear forces.[6]

The Soviet proposals added fuel to the anti-nuclear campaign in the United States. By the spring of 1982, resolutions supporting a freeze of nuclear arsenals had been adopted by 320 city and 56 county councils in the country. In June 1982, one million people descended onto the streets of Manhattan to protest the continuing nuclear arms race. The National Council of Churches took a position in favor of the freeze, as did the Synagogue Council of America. The Catholic bishops were still debating, as were the member churches of the National Association of Evangelicals. The latter became Reagan's last hope.[7]

On March 8, 1983, Reagan appeared at the Evangelical Convention in Orlando, Florida, facing a difficult task. The group was divided on the freeze issue, with Quakers and Evangelical Mennonites supporting the pro-freeze protesters.

Reagan began his address to the 1,200 delegates by thanking them for their warm welcome and for their prayers. He then broke the ice by telling a joke about a politician who got to heaven and whom St. Peter treated much better than priests. When the politician asked why that was so, St. Peter responded that there were thousands of righteous clergy, but he was the first politician to have made it to paradise. The joke was met with laughter. Reagan went on to remind the audience about his efforts to restore public prayer in schools and assure them that his administration would defend in court its decision not to allow the prescription of contraceptives to teenage girls without the consent of their parents.

As Reagan transitioned from domestic to foreign policy, he embraced the language of the Christian crusade against communism. The word "evil," used in both segments of the speech, tied them together. Reagan spoke about his policy toward the Soviet Union as a life-and-death struggle against an atheistic regime that rejected all moral principles. He called on his audience to pray for the salvation of those who were in totalitarian darkness, asserting that unless they discovered God, they would remain the "focus of evil in the modern world." His next story hinted at nuclear holocaust as a price not too high to be paid for preserving faith in God. Reagan cited a young father who had allegedly told him: "I would rather see my little girls die now, still believing in God, than have them grow up under communism and one day die no longer believing in God." Back in 1957, Mao was prepared to sacrifice half of China's population in the nuclear holocaust in the name of communism. Reagan seemed to be ready to sacrifice his own countrymen in the name of the Almighty.[8]

The speech became known as the "evil empire" speech for the name Reagan gave in it to the Soviet Union. The term did not come from Scripture, despite Reagan's numerous references to religion, but from popular culture. The principal author of the speech, a thirty-four-year-old former Pulitzer Prize–winning journalist named Anthony R. Dolan, borrowed the term from the Hollywood blockbuster *Star Wars*, whose third episode was about to be released in May 1983. The "evil empire" of the film was the Galactic Empire, which had first appeared in an earlier episode of the *Star Wars* series, *The Empire Strikes Back* (1980). Its flag and symbols resembled those of the Nazi Reich. Now the term was being applied to the Soviets.[9]

The *Star Wars* imagery evoked in the "evil empire" speech of March 8, 1983, followed the president into his next major speech, delivered shortly thereafter, on March 23. The televised speech to the nation dealt specifically with issues of national security. Reagan proposed the adoption of a Strategic Defense Initiative (SDI) to develop a new anti-ballistic missile system. Given its reliance on placing parts of the system in outer space, Reagan's critics almost immediately dubbed it the "Star Wars" speech. Under that name it entered the annals of public memory and history, although stars and outer space were not mentioned even once in the text. Instead, Reagan talked about "a program to counter the awesome Soviet missile threat with measures that are defensive."

Reagan, always an effective salesman, was eager to impress his listeners with the degree of safety that the new program would bring them. "What if free people could live secure in the knowledge that their security did not rest upon the threat of instant U.S. retaliation to deter a Soviet attack, that we could intercept and destroy strategic ballistic missiles before they reached our own soil or that of our allies?" With this question Reagan not only relegated the MAD strategy to the "ashbin of history" but also destabilized the world order based on it. The speech heralded a major departure from American policy of the previous decade and a half. Reagan was prepared to scrap MAD, although he admitted that nuclear deterrence based on that strategy had worked in the past and prevented nuclear war. But he wanted more.[10]

White House staffers explained to journalists that the new technology on which the anti-ballistic missile system was to be based included lasers, microwave devices, and particle and projective beams directed from satellites, airplanes, and ground installations to intercept Soviet missiles. Admittedly, the technology did not yet exist and would take decades to develop. "I call upon the scientific community in our country, those who gave us nuclear weapons, to turn their great talents now to the cause of mankind and world peace, to give us the means of rendering these nuclear weapons impotent and obsolete," declared Reagan.[11]

Among those invited to the White House to attend a reception organized in conjunction with Reagan's big announcement was the father of the hydrogen bomb, Edward Teller, who had done most to convince the president to invest his trust and the nation's resources in the new program. Back in 1967, Reagan's first year as governor of California, Teller had led the future president on a tour of the Lawrence Livermore National Laboratory and used the opportunity to introduce him to the latest research on the nuclear-armed anti-ballistic missile system.

The idea of an ABM system appealed to Reagan, who had never reconciled himself to the MAD logic. In January 1982, when Reagan's advisers proposed a space-based anti-ballistic missile system, they relied on Edward Teller to convince him. At the center of the SDI vision was the X-ray laser, a weapon that, according to Teller, would be able to shoot down multiple missiles with a single beam. The laser, which was to be powered by the explosion of a small hydrogen bomb, was still under development at the Livermore Laboratory. The scientists on staff were not sure whether it could be produced before the end of the century, if at all, but Teller convinced Reagan's advisers and eventually the president himself to accept his vision and replace "mutually assured destruction" with what he called "assured survival."[12]

Teller was eager to sell his laser-based anti-ballistic missile system to the White House not only as a military item but also as a propaganda weapon with which the administration could successfully attack the freeze movement while holding the high moral ground. The tactic worked, and once the chiefs of staff agreed to support the development of the new technology as a way of switching from offensive to defensive conduct of a future nuclear war, Reagan was happy to employ the concept as both a future military strategy and an immediate political weapon.[13]

Reagan embraced the SDI idea with the zeal of a neophyte. It offered a solution to his old problem of opposing MAD on grounds of reason and morality without being able to offer anything to replace it. His actor's imagination allowed him to take a leap into the future promised by Teller's X-ray laser that most scientists and politicians were unable or reluctant to make. Among those caught by surprise and astonished by the announcement of a program not yet proved feasible were many Pentagon staffers. Senator Edward Kennedy, one of the leaders of the "freeze" movement, was the first to refer to Reagan's Strategic Defense Initiative as a "reckless Star Wars scheme." The "evil empire" of Reagan's March 8 speech now became associated with the "Star Wars" initiative of his March 23 address.[14]

While others attacked or doubted Reagan's belief in the feasibility of space-based laser defenses, or ridiculed it as mere science fiction, his embrace of the concept immediately made it a political and geostrategic reality that the Soviet Union had no choice but to deal with. Yurii Andropov wasted no time in responding to Reagan's "Star Wars" speech. Four days later, in an interview with the leading Soviet newspaper, *Pravda*, the new general secretary rejected Reagan's claim that the United States was lagging behind the Soviet Union in nuclear armaments, asserting that the number of nuclear warheads on strategic missiles had grown from four to more than ten thousand. He admitted that the Soviet arsenal had grown as well but declared that the Soviets had simply reached parity with the United States, making further nuclear blackmail impossible.

Referring to SDI, Andropov declared that the Americans were using it as a cover to continue building up their offensive strategic forces with the aim of delivering a first strike against the Soviet Union. The United States, he claimed, was rejecting the principle of interconnection between offensive strategic weapons and anti-missile defenses that had served as the basis of the SALT I treaty signed in 1972 by Brezhnev and Nixon. The realization of SDI, stated Andropov, "would open the floodgates of a runaway race of all types of strategic arms, both offensive and defensive."

Andropov pledged that the Soviet Union would never permit American strategic superiority. The Soviet defense minister, Marshal Dmitrii Ustinov, backed by Soviet scientists eager to prove their worth, was pushing for a Soviet anti-ballistic missile system as a response to SDI. But Andropov knew that, given the sorry state of the economy he had inherited from Brezhnev, a new arms race would bankrupt the Soviet Union. Like Reagan, Andropov had no money to invest in his own SDI. The difference was that Reagan could borrow, while Andropov could not, as the Soviet economy was in permanent recession. He called on Washington to join the USSR in preventing a "nuclear catastrophe."[15]

Andropov's interview demonstrated the degree of concern that Reagan's initiative had produced in Moscow and offered an insight into the Soviet leadership's thinking about the Americans. The Soviets saw Reagan as an individual who could indeed start a nuclear war. "Is he just playing his game and being a hypocrite, or does he really realize that for all our ideological disagreements, you just cannot bring about a confrontation in the nuclear age?" Andropov asked the Soviet ambassador to the United States, Anatolii Dobrynin. The ambassador had no good answer to that question. Andropov stressed the need to keep working with Reagan but added: "We should be vigilant, because he is unpredictable."[16]

Although American politicians and intelligence services did not know it at the time, Andropov truly believed that the American nuclear buildup, including the deployment of Pershing II and cruise missiles in Europe, could be part of a plan to prepare the United States for a first nuclear strike against the USSR. In 1981, when Andropov was still head of the KGB, he instructed his subordinates "not to miss the military preparations of the enemy, its preparations for a nuclear strike, and not to miss the real risk of the outbreak of war." Operation RYaN—*raketno-yadernoe napadenie*, or nuclear missile attack—was born, with the KGB and Soviet military intelligence (GRU) alerted to look for signs of impending nuclear war.[17]

The CIA later concluded that Operation RYaN was triggered by American "psychological warfare operations" launched after Reagan's arrival in the White House: American airplanes and naval vessels took air and water samples close to the Soviet borders, approached Soviet missile installations, and tested Soviet early warning systems near naval and military bases. The plan was to put the Soviets on constant alert, making them guess what would come next. Their guess was a possible nuclear attack. In February 1983 Operation RYaN was assigned priority status by Moscow—a matter of "particularly grave importance." Soviet intelligence operatives throughout the world were instructed to keep "constant watch" for signs of a coming nuclear attack on the USSR.[18]

On September 1, 1983, Soviet air defenses in the Far East went on the alert as radar detected a Boeing 747 flying above the Kamchatka Peninsula. American airplanes had probed Soviet early warning systems before, but now they had clearly crossed the line. A plane flying over Kamchatka, with its missile installations, and then over Sakhalin Island, testing Soviet early warning systems, was more than Soviet commanders could tolerate. The Boeing was about to leave Soviet airspace, presumably with loads of data on Soviet missile installations and air defenses, when the order came to shoot down the intruder. The Soviet Sukhoi S-15 supersonic interceptor that followed the Boeing fired two air-to-air missiles. The wreckage of the Boeing fell into the sea a few dozen kilometers from Sakhalin Island.[19]

The Boeing 747 shot down over Sakhalin was in fact Korean Air Lines (KAL) flight 007 on its way from New York to Seoul via Anchorage. It had lost its way because of a navigational error and was flying on autopilot up to 500 kilometers (31 miles) off its expected route. Two hundred sixty-nine people, including the crew and the passengers, died in the missile attack on the plane. Among the victims was the forty-eight-year-old congressman Larry McDonald of Georgia, who happened to be the president of the anticommunist John Birch Society. The Soviet government issued a statement saying that an unidentified plane had crossed into Soviet airspace and was followed by Soviet interceptors but had subsequently disappeared from the radar screen, flying "in the direction of the Sea of Japan."

Yurii Andropov was as surprised as everyone else. Discussing the situation with the Soviet ambassador in Washington, Anatolii Dobrynin, he told the ambassador: "Our military made a gross blunder by shooting down the airliner." He cursed "those blockheads of generals who care not a bit for grand questions of politics." But he still believed that the plane had been on a spy mission and was angry with Reagan, who had called the shooting down of the plane a "crime against humanity." Ten days later Reagan revoked the right of the Soviet state airline, Aeroflot, to operate flights to and from the United States, making direct air connections between the two countries impossible.[20]

It turned out that Andropov had been too quick to tar all generals with the same brush. A few weeks after the KAL 007 accident, General Yurii Votintsev, the commander of the Soviet Air Defense Missile Units, refused to reprimand his subordinate, Lieutenant Colonel Stanislav Petrov, for not jumping the gun and launching a missile attack on the United States in response to what the Soviet early warning system had erroneously identified as an American nuclear missile attack—the Operation RYaN that Soviet intelligence was so concerned about. Petrov had been on duty on September 25, 1983, at the control center of the space-based system for the interception of missile attacks, approximately 100 kilometers (62 miles) from Moscow, when the air defense early warning system sounded the alarm—five missiles had apparently been launched from American territory and were approaching the Soviet Union.

According to instructions, Petrov had to report the missile attack to his superiors. The decision on how to respond would rest with Andropov if there was time for the news to reach him. Petrov had little doubt that the Americans could attack but doubted that a first strike would involve only five missiles. The strike had to be massive. He also suspected a possible malfunction of the satellite-based early warning system—he knew its strong and weak points better than anyone else—and actually wrote instructions for the officers on duty. He decided not to report the "attack," as those empowered to make decisions had no clue how unreliable the system could be. As it turned out later, sunbeams penetrating high-altitude clouds over North Dakota had made the Soviet space-based system malfunction. Petrov would be hailed as a hero, but only much later, after the end of the Cold War. The first to tell the story of what happened on the night of October 25 was Petrov's commander, General Votintsev, who refused to reprimand his subordinate after first hearing about the accident.[21]

While the incident involving Petrov remained a state secret, the shooting down of KAL 007 was central to the rapid deterioration of

Soviet-American relations. On September 29, 1983, three days after the Petrov incident, *Pravda* published Andropov's statement calling the KAL 007 flight a "sophisticated provocation organized by US special services and using a South Korean airplane" and "an example of extreme adventurism in policy." Later that month the Soviet foreign minister, Andrei Gromyko, met with his American counterpart, George Shultz. The meeting, which focused on KAL 007, became not just tense but openly hostile. Gromyko remembered it as the "sharpest exchange" that he had ever had with an American secretary of state. He added: "and I had talks with fourteen of them." [22]

On November 19, 1983, *Pravda* editors gave special attention to Soviet Missile Forces and Artillery Day, a little-known military holiday that had now acquired a special meaning. The Soviet defense minister, Marshal Ustinov, contributed an article to the *Pravda* issue that received its imprimatur directly from the Politburo. "The White House has openly set a course for the achievement of military superiority," wrote Ustinov. According to him, the ultimate goal was the elimination of the socialist system and national-liberation movements around the globe. Ustinov restated the Soviet position that the USSR would not launch a first strike but was prepared to respond in kind if attacked. In response to the deployment of American missiles in Europe, he threatened to end the moratorium on the deployment of the SS-20 Sabers.[23]

Ustinov also promised other measures to be undertaken in coordination with Soviet allies. He did not specify what those measures might be, but it was not difficult to imagine another Cuba scenario—the deployment of Soviet nuclear missiles in East European states, which the Kremlin had never tried. The MAD logic killed by Reagan's announcement of the Strategic Defense Initiative had stopped working.

Chapter 24

THE FALL OF THE NUCLEAR COLOSSUS

President Reagan understood the world around him first and foremost in moral, ethical, and emotional terms. That was how he came to realize the danger that his uncompromising rhetoric, provocative American behavior on the Soviet borders, and the arms buildup presented to his country and the world. What led him to reconsider and slowly begin changing his policies was the fear that nuclear weapons and the prospect of nuclear war had always inspired in him, and the growing realization that the Soviets feared he would initiate such a conflict.

The change in Reagan's attitude demonstrated itself on Columbus Day (October 10), 1983. After some rest and horse riding at Camp David, the president loaded his video player with a brand-new Hollywood movie that was scheduled for general release the following month. In the film, titled *The Day After*, tension between NATO and the Warsaw Pact resulted in a nuclear conflict. It showed the city of Lawrence, Kansas, devastated by a nuclear strike, leaving the survivors to struggle for whatever semblance of life they might manage to recover. "Unflagging nightmare of a program," commented a BBC reporter.

The film made a strong impression on Reagan. "Kansas wiped out in a nuclear war with Russia," he wrote in his diary that day. "It is powerfully done—all $7 mil. worth. It's very effective & left me greatly depressed." The film made Reagan reflect on his policies at home and abroad. "Whether it will be of help to the 'anti-nukes' or not, I can't say," continued Reagan. "My own reaction was one of our having to do all we can to have a deterrent & to see there is never a nuclear war." He was counting on the Strategic Defense Initiative to provide a shield against Soviet missiles and worried that the anti-nuclear movement might prevent him from making that shield a reality. "We know it's 'anti-nuke' propaganda but we're going to take it over & say it shows why we must keep on doing what we're doing," wrote Reagan on November 18, two days before the official release of the movie.[1]

Around the same time, the head of the KGB station in London, Lieutenant Colonel Oleg Gordievsky, who was in fact a double agent working for the KGB by day and for MI6 by night, reported to his British handlers about Soviet paranoia caused by a NATO military exercise called "Able Archer." The exercise ran from November 7 to 11, 1983, and simulated military operations under conditions of nuclear war. The Soviets were on the lookout for a nuclear attack: their Operation RYaN was in full swing, and they were afraid that the exercise was a cover for a first nuclear strike.

Gordievsky's report made it all the way to Margaret Thatcher, and eventually to Ronald Reagan. Whether or not there was a war scare in the Soviet Union, as Gordievsky suggested, it became a reality in the eyes of the Western leaders. "I don't see how they could believe that—but it's something to think about," Reagan told the CIA chief, William Casey, upon receiving his report about Gordievsky's revelations and other pieces of intelligence that supported them.[2]

Reagan showed he was concerned about how Soviet leaders saw his actions, when he delivered a speech on Soviet-American relations on January 16, 1984. There was a new tone in the speech, which declared that the United States was "in the strongest position in years to establish a constructive and realistic working relationship with the Soviet

Union." The idea was that the military buildup undertaken during the first years of Reagan's administration had created preconditions for the United States to negotiate with the Soviets from a position of strength. There was no more talk about a crusade against the Soviet Union or attacks on the Soviet leaders' ideology or moral principles.

Instead, in parts of the speech written by Reagan himself, he called for peace while appealing to the common concerns of average people on both sides of the Cold War divide. "Just suppose with me for a moment that an Ivan and an Anya could find themselves, oh, say, in a waiting room or sharing a shelter from the rain or a storm with a Jim and Sally," suggested Reagan. "And there was no language barrier to keep them from getting acquainted. Would they then debate the differences between their respective Governments? Or would they find themselves comparing notes about their children and what each other did for a living?" He went on: "If the Soviet Government wants peace, then there will be peace." After suggesting that Moscow and Washington could "strengthen peace and reduce the level of arms" by working together, he concluded with a call to action: "Let us begin now."[3]

The change in rhetoric also meant Reagan's attitude changed. He was not prepared to give up on what he considered to be the fully justified and long overdue buildup of the American nuclear and missile arsenal or the "Star Wars" that so frightened his opponents. But he was prepared to talk to the Soviet leaders about the implications of his tacit admission that the logic of mutually assured destruction was untenable. "Star Wars" undermined MAD, but MAD, or the balance of terror, had to be addressed by negotiations on the control or reduction of the two countries' nuclear arsenals. The Soviet leaders could understand such realpolitik.

The problem was that they kept dying on him—first Andropov in February 1984, then Konstantin Chernenko in March 1985. But then, unexpectedly, there came a glimmer of hope: the Central Committee in Moscow elected an energetic fifty-four-year-old successor to Chernenko, a former Central Committee secretary in charge of agriculture, Mikhail Gorbachev.

Reagan and Gorbachev met for the first time in Geneva in November 1985, eight months after Gorbachev's election to the top party office. "I have met a caveman," Gorbachev told his aides after his first conversation with Reagan. The enterprising Soviet leader was not referring to Reagan's age but to the depth of his anticommunist views. At the top of the summit's agenda was the nuclear arms race, but no agreement was signed and, more importantly, there was no expectation of such an agreement. On the eve of the summit, anticipating that Gorbachev would demand an end to his Strategic Defense Initiative, Reagan wrote in his diary: "Well, this will be a case of an irresistible force meeting an immovable object."[4]

"The question of ending the arms race is of critical importance," was the key message of Gorbachev's original pitch to Reagan. The Soviet leader wanted to stop it and had ideas on how to accomplish his goal. But Reagan was not about to rush into negotiations. "We don't mistrust each other because we are armed, we are armed because we mistrust each other," was his leitmotif. For Reagan, the summit was a trust-building exercise. Trust could develop if the Soviets agreed to withdraw from Afghanistan and stop supporting revolutions in the Third World.

As expected, Gorbachev pushed for the abolition of SDI, which he did not call "Star Wars" but characterized as a step toward the weaponization of space. "Space weapons will be harder to verify and will feed suspicions and mistrust," argued Gorbachev. He was prepared to reduce his nuclear arsenal by half if the Americans did the same and gave up SDI. Reagan, as he had done publicly before the summit, emphasized SDI as a defensive weapons system that could not harm anyone. "I am talking about a shield," he argued. "If a defensive system is found, we would prefer to sit down and get rid of nuclear weapons and, with them, the threat of war."[5]

Reagan was prepared to do something that Truman had balked at—to share new, in this case not nuclear but anti-missile, technology with the Soviets once it was developed. Many of Reagan's advisers found that offer controversial: some of them wanted an arms race that would bankrupt the Soviets, while others wanted to use SDI as a means of getting concessions from them. Gorbachev was also unreceptive, but for reasons different from those advanced by Reagan's aides.

"Why don't you believe me when I say the Soviet Union will never attack?" Gorbachev asked Reagan. The president explained that "no one can say to the American people that they should rely on personal faith rather than sound defense." In effect, he was stepping into Gorbachev's trap. "Why should I accept your sincerity on your willingness to share SDI when you do not even share your advanced technology with your allies?" responded Gorbachev, asking another question. According to a different version of the conversation, Gorbachev simply laughed and pointed out American unwillingness to share technical data about milking machines with the Soviets.[6]

The two leaders parted ways without reaching understanding on anything. Gorbachev was anything but optimistic. When his close adviser and ally Aleksandr Yakovlev quoted the famous British novelist Graham Greene's assessment of Reagan as a clown, Gorbachev responded: "This is the danger!" He needed a leader with whom, in the words of Margaret Thatcher about himself, he could "do business." Gorbachev was not sure that he had found one in Reagan, but he also knew that he had no choice but to deal with him. "We face a very hard struggle," Gorbachev told the Politburo in May 1986. "We have to negotiate with a 'band' [of thugs] but negotiate nevertheless." Publicly he spoke about the "spirit of Geneva," meaning readiness to continue the dialogue.[7]

Faced with Reagan's resumption of the arms race, Gorbachev told his close ally, the new Soviet foreign minister, Eduard Shevardnadze, that he had two choices: to cut consumption and follow the Americans into a new nuclear buildup or seek to ease international tensions. With the economy in recession and the living standards of the Soviet population in decline, the internal threat to the regime loomed larger than the external one in Gorbachev's eyes. If he could manage to stop the arms race and begin disarmament, then the savings would improve the economy and the standard of living, allaying popular discontent.

The economy, which had not been doing well for years, was now in steep decline. According to official figures, between 1980 and 1985 Soviet GDP had been growing at a rate of less than 4 percent—the lowest rate in Soviet history since the 1920s. The CIA estimated the growth rate at closer to 2 percent, with real growth amounting to less than 1 percent. Military spending, concealed in allocations and subsidies to the numerous ministries and agencies that constituted the Soviet military-industrial complex, amounted to 40 percent of the state budget. The fall of world oil prices had cost the Soviet budget thirteen billion rubles. An even bigger hit came from Gorbachev's attempt to improve discipline and stop the physical and moral degradation of Soviet society by introducing an anti-alcohol campaign. There were 4.5 million alcoholics in the USSR. The campaign cost forty-three billion rubles in lost revenue, while doing little to improve health standards or discipline, as the population switched to homebrew.[8]

In March 1986 Gorbachev called on the Politburo to "make every effort not to bankrupt the country through defense." He wanted to save money on the arms race and make the highly technological military sector produce goods for public consumption, thereby generating income to support itself. "In order to solve the problem, we have to change the military," Gorbachev told the Politburo on April 24, 1986, as his colleagues discussed the lack of consumer goods produced by "light industry." He added: "And, by the way, where is defense going to find money?" The idea was that profits obtained by the military-industrial complex from the production of civilian goods could be used to develop arms programs. But such economic calculation within a single enterprise or even the entire military complex was hardly more than a pipe dream.[9]

As Gorbachev spoke those words, the operators at reactor no. 4 of the Chernobyl (Chornobyl) nuclear power plant in Ukraine were getting ready to run a safety test at their workplace. The plant was an example of cooperation between the military-industrial complex and the consumption-oriented civilian economy that might very well have served as an inspiration, if not a model, for the conversion of part of the military economy toward civilian use proposed by Gorbachev.

The Chernobyl plant used high-powered RBMK channel–type reactors, with water as a coolant and graphite as the moderator of the nuclear reaction, which could produce plutonium. Originally based on the American Oak Ridge model of a graphite reactor stolen by Soviet spies, nuclear reactors of that type had been adapted to the production of electricity. In fact, they were dual-use reactors that could be reconverted to plutonium production if required. The design of the reactor was top secret, and not all its characteristics were known even to the operators charged with running it.

To conduct the safety test of the reactor's turbine generator, the operators reduced power to a level at which the reactor became unstable. At 1:23 a.m. on April 26, 1986, when they pressed the "scram" button in an attempt to shut down the reactor, it exploded. Anywhere between 50 and 200 million curies of radioactivity were released from the opening in the reactor's core during the next few weeks. The Soviets evacuated approximately 50,000 people from the nearby city of Prypiat. They were followed by another 90,000 people from the 10-kilometer (6-mile) and then 30-kilometer (19-mile) exclusion zone, as well as other areas affected by the fallout. The army was thrown into the exclusion zone to measure radiation levels, conduct decontamination, help move people, and eventually build a concrete shelter or sarcophagus above the damaged reactor. Altogether 600,000 men and women mobilized through the army reserve passed through the Chernobyl exclusion zone, absorbing radiation and returning home after fulfilling their duty to the state.

For months, the top Soviet leadership could hardly focus on anything other than dealing with the consequences of what became the world's largest nuclear catastrophe. The economic damage was enormous, counted in hundreds of billions of Soviet rubles. Prospects of doubling the number of Soviet nuclear power stations in the course of the next five years were scrapped. Western criticism of the Soviet refusal to make a timely announcement about the accident, together with popular demand within the USSR to tell "the truth about Chernobyl," meaning information about the consequences of the disaster, forced the regime to loosen its control over the media. The policy of *glasnost* or "openness" was born, allowing Soviet citizens to discuss Soviet economic and social problems in public. The party itself would soon be under attack. The Chernobyl explosion jumpstarted the democratization and subsequent demise of the Soviet political system.[10]

The contamination caused by the Chernobyl explosion, in which only part of the plant's nuclear fuel escaped the reactor, showed the Soviet leadership what a nuclear war could look like. It was a frightening picture. Arguably, that real-life drama made a stronger impression on Gorbachev than *The Day After* had made on Reagan. On May 8, 1986, twelve days after the accident, Gorbachev told the Politburo that he was prepared to negotiate the reduction of nuclear weapons even with a "gang." "Otherwise, what?" asked Gorbachev. "There it is—the Chernobyl accident. Just a breath, and we felt what nuclear war actually is."[11]

In September, with the Chernobyl situation finally under control and the sarcophagus above the damaged reactor nearing completion, Gorbachev wrote to Reagan, suggesting a meeting to break the deadlock in the arms control negotiations conducted by diplomats in Geneva. The two leaders met on October 11 and 12, 1986. Some believed that Gorbachev had lured Reagan into a trap, as the meeting had no clear agenda and was organized as a discussion between the two leaders, with their foreign policy teams relegated to secondary roles. One on one, Gorbachev believed, and many Americans feared, he could emerge victorious in the nuclear duel with Reagan.

To different degrees, the two leaders were averse to the very idea of nuclear weapons and found in that aversion a platform on which they could agree and build. They met in Reykjavik, Iceland. When Gorbachev proposed his 50-percent cut of the arsenals of the two nuclear superpowers, Reagan accepted the proposal. They agreed to halve their "strategic offensive weapons" in five years and eliminate "offensive ballistic missiles" altogether in the subsequent five years. Reagan's advisers were terrified when the president agreed to substitute "weapons" for "missiles" in the text of the treaty proposed by Gorbachev. "Weapons" included not only the missiles over which the Americans were prepared to negotiate but also strategic bombers, in which the United States had superiority, and which had not been included in the earlier discussions. It was a revolutionary breakthrough. It looked as though Gorbachev's gamble had worked. There was only one item to be decided for the deal to work. That was SDI.

As Gorbachev kept pushing for the elimination of SDI, Reagan agreed not to deploy it for the next ten years, limiting the program to the "laboratory" and testing. It sounded like a major concession. In the course of ten years they were supposed to eliminate nuclear weapons altogether, so what difference would SDI make? But Gorbachev saw it differently: by the end of the ten-year period, his country would have nothing, while the American SDI technology would be complete. If the arms race were to begin anew following the decade of elimination, the USSR would be at the mercy of American missiles that could reach it without obstacles, while Soviet missiles would be destroyed in space by SDI. Gorbachev's offer to reduce or even eliminate the nuclear arsenals was conditioned on scrapping SDI completely. Reagan refused.[12]

There was no deal. The agreement of the century fell apart over "one word," as Reagan suggested to Gorbachev at the end of the summit. The two leaders left Reykjavik in a gloomy mood, with Reagan not joining Gorbachev for a press conference nor holding one of his own. Speaking a few days later to the Politburo, Gorbachev stated: "Some attempted to depict the drama in Reykjavik (and the situation developed dramatically indeed) as if everything depended on one word. No, it depended on principle. We made certain concessions, but we cannot make a concession that would endanger the security of our state." Nevertheless, Gorbachev declared a partial victory. "Reykjavik is the new start of our mass peace offensive," Gorbachev told his colleagues.[13]

Like Khrushchev on the eve of his Vienna summit with Kennedy, Gorbachev at Reykjavik hoped to surprise his American counterpart and achieve the results he desired. But in both cases, the Soviet reaction was retreat. Khrushchev had hidden behind the newly constructed Berlin Wall, soon dropping his demand for an American withdrawal from West Berlin. Gorbachev decided to back away from his demand to stop the development of SDI. Instead, in February 1987 the Politburo supported his proposal to "unbundle" the negotiation package. The Euromissiles were assigned to a separate category unrelated to SDI. The Soviet military agreed to trade their SS-20s deployed in Europe for the American Pershings—both were supposed to be withdrawn from Europe.[14]

This retreat from positions developed under Khrushchev and Andropov, and held originally by Gorbachev, was presented as "new thinking." Gorbachev was prepared to make concessions in order to avoid a new arms race and reduce the number of nuclear weapons, freeing resources for economic development. In the realm of nuclear weapons, the new approach was called "strategic parity." Gorbachev told the Politburo: "Strategic parity gives us a reliable guarantee of the defense of our country, and our opponent will not attack us, otherwise he will sustain an unacceptable retaliatory strike." What Gorbachev called "strategic parity" was characterized by some Politburo members as a "sound sufficiency." In effect, Gorbachev had given up on the arms race with the United States. His thinking was now more in line with that of the leaders of Britain, France, and China. More than ever before, Soviet nuclear arms were becoming weapons of deterrence.[15]

On November 1, 1987, HarperCollins published an English translation of Gorbachev's new book, *Perestroika: New Thinking for Our Country and the World*. There the Soviet leader presented his new vision of Soviet foreign policy. "We are all passengers aboard one ship, the Earth, and we must not allow it to be wrecked," wrote Gorbachev. "There will be no second Noah's Ark." Gorbachev claimed that he "was against SDI, because we are for complete elimination of nuclear weapons." But he was no longer prepared to treat the issue as an impassable barrier to improved Soviet-American relations. The publication was timed to coincide with preparations for a new summit between Gorbachev and Reagan to be held in Washington in December of that year.[16]

On December 7, 1987, when Gorbachev arrived for the meeting, he was welcomed in the streets of the American capital by many who had protested Reagan's "Star Wars." The two leaders signed the Intermediate-Range Nuclear Forces (INF) Treaty, putting an end to the Euromissile crisis that had begun in the late 1970s. The basis of the treaty was Reagan's 1981 "zero-zero option," which Gorbachev's predecessors had rejected. For the first time, the two sides were not just restricting their arsenals but eliminating entire classes of weapons. Before the end of the Cold War, 1,692 missiles were eliminated on the basis of the agreement.[17]

The INF Treaty became possible because of Gorbachev's readiness to make one more concession, allowing American on-site inspections of Soviet nuclear missile installations. This obstacle to verifiable treaties, which had been in place since the days of Khrushchev, was now gone. Khrushchev and his successors had opposed inspections not because they were concerned about American "spying" but because they did not want to reveal the inferiority of their nuclear armaments to the other superpower. Gorbachev did not care about the Americans discovering the actual state of his nuclear sword and shield. In his mind, the latter was strong enough to serve as a deterrent.

Ronald Reagan had every reason to feel triumphant. "It was a good meeting & it's plain he really wants more reduction of nuclear weapons," recorded the president in his diary entry for December 8, 1987, the day on which he and Gorbachev signed the INF Treaty. Not only was the treaty based on Reagan's "zero-zero option," originally rejected by Andropov and then Gorbachev, but his beloved SDI also remained intact. On the following day Gorbachev made his last attempt to kill SDI. He proposed that for ten years the two sides should abide by the Anti-Ballistic Missile Treaty of 1972 and only then negotiate whether SDI should be deployed or not. Reagan rejected the offer. In May 1988 he flew to Moscow to ratify the INF Treaty. While strolling on Red Square, he was asked whether he still stood by his definition of the Soviet Union as the "evil empire." He responded in the negative: "No, I was talking about another time, another era."[18]

Despite the positive Western public response to Gorbachev's "new thinking," things were not going well for him either at home or abroad. Allies and clients were abandoning him with frightening speed and in increasing numbers. In December 1988, Gorbachev pledged from the podium of the United Nations that he would not use force to keep the East Europeans in the Soviet camp. He believed that they would choose socialism on their own. Instead, they rebelled, and by the end of 1989 communist regimes had been overthrown in all the Eastern Bloc countries except Albania.[19]

In the Soviet Union, Gorbachev's reforms, known as *perestroika* or restructuring, exacerbated the economic crisis already under way. The old command economy was being destroyed from the top, while a new one based on private enterprise refused to grow from below. Gorbachev hoped that the introduction of electoral democracy would give the political system inherited from Joseph Stalin a new lease on life. Instead, the first competitive elections since 1917, held in the spring of 1990, brought pro-independence forces to power in the Baltic republics and made them a powerful force in several others, including Ukraine. Gorbachev's former protégé and now rival, Boris Yeltsin, declared the sovereignty of the Russian Federation, the core of the Union.

As Gorbachev tried to negotiate a looser Union with the rebellious republics, some of his own aides staged a coup in August 1991 while their boss was vacationing in the Crimea. Yeltsin led the resistance to the coup, which collapsed without a strong leader. By the time Gorbachev returned to Moscow, Yeltsin was well established in power. It was the final countdown for the USSR.

George H. W. Bush, who succeeded Reagan in 1989, managed to get one more nuclear arms deal from Gorbachev before the Soviet leader lost all his power. Bush visited Moscow in late July 1991. On the agenda was the signing of the Strategic Arms Reduction Treaty (START), which the two sides had begun to negotiate in the early 1980s. It was the most complex but also the most radical arms reduction agreement ever worked out. The number of nuclear vehicles—ICBMs and strategic bombers—was limited to 1,600 on each side, while the number of warheads was limited to 6,000. Those were extremely significant reductions. As a result, the two sides cut their nuclear arsenals by 80 percent.[20]

Once again, Gorbachev was making concessions. Strobe Talbott, who would later serve as President Bill Clinton's point man on Russia, wrote in *Time* magazine on the eve of the Bush-Gorbachev summit: "On almost every major question in START, the US demanded and got its way. . . . Gorbachev is tacitly accepting a position of overall inferiority, at least in the near term, since he is giving up right away much of the USSR's principal strength, which is land-based ballistic missiles, allowing the US to keep its advantages in bombers, cruise missiles and submarine weapons." There was no mention of Reagan's SDI. Gorbachev considered making a unilateral statement threatening a Soviet exit from the START treaty in case of the deployment of SDI but eventually dropped the idea.[21]

On Christmas Day, December 25, 1991, a somber and clearly shaken Mikhail Gorbachev faced television cameras to deliver his final address as president of the Soviet Union. The superpower had been dissolved a few weeks earlier by decision of the leaders of its core republics, Russia, Ukraine, and Belarus. The rest of the republics still belonging to the USSR followed suit. "We live in a new world," declared Gorbachev in his farewell address. "The Cold War has ended, the arms race has stopped, as has the insane militarization which mutilated our economy, public psyche and morals. The threat of a world war has been removed." Soon after he ended his speech, his briefcase with codes required for the launching of nuclear weapons was passed to the new master of the Kremlin, Boris Yeltsin.[22]

In the field of nuclear arms, Gorbachev's new thinking amounted to the Soviet Union's adoption of a posture akin to that previously assumed by Britain and other countries, which regarded nuclear weapons exclusively as a deterrent against a more powerful enemy. Moscow remained equal with Washington in terms of its nuclear arsenal but lost parity in its ability to deliver its warheads to the target. Gorbachev's shift became an important precursor of the end of the Cold War. The nuclear competition was over, but the race to stop the proliferation of nuclear weapons had just entered its new and most dramatic stage. The Soviet nuclear arsenal did not dissipate into thin air.

The fall of the USSR produced the single greatest act of nuclear proliferation in world history. Instead of one country, four states possessing former Soviet nuclear missiles appeared on the world map. Apart from Russia, they were Ukraine, Belarus, and Kazakhstan. Three of them would give up their nuclear arsenals before the end of the decade, reversing the proliferation process started by the Soviet collapse.

Chapter 25

GIVING UP THE BOMB

In January 1993, George H. W. Bush scored his last major victory in the international arena. Soon after New Year's Day, he boarded a plane to Moscow. The treaty that he signed on January 3 entered the annals of nuclear history as START II. Bush had begun negotiating the treaty with Gorbachev in the fall of 1991, and now he signed it with Gorbachev's rival and successor, Boris Yeltsin. START II mandated the elimination of nuclear-armed MIRV missiles, which were considered first-strike weapons and particularly important in that regard. Eliminating them meant greatly reducing the probability of a first strike from either side.[1]

The end of the Cold War and Soviet-American geopolitical rivalry produced much more than the end of the nuclear arms race between the two superpowers. The 1990s became the denuclearization decade par excellence. Two countries, Brazil and Argentina, agreed to stop the development of their nuclear programs, and four countries agreed to give up their nuclear arsenals. Among them was the post-Soviet republic of Ukraine, which was born nuclear when it proclaimed independence in 1991, coming into possession of one of the world's largest nuclear arsenals—third in size after those of the United States and now independent Russia. Belarus and Kazakhstan also gave up their share of the Soviet nuclear arsenal. Outside the region, South Africa turned in its nuclear weapons.

The political change and collapse of the authoritarian regimes triggered first by the relaxation of the tensions between East and West and then by the end of the Cold War produced the global peace dividend that turned into a nonproliferation bonanza. For many it looked like the nuclear arms race, if not the nuclear age itself, was finally over.[2]

The first country not to control or limit its nuclear stockpile but to eliminate it altogether was the Republic of South Africa, whose white supremacist government built a secret nuclear arsenal to protect itself ostensibly against the Soviet Union and its allies, but in reality against the black majority in the country and beyond its borders. The end of the Cold War spelled the end not only of apartheid but also of South Africa's nuclear status. The two processes were closely interconnected.

South African involvement in all things nuclear began in 1944 as part of the British nuclear program, when Winston Churchill asked the head of the South African government, Jan C. Smuts, to make an inventory of the country's uranium deposits. Smuts agreed, and it was soon discovered that the republic's deposits were considerable. By 1948 the South African Energy Board was created to oversee uranium mining, which got a boost with the opening of the country's first uranium-extraction plant in 1952. In three years there were sixteen mines producing uranium for American and British reactors and bombs. But in 1957, taking advantage of the Atoms for Peace program, South Africa signed a fifty-year deal with the United States whereby the Americans supplied a 20 Mw reactor called Safari-1 and 90-percent enriched uranium.

The reactor went critical in 1965. By that time the first generation of South African nuclear scientists, who had been sent for study abroad in the 1950s, were back home, running the reactor. In 1969 they started a uranium enrichment project that became operational in the early 1970s. The Plowshares program in the United States, which utilized nuclear explosions for peaceful purposes, attracted a good deal of attention in South Africa as a means of developing new technology for the country's mining industry. Research began in 1971, proceeding by 1974 to the development stage, which involved drilling shafts and building a test site for what was defined as peaceful nuclear explosions.[3]

Soviet satellites located the site in 1977, causing an international scandal and forcing the South African government to abandon its nuclear test site. In the following year Pretoria decided to repurpose the program in order to build a bomb. The government adopted a three-phase approach to the use of the future bomb as a deterrent. In the first phase, the authorities' posture would be one of strategic ambiguity, neither confirming nor denying possession of the bomb; if threatened, they would proceed to phase two, revealing the existence of the bomb to South Africa's allies, Washington in particular. In phase three, should the military situation require it, the government would announce its possession of the bomb, possibly with a nuclear test. There was officially no phase four, but it is not hard to discern what it would have entailed—the use of the bomb as a weapon of last resort.[4]

The official justification of this approach, given decades later, was the need for "limited deterrence" against possible Soviet aggression. But the decisions were made against the background of the growing anti-apartheid movement in South Africa and the rise of anticolonial movements on its borders. Soon after World War II, the South African government established the system of apartheid—a codified system of racial segregation and discrimination that allowed the white minority of 20 percent to control the black majority of 80 percent. Opposition to apartheid began almost immediately, with most South African religious organizations condemning the government policy—alas, to no avail. In the 1950s the African National Congress (ANC), the country's largest African political party, launched nonviolent campaigns of defiance and protest.

Opposition to apartheid intensified and turned violent after the Sharpeville massacre of 1961, in which the police killed sixty-nine black protesters and wounded hundreds more. The ANC was banned but continued to engage in armed struggle in the underground. Nelson Mandela, an ANC leader, arrested in 1962 for his role in the sabotage campaign against the government and sentenced to life behind bars, became the symbol of the anti-apartheid movement throughout the world. A member of the clandestine Communist Party, he also embodied the connection established in the minds of both supporters and opponents of apartheid between the anti-apartheid struggle and the ideological and geopolitical divides of the Cold War. Washington regarded the white supremacist government in Pretoria as an ally, while Moscow saw the communist-led ANC as a source of support.[5]

The treatment of the black majority produced worldwide protests, making South Africa a pariah state. In 1961 it was forced to leave the British Commonwealth because of criticism of the apartheid policy by the member states. From the mid-1960s, South Africa was involved in the so-called Border War on the territory of Namibia, preventing that future country from becoming independent of South Africa and its government. The conflict was internationalized, with the Soviet Union and China backing the rebels. In 1975, the South African armed forces began military operations in Zambia and Angola. The latter country was descending into a bloody civil war involving not only the Soviets and the Americans but also the Chinese and the Cubans.[6]

By the 1970s it appeared that the South African leaders shared the nightmare that had afflicted Prime Minister David Ben-Gurion of Israel in the 1950s and 1960s: a united front of neighboring states and non-state movements against what they perceived as European settler colonialism. While Israel was criticized for its policies toward the Palestinian minority and the occupation of territories of neighboring states, the apartheid government in Pretoria was subjected to a UN Security

Council weapons embargo, recommended but not imposed economic sanctions, and suspension of its membership in UN bodies, including the IAEA. Israel and South Africa found themselves in comparable situations and opted for the same solution: to build the bomb. The two countries would keep their capabilities secret, hiding them as much from their Western supporters as from their regional enemies. They would also cooperate with each other on the development of their missile capabilities, with Israel helping South Africa to increase the range of its missiles and develop its ballistic missile program.[7]

The plutonium enrichment plant in Valindaba outside Pretoria produced its first highly enriched uranium in January 1978. Before the end of the following year, South Africa had its first nuclear explosive device, the entire project involving fewer than 1,000 people and costing the government less than 1 percent of its military budget. South Africa's participation in the Atoms for Peace and Plowshares programs, which had helped the country develop expertise and train cadres, was clearly paying off. The weaponization of the South African atom continued, and by 1982 the government had its first nuclear bomb. Five more were added before the end of the decade.[8]

Ironically, it was just as Pretoria achieved its dream of becoming a full-fledged nuclear state that the country's rulers saw the need for a bomb disappear. The South African Border War came to an end in March 1990 with the declaration of Namibian independence. It was preceded by an agreement in 1988 between the governments of Angola, South Africa, and Cuba to end foreign involvement in the Angolan civil war. The deal that saw the withdrawal of Cuban and South African troops from Angola was brokered by the United States with the participation of the Soviet Union, which by then was busy trying to reenter the world community on the new terms defined by Gorbachev.

The disappearance of the Soviet threat to American interests in Africa removed the key reason for US backing of the apartheid government, which was already subject to congressional sanctions. The United States wanted a transition to democracy, and in 1990 Pretoria made the first step in that direction, lifting the ban on the activities of the African National Congress. Nelson Mandela was released after spending twenty-seven years in prison. White rule in South Africa was coming to an end, and no nuclear weapon could preserve it. The survival of the old political and business elites depended on finding a modus vivendi with the black majority and reentering the world community of nations.

The crucial preparations for dismantling the South African nuclear arms program had been made in 1989 with the election of Frederik Willem de Klerk as the country's new president. Early in 1991 the uranium enrichment facility was shut down, and by summer the disassembly of the country's first atomic bomb had begun. Nuclear disarmament was conducted as secretly as armament before it, the main reason being the government's desire not to be seen as a violator of international rules at a time when it wanted to clean up its act and have the sanctions removed. In July 1991 South Africa joined the Nuclear Non-Proliferation Treaty as a non-nuclear state. By that time its nuclear program had been completely dismantled. De Klerk, the last white president of South Africa, publicly admitted the existence of the program and its termination in 1993.[9]

Like the Soviet Union, South Africa dramatically revised its nuclear strategy in response not so much to foreign as to domestic changes. But the South African metamorphosis turned out to be much more dramatic than the Soviet one. If, as a result of the introduction of democratic elections and change of regime, Russia lost its empire but kept a reduced nuclear arsenal, Pretoria kept the country together but gave up its entire nuclear stockpile. In that sense, the South African situation was closer to that of the non-Russian Soviet republics that had inherited part of the Soviet nuclear arsenal. One of the basic reasons why South Africa and the former Soviet republics of Ukraine, Belarus, and Kazakhstan decided to give up the bomb was the desire of their governments to become fully recognized members of the international community, now closely adhering to the nonproliferation regulations enforced by the major nuclear powers.

The White House looked at the accession of the Soviet republics to the nonproliferation regime as a precondition for international recognition of their independence and economic assistance even before the USSR fell apart.

In early September 1991, soon after the failed coup against Mikhail Gorbachev in Moscow, Secretary of State James Baker came up with a set of principles to guide American policy toward the new republics which, like Ukraine, had declared their independence of Moscow. Apart from peaceful political transition, inviolability of borders, and commitment to human rights, the new governments were expected to honor the Soviet Union's international obligations. This meant that the START agreements would now apply to any new republic with nuclear weapons on its territory.

But Washington was not rushing to embrace pro-independence tendencies in the republics. The Americans were concerned, and their concern was heightened by Gorbachev, that the Soviet Union could become another Yugoslavia, which was then descending into the abyss of civil war—only with nuclear weapons. That frightened even the most hardened politicians. The Americans would have preferred that Gorbachev stay in power with the Soviet nuclear arsenal under his control. Seeing the writing on the wall, however, they were eager to cash in their chips with Gorbachev before anarchy could ensue. In September and October, Bush engaged Gorbachev in new arms reduction talks that would eventually lead to his signing of the START II agreement with Boris Yeltsin in January 1993.[10]

The Ukrainian referendum, held on December 1, 1991, effectively put an end to the existence of the USSR. More than 90 percent of the voters endorsed the independence of their country. A week later Yeltsin, together with the newly elected Ukrainian president, Leonid Kravchuk, and the head of the Belarusian parliament, Stanislav Shushkevich, dissolved the nuclear superpower. From their meeting place in a Belarusian hunting lodge, the three leaders called Bush. Yeltsin informed him that the leaders of the three countries—all had nuclear weapons on their territory—had agreed to bring them under joint control. Yeltsin was confident that the leader of the fourth nuclear republic, Nursultan Nazarbayev of Kazakhstan, would support their decision. The three leaders agreed to honor Soviet international obligations: START would live on.[11]

Faced with a united front of the key Soviet republics, Bush did not object. Before the end of December, Gorbachev resigned, and the USSR disappeared from the map of the world. For Washington, the question was what to do with the Soviet nuclear arsenal. Bush's preferred solution was to transfer all nuclear weapons from Ukraine, Belarus, and Kazakhstan to Russia, hoping that Moscow would be a more reliable partner in nonproliferation efforts than the governments of the newly independent states, which had no experience of handling nuclear arsenals.

In May 1992 James Baker gathered the foreign ministers of the newly independent nuclear states of Russia, Ukraine, Belarus, and Kazakhstan in Lisbon to sign what became known as the Lisbon Protocol. The new countries with nuclear weapons on their territory became cosigners of the START II treaty, negotiated less than a year earlier by George Bush and Mikhail Gorbachev. The trick was that by signing the Lisbon Protocol the foreign ministers of Ukraine, Belarus, and Kazakhstan were also pledging to remove strategic nuclear missiles from their territory as soon as possible and join the Non-Proliferation Treaty as non-nuclear states. Their participation in START was thus temporary and would lead not to the limitation of their nuclear arsenals but to their complete liquidation. The formula embedded in the Lisbon Protocol was strongly supported by Russia, which would now become the only post-Soviet nuclear state.

Responding to American and Russian demands, Ukraine, Belarus, and Kazakhstan signed the Lisbon Protocol. James Baker and the Russian foreign minister, Andrei Kozyrev, had every reason to celebrate a joint victory, but there were also clouds on the horizon. Ukraine submitted a note indicating that it considered itself the rightful owner of the nuclear weapons on its territory and, while it was prepared to remove them, it would need security guarantees from the nuclear states. That was a sign of trouble to come. The protocol had to be ratified by the parliaments of the future non-nuclear states, and, while those of Kazakhstan and Belarus raised little challenge, the Ukrainian parliament refused to follow suit. An influential group of parliamentarians that had first manifested its opposition to nuclear disarmament in the spring of 1992 now organized itself as a "working group" on the ratification of the Lisbon Protocol. They refused to support ratification unless Washington and Moscow met the Ukrainian demands.[12]

Ukraine's procrastination presented a major challenge to the nonproliferation efforts of both Washington and Moscow. At that moment, Ukraine possessed the third-largest nuclear arsenal in the world. Russia was in possession of almost 70 percent of Soviet nuclear weapons, while Kyiv inherited 17 percent of the Soviet arsenal. The rest was divided between Kazakhstan and Belarus. With 1,800 strategic warheads, 176 long-range ballistic missiles, and 42 strategic bombers, Ukraine found itself right behind the United States and Russia. Its nuclear arsenal was larger than that of the United Kingdom, France, and China combined.[13]

Few people could imagine that Ukraine, whose name had become almost synonymous with the Chernobyl nuclear disaster, would present any challenges to the nonproliferation process when the country's parliament made its first step toward independence in July 1990 and declared the sovereignty of the republic. Scarred by the experience of Chernobyl, the Ukrainian deputies were eager to declare their commitment to their country's non-nuclear status. "The Ukrainian SSR solemnly declares its intention to become in future a perpetually neutral country, taking no part in military blocs and abiding by the three non-nuclear principles: not to accept, produce, or acquire nuclear weapons," read the text of the declaration.[14]

The Chernobyl disaster was not the only reason why Ukrainians did not want the Soviet nuclear weapons. The declaration postulated Ukraine's right to have its own armed forces, and implied that the country was prepared to leave the Soviet Union and wanted Soviet troops, missiles, and nuclear bombs out of its territory. Continuing presence of the nuclear weapons on Ukrainian territory meant at that time continuing presence of the Soviet troops. The situation did not change much after the Ukrainian parliament declared full independence from the Soviet Union on August 24, 1991. Viacheslav Chornovil, the head of the country's pro-independence movement, Rukh, claimed Ukraine's ownership of all Soviet assets, including the nuclear weapons on its territory. But he proposed to trade them in return for the right to create Ukraine's own armed forces and international recognition of Ukraine as an independent state.[15]

To gain legitimacy as an independent country, Ukraine had to get rid of its nuclear weapons—that was the message to Kyiv from Washington. In October 1991, parliament responded by adopting a resolution on the non-nuclear status of Ukraine, committing the yet unborn country to divest itself of nuclear weapons, observe the conditions of START I, and accede to the Non-Proliferation Treaty as a non-nuclear state. The driving force behind the resolution was Ukraine's concern that unless the nuclear weapons were removed, Russia would invade under the pretext of safeguarding them. But there was a caveat in the document: "Ukraine insists on its right to control the non-use of the nuclear weapons located on its territory." This was a claim to at least partial control of the weapons.[16]

The immediate question that emerged after the dissolution of the Soviet Union was who would control the Soviet nuclear arsenal once the warheads were removed from the non-Russian republics. Moscow and Kyiv clashed over the issue in March 1992, when President Leonid Kravchuk ordered a stop to the redeployment of tactical nuclear weapons from Ukraine to Russia. "We want guarantees that they can't be used anywhere. I don't want to make anybody else stronger," declared Kravchuk. Behind Kravchuk's decision were Ukraine's worsening relations with Russia. In February 1992, the two countries differed over the issue of funding the transportation of nuclear weapons to Russia. There were also increasing tensions over the future of the Black Sea Fleet, which Russia considered part of its strategic forces, while Ukraine counted it as part of its Soviet inheritance. The issue of the fleet was closely associated with the ownership of the naval base in Sevastopol and ultimately involved control over the Crimean Peninsula as a whole.[17]

In early April 1992 Kravchuk met with the Defense Council of Ukraine to discuss nuclear disarmament. Among those present was Stanislav Koniukhov, the chief of the "Southern" designer group in Dnipropetrovsk that had developed most of the Soviet strategic missiles, including the SS-4 "Sandal," the SS-5 "Skean," the SS-7 "Saddler," the SS-8 "Sasin," the SS-9 "Scarp," the SS-15 "Scrooge," the SS-17 "Spanker," and the latest weapon, the SS-24 "Scalpel." Koniukhov argued that Ukraine had the expertise and facilities required to dismantle nuclear weapons on its own. He favored gradual nuclear disarmament synchronized with other countries; otherwise Ukraine would have no deterrent against possible attack. He was prepared to get rid of Ukraine's 130 SS-19 missiles designed and produced in Russia but wanted to keep the SS-24s designed by his office. They were armed with ten MIRV warheads, and their deployment had begun in 1987. Koniukhov also informed Kravchuk that Ukraine had leverage over Russia: only his people were trained to provide maintenance for the Ukraine-built missiles now stationed in Russia.[18]

Within a week after the Defense Council meeting, Kravchuk took control of the Forty-Third Missile Army located in Ukraine, whose missiles and personnel had formed the backbone of the Soviet missile forces in Cuba in 1962. Meanwhile, the Ukrainian parliament accused Russia of failing to ensure the non-use of nuclear weapons on Ukrainian territory and instructed the government to establish "technical control" over them. Ukraine insisted on international control over the destruction of such weapons in Russia. Russia refused. Only after the

Ukrainians were allowed to monitor the destruction of their weapons on Russian territory, the transfer of tactical nuclear weapons to Russia was resumed and the immediate crisis resolved. But a bigger one over strategic nuclear weapons was looming on the horizon.[19]

The signing of the Lisbon Protocol in May 1992 committed Ukraine to non-nuclear status. It was a concession to the Americans, but growing worries about Russia made the Ukrainian parliamentarians postpone ratification of the protocol. In November 1993, after long delays and debates, the Ukrainian parliament passed a resolution accepting the START I conditions and thus taking on part of the responsibilities of the now defunct USSR but refusing to join the Non-Proliferation Treaty as a non-nuclear state. Instead, Ukraine agreed to reduce its nuclear arsenal in accordance with the provisions of the START I Treaty but insisted on leaving all nuclear weapons not covered by the treaty under its control. Given that the control was physical but not operational, Kyiv declared that Ukraine was not a nuclear state.

In Washington, Bill Clinton and his administration took over the White House in January 1993, eager to seek new approaches to the Ukraine nuclear problem. Ukraine's security concerns were legitimate, admitted Clinton's university friend Strobe Talbott, now a special envoy to the post-Soviet states. He was not alone in his view. In the summer of 1993 John Mearsheimer, one of the most respected American foreign policy experts, declared himself in favor of Ukraine's nuclear status. "Ukraine cannot defend itself against a nuclear-armed Russia with conventional weapons, and no state, including the United States, is going to extend to it a meaningful security guarantee," wrote Mearsheimer in *Foreign Affairs*. "Ukrainian nuclear weapons are the only reliable deterrent to Russian aggression. If the U.S. aim is to enhance stability in Europe, the case against a nuclear-armed Ukraine is unpersuasive."[20]

Clinton and Talbott were not prepared to go so far as to recognize or encourage a nuclear-armed Ukraine, but they were prepared to recognize Ukraine's claim to ownership of the weapons on its territory and its entitlement to security assurances and compensation. In January 1994, with the Ukrainian parliament still refusing to ratify the Lisbon Protocol, Clinton convinced President Kravchuk to sign a trilateral agreement including President Boris Yeltsin of Russia. Kravchuk, facing international isolation and a mounting economic crisis at home, agreed to sign a deal that promised Ukraine unspecified security assurances, technical support for dismantling the missiles, and compensation for enriched uranium from Ukrainian nuclear warheads.[21]

This arrangement was not what Kravchuk and Ukrainian officials had hoped for, but they were in a poor negotiating position. Ukraine was in the midst of a multifaceted crisis: a separatist movement in the Crimea supported by Russia and conflict over the Black Sea Fleet were only part of the story. The Ukrainian economy was in a tailspin. "Economic Collapse Leaves Ukraine With Little to Trade but Its Weapons," ran the title of an article in the *New York Times* published on January 13, 1994, the day before the three leaders met in Moscow to sign the agreement. In 1994 Ukraine lost 23 percent of its GDP as compared with the previous year, and inflation was out of control, reaching 891 percent at the end of the year.[22]

In the previous autumn, parliament had lifted the moratorium on the construction of new nuclear power plants that it had enacted in the heat of anti-Chernobyl mobilization during the last years of Soviet rule. Now, to the horror of Europeans, Chernobyl was free to operate. The Ukrainians needed energy and could not afford to buy it from Russia. Both at home, by lifting the ban on nuclear power, and abroad, by agreeing to remove their missiles in exchange for vague security assurances and compensation, the Ukrainians were taking nuclear risks to save their flatlining economy and, ultimately, their statehood.[23]

Badly in need of American political support vis-à-vis Russia and economic assistance, the Ukrainian political elite eventually gave in to Washington's pressure. In February 1994 parliament ratified the Lisbon Protocol. In November of that year it voted for Ukraine to join the Non-Proliferation Treaty as a non-nuclear state. In December, the new

Ukrainian president, Leonid Kuchma, formerly the director of Pivdenmash, the largest missile factory in Europe, and in the past a nuclear hawk in parliament, went to Budapest to sign what became known as the Budapest Memorandum. Signed by the United States, Russia, and Great Britain and later joined by two other nuclear states, China and France, the memorandum provided security assurances to Ukraine, Belarus, and Kazakhstan in exchange for their accession to the Non-Proliferation Treaty.[24]

Ukraine had become a non-nuclear state de jure, but the removal of the warheads and dismantling of the missiles and nuclear arms infrastructure, in particular forty-six silos for intercontinental missiles, would continue for the rest of the decade. The last nuclear warheads and missiles left its territory in 1996. In 2002, the last silo was demolished to prevent future use, and one of the key missile complexes became a museum. Ukrainians did as they were told, but the deal they got did not solve their main problem—Russia. "If tomorrow Russia goes into the Crimea, no one will even raise an eyebrow," stated former president Kravchuk, predicting the situation that occurred almost twenty years later. Russia annexed the Crimea in the spring of 2014, while the signatories of the Budapest Memorandum failed to intervene to stop the Russian aggression and secure the territorial sovereignty of Ukraine ensured by the memorandum.[25]

The denuclearization of Ukraine opened the door to Russia's invasion two decades after the signing of the Budapest Memorandum. It also led to the annexation of part of Ukraine's territory—the first such occurrence in European history since World War II. Furthermore, it dealt a major blow to the processes of denuclearization and nonproliferation by showing that the security assurances offered by nuclear states in return for giving up nuclear weapons could not replace the power of the weapons themselves as a deterrent. Ukraine had succeeded in claiming ownership of nuclear weapons but failed to acquire sufficient political, legal, and military guarantees to replace the weapons as a deterrent to foreign invasion. John Mearsheimer turned out to be right in asserting that no state was prepared to give Ukraine such guarantees.

The Ukrainian case demonstrated the multiplicity of roles that nuclear weapons can play in advancing a state's domestic and foreign agendas. By denouncing the country's nuclear status in 1990 and 1992, the Ukrainian parliament responded to the anti-nuclear sentiment produced in Ukrainian society by the Chernobyl nuclear disaster and manifested the country's desire to leave the Soviet Union and its military-political infrastructure. The endorsement of Ukraine's non-nuclear status helped to secure American consent for the disintegration of the USSR and the international community's recognition of Ukraine and other nuclear republics. But once independent, Ukraine did its best to reclaim the nuclear weapons on its territory as a means of dealing with resurgent Russia and its claims to the Black Sea Fleet and, later, to the Crimean Peninsula.

Ukraine eventually gave up physical control over nuclear weapons under joint diplomatic pressure from the United States and Russia and in exchange for financial compensation and badly needed economic assistance. The desire for international diplomatic recognition, combined with Ukraine's self-image as a non-nuclear country, contributed to that decision. Last but not least, Ukraine's close ties with Russia, which included not only economic and military infrastructure but also history, culture, and identity, made many in the Ukrainian establishment dismiss the very idea of a future conflict with Russia, especially one in which, given the proximity of the two countries, nuclear arms could be used either as a deterrent or as offensive weapons.[26]

The Budapest Memorandum demonstrated an unprecedented level of cooperation between the two countries most interested in its signing, the United States and Russia. Never before or after did the two nuclear superpowers work so closely together in promoting the nonproliferation regime. While their long-term visions of the future did not coincide, their immediate goals with regard to nonproliferation turned out to be the same. The United States forced the Non-Proliferation Treaty on newly independent nations, making sure that continuing tensions between the post-Soviet states would not lead to nuclear war. Russia disarmed three states that belonged to what the Kremlin considered its sphere of influence, making future interference in the internal affairs of those states and even territorial annexations much easier.

Throughout the 1990s, the joint efforts of Washington and Moscow boosted nuclear nonproliferation initiatives around the world. One such initiative was the creation of nuclear-free zones. It turned out to be more successful in Africa than in Europe. In 1996 Ukraine and Belarus, no longer having a nuclear arsenal to lose, attempted to declare a nuclear-free zone in central and eastern Europe. There were no takers. The countries of the region, including Poland, Hungary, and now-divided Czechoslovakia, were aiming for membership in the nuclear-armed NATO. If they did not intend to join the alliance at the time, then their plan, as in the case of Poland, was to develop their own nuclear arsenals. But similar initiatives in other regions were more successful. The African Nuclear Weapons–Free Zone was launched in April 1996, with South Africa joining the treaty on the day it was opened for signing and ratifying it in less than two years.[27]

Chapter 26

THE RETURN OF FEAR

In September 1996, the United Nations General Assembly adopted a historic Comprehensive Nuclear Test Ban Treaty that prohibited nuclear testing for both military and peaceful purposes in all spheres. Underground tests were now banned as well. Two-thirds of the UN member nations voted in favor of the treaty. The dream of generations of anti-nuclear activists since the Castle Bravo test of 1954 finally came within reach. Among the signatories were the United States, Russia, the United Kingdom, France, China, and Israel.[1]

Conspicuously absent from the signatories of the new landmark treaty were India, Pakistan, and North Korea, which were working on their own nuclear arms programs. They needed tests, even underground ones, to enter the atomic weapons club. All three would join it in the next ten years. The decade of nuclear arms control, disarmament, and nonproliferation that began with the Gorbachev-Reagan agreement on intermediate-range nuclear-armed missiles in 1987 was to be followed by a decade of massive proliferation. The Indians and Pakistanis would test their atomic bombs in 1998, and the North Koreans would follow suit in 2006.

The nuclear tests conducted by the three non-signatories of the treaty would turn the years between 1998 and 2006 into the period of greatest nuclear arms expansion since the 1960s, when France, China, and Israel acquired their atomic bombs. The nuclear disarmament initiated by the two superpowers in the late 1980s led to the end of the Cold War, but that did not dispel international tensions and regional rivalries elsewhere. Left to their own devices, India, Pakistan and North Korea took the initiative to ensure their security.

The new atomic states had a significant feature in common. All three were new formations on the world map, carved out of former colonial possessions by outside forces in ways that left them feeling deeply insecure. They came into the world with a host of unsettled ethno-national, religious, ideological, geostrategic, and territorial issues. Their vulnerabilities were caused by various factors and sometimes defined in different terms, but all three countries decided to guarantee their security by obtaining an atomic bomb.

The first country to acquire nuclear weapons in the post–Cold War era was India. The nuclear explosive device intended for peaceful purposes that India tested in 1974 was not a nuclear bomb, and the government decided not to build one at the time. Nor did India test any nuclear explosive device for the rest of the 1970s or the 1980s.

As the Cold War came to an end, it looked for a while as though India would stay away from the nuclear weapons indefinitely. The concerns that many had about Pakistan's possible acquisition of a nuclear bomb faded into the past as the United States took a strong stand against the building of a nuclear weapon by its traditional ally in the region and even imposed short-lived sanctions on Islamabad. But India and Pakistan were involved in a seemingly endless conflict over Kashmir that produced mutual suspicion and anxiety. The balance of terror between them was based not on the possession of a nuclear weapon but on the possibility that either state might acquire one at short notice. Since a political settlement appeared unlikely and tensions remained high, each country decided to gain an advantage by building a nuclear weapon.

Paradoxically, it was the success of the nonproliferation policies pushed by the United States and backed by Russia after the Cold War that prompted India and Pakistan to choose the nuclear option. Those new measures, promoted not only by the two nuclear superpowers but also by the other nuclear "haves," seemed likely to make it permanently impossible for non-nuclear states to obtain atomic weapons. That was the main reason why the two governments balked at signing the Comprehensive Nuclear Test Ban Treaty. Since it banned all nuclear testing, the treaty would prevent any non-nuclear state from building a bomb and ensuring that it would work. Next on the agenda was a treaty on the production of fissile materials for the nuclear weapons, making such activity illegal. Not only Pakistan but also India refused to sign the agreement, which, ironically enough, had been advocated by its founding father, Jawaharlal Nehru.[2]

In May 1998 India surprised the world, including American diplomats and spies, by initiating a series of tests not of explosive devices intended for peaceful use but of actual bombs. The decision was made by the new prime minister, Atal Bihari Vajpayee. He and his nationalist Bharatiya Janata Party came to power in March 1988, promising to make India a nuclear state and to bring the entire Kashmir region under Indian control. The two objectives were connected, especially given that India's opponent in Kashmir was not only non-nuclear Pakistan but also a fully nuclear China, which controlled part of the region.

The set of nuclear tests ordered by Vajpayee, called Pokhran II, took place on the same proving ground as the Pokhran I or "Smiling Buddha" test of 1974. One fusion and two fission bombs were exploded in underground tests on May 11, 1998. Two more fission bombs were detonated on May 13. Vajpayee proudly declared that India now had "a proven capability for a weaponized nuclear programme." Indeed, after the twenty-four-year lull between Pokhran I and II, India had exploded not just a nuclear device but a set of bombs. India's first nuclear test had become known as "Smiling Buddha"; the tests of 1998 were code-named "Shakti-98." In Hinduism, *Shakti* stands for the primordial cosmic energy that moves through the universe. Cosmic or not, the political shockwave created by the tests not only took the world by surprise but struck neighboring Pakistan with overwhelming force.[3]

Because of its role in countering the Soviet occupation of Afghanistan, Pakistan had emerged unscathed by the brief period of American sanctions imposed on it in 1979 and used the lull in the Indian nuclear arms program to catch up with one of the region's hegemons. To that end Pakistani scientists, spies, and diplomats promoted and took full advantage of the worldwide underground traffic in nuclear technology and materials. A "cold test" of a nuclear device containing a subcritical mass of fissile material that produced no yield to speak of was conducted in 1983. More such tests followed in the 1980s and then the early 1990s.

The administration of President George H. W. Bush reimposed sanctions on Pakistan, but they had little impact, because Islamabad did not believe that the Americans would continue their economic assistance after the end of the Cold War, bomb or no bomb. In fact, economic assistance continued, although it was now delivered by the United Nations. There was no consistency in American policy either, as the administration of President Bill Clinton lifted the sanctions imposed by his predecessor. The military government that came to power in 1996 lacked political legitimacy and tried to acquire it by flaunting its nuclear program, which was popular with the public. Pakistan successfully developed its nuclear capability but technically remained a non-nuclear weapons state until May 1998.[4]

On May 28, Islamabad responded to news of the Indian nuclear tests earlier that month with news of its own: Pakistan exploded five nuclear bombs with a combined yield of 40 kilotons of TNT. Two days later another announcement was made in Islamabad: one more atomic bomb went off, with a yield of 12 kilotons. Skeptics said that Pakistan was bluffing about the number of bombs in its possession, but no one questioned that it had exploded at least one atomic bomb. The Indian and Pakistani tests were all conducted underground and did not violate the Partial Nuclear Test Ban Treaty, to which both countries were signatories. But as neither country signed the Comprehensive Nuclear Test Ban Treaty, they were not violating any treaty of which they were part. When it came to the response of the international community, that did not matter much.[5]

The United States imposed sanctions first on India and then on Pakistan, citing the Arms Export Control Act, which allowed the American president to sanction non-nuclear states once they acquired "nuclear explosive devices." But those measures proved temporary, as the geopolitical situation changed once again. Like the sanctions of 1979 on Islamabad, those of 2001 on India and Pakistan were lifted because of another crisis in Afghanistan. This time it was the 9/11 attacks, which led the United States to invade Afghanistan. The sanctions were lifted less than two weeks after the attacks. Geopolitics had trumped nonproliferation efforts and concerns. Two new nuclear states had joined the exclusive club of the five original nuclear powers.[6]

The Indian nuclear test produced a limited but still chain reaction of nuclear proliferation. No other country but Pakistan decided to acquire nuclear weapons because India had done so. But Pakistan caused proliferation in a different way. To build an atomic bomb, the country's political and scientific authorities had to create an illegal system giving them access to scientific knowledge, raw materials, and technology that they had not previously possessed. That network was now available for use by other aspiring nuclear states. The first in line was North Korea.

In the 1990s North Korea, a country of more than twenty million people, was one of only four communist regimes remaining in the world. They all happened to be outside Europe. Aside from Castro's Cuba in the Caribbean, communists still held power in North Korea, Vietnam, and China. The 1990s turned out to be a period of growth for China and Vietnam and of decline for Cuba and North Korea, owing largely to the collapse of the Soviet Union and the end of Soviet subsidies. While the Cubans struggled, North Koreans died en masse—the famine that hit the country in 1994 lasted several years and killed millions of people. Yet the regime never gave up on its nuclear weapons program.[7]

The North Korean nuclear program began in the 1960s as a direct response to the deployment of American nuclear weapons in South Korea. In 1957, the Eisenhower administration decided to place Jupiter medium-range nuclear ballistic missiles in Italy and Turkey; the deployment of nuclear weapons in South Korea was approved about the same time. If in Italy and Turkey the nuclear weapons were under joint American and Italian or Turkish control, with the NATO allies operating the missiles and the United States maintaining control over the warheads, in South Korea the nuclear weapons were placed under the exclusive authority of United States Forces Korea. The weapons in question, atomic cannons and Honest John atomic rockets, were the first-generation nuclear-capable weapons, difficult to operate, that could not reach Soviet territory from anywhere in Europe.

The rationale for placing nuclear weapons in South Korea had little to do with the Soviet Union, China, or North Korea per se and was heavily influenced by a change in American thinking about nuclear deterrence. In 1956 President Eisenhower told the National Security Council that spending $800 million per year on the upkeep of US forces in South Korea was a problem for the national budget. The troops had to be reduced in numbers—a prospect that terrified the South Korean president, Syngman Rhee. To calm his nerves and keep the South Korean ally happy, Eisenhower was prepared to deploy nuclear weapons on the peninsula. They were supposed to demonstrate continuing American support of the South Korean regime, and achieved their purpose. In June 1957 Syngman Rhee expressed his gratitude to Eisenhower for the decision.

There was no way to keep the nuclear deployment secret from the North Koreans. Under the armistice agreement that ended the Korean War, neither side was allowed to introduce new weapons of any kind. Accordingly, in June 1957, the Eisenhower administration suspended the paragraph of the armistice agreement prohibiting the introduction of new weapons. North Korea immediately accused Washington of planning to turn South Korea into an American nuclear base. The Soviet Union and Czechoslovakia made similar protests in the United Nations, but such protests had little impact on Eisenhower. The decision had been already made, and the first nuclear weapons were installed in South Korea in 1958.

In North Korea, the anxious Kim Il-sung, founder of the Kim communist dynasty, began to prepare his country for nuclear attack. He ordered that all new military and major economic facilities in areas close to the demilitarized zone be constructed underground. The militarization of the country, which had been doing better economically than South Korea and had a significantly smaller army, had begun. In 1963 Kim Il-sung turned to the Soviet Union for help in starting the development of his own nuclear weapons. The Soviets, burned by their assistance to the Chinese in the 1950s, refused. Instead, they offered to build a nuclear research reactor. The offer was accepted, but in 1964, when the Chinese tested their own atomic bomb, Kim Il-sung turned to Mao Zedong with the same request. Once again, he was turned down. For the time being he had to rely on his underground structures to deal with possible nuclear attack from the south and limit his nuclear ambitions to the Soviet-built nuclear research reactor.[8]

Kim Il-sung was not the only Korean leader dreaming about nuclear weapons of his own. So was President Park Chung-hee of South Korea, in power from 1962 until his assassination in 1979. In 1970, in the middle of the Vietnam War, Washington informed Seoul about its desire to withdraw 20,000 of the 63,000 American troops remaining on the peninsula and did so in the following year despite South Korean objections. Richard Nixon's opening to China, the discovery of underground infiltration tunnels from North Korea to the south, and the murder of President Park's wife by a pro–North Korean assassin added to the sense of insecurity in Seoul.

In 1974 American diplomats and intelligence officers began to suspect that Park Chung-hee was developing a nuclear arms program of his own. He had allegedly tasked his scientists to build a bomb by 1977. American experts expected a South Korean bomb by 1980. With concerns about nonproliferation heightened by the Indian nuclear explosion of May 1974, Washington did its best to disrupt Seoul's plans to purchase nuclear technology abroad. By that time, South Korea had a signed deal with the Canadians, who agreed to supply one of their CANDU 6 heavy water reactors. To get that deal going, South Korea even joined the Non-Proliferation Treaty as a non-nuclear state and ratified it in April 1975. Another deal was made with the French, who agreed to supply a reprocessing plant.[9]

In June 1975 Henry Kissinger, who had survived Richard Nixon's resignation the previous year and remained secretary of state under Gerald Ford, instructed the US embassy in Seoul to initiate a "more open discussion" with the host government about American concern over the South Korean nuclear program. He was particularly concerned about the plans to acquire a French reprocessing plant. "It will in fact increase their nuclear weapons potential," wrote Kissinger. "Normal reactor grade plutonium could be used in sophisticated bomb designs or, at the cost of some uncertainty in yield, even in less sophisticated weapons," he continued.

On a visit to Paris, Kissinger managed to persuade the French to cancel the deal. The French government went along, reimbursing its companies supplying equipment to South Korea for the lost profits. The South Koreans had no choice but to acquiesce. Also canceled under American pressure were negotiated deals for the purchase of the research reactor from Canada and a mixed-oxide (MOX) fuel fabrication plant from Belgium. The South Koreans did not abandon their nuclear program entirely but had to collaborate with the Americans on non-weapons research programs.[10]

The Americans continued to keep an eye on their ally. In May 1977 they were alerted to an active discussion of the nuclear option in the government-controlled media. Nevertheless, the CIA was unable to detect any steps toward the renewal of the nuclear weapons program and concluded that the media discussion was a reaction to President Carter's decision to prepare a plan for withdrawing American nuclear weapons from the peninsula. It was a means of reassuring the South Koreans that their government would protect them with or without American nuclear weapons. The Americans eventually decided not to withdraw their missiles. South Korean nuclear ambitions sustained a huge blow on October 26, 1979, when President Park, the chief proponent of the program, was assassinated by his own chief of intelligence.[11]

Seoul's public flirtation with nuclear weapons in the 1970s did not go unnoticed in Pyongyang. In 1974, alerted by the first signs that South Korea was working on its own bomb, the North Koreans turned to the Chinese not for the bomb itself but for help in building their own nuclear weapon. Once again they were turned down, but again they refused to give up. By the end of the 1970s, as South Korea abandoned its hopes for nuclear status, the North Koreans launched their own nuclear arms program. With two of its neighbors, China and the Soviet Union, having nuclear weapons, and the third, South Korea, hosting American missiles and by all accounts building a bomb of its own, the pressure on the reclusive North Korean regime to acquire an atomic bomb was growing. It was exacerbated by the struggling North Korean economy and challenges it presented to Pyongyang's defense capabilities.

Since the mid-1950s, the North Korean regime had developed its own brand of communism. It included the cult of the leader, the founder of the North Korean state and party Kim Il-sung, economic self-sufficiency (*Juche*), and an independent foreign policy, even toward the USSR and China. Its outcome was the militarization of the North Korean economy and society in the atmosphere of a besieged fortress. The government of South Korea was perceived as the main threat to the regime, as it had been ever since the end of the Korean War, which the North Korean communists had started and lost.[12]

Until the 1970s, both North and South Korea were doing poorly in economic development. If anything, the North was ahead of the South. But in the course of the 1970s the situation changed dramatically. South Korea forged ahead with the help of American and Japanese investments, while North Korea lagged, its economy strangled by excessive centralization and lack of investment, with the regime's two main sponsors, the USSR and China, offering little help. The poorly performing economy could not provide enough resources to build expensive aircraft, modern tanks, and artillery. So Pyongyang decided to cut defense expenses by building nuclear weapons. By the early 1980s, US intelligence had begun to pick up signs suggesting the existence of a nuclear research program in the country.

In 1979, North Korea began the construction of its first reactor, modeled on the 1950s British dual-purpose Magnox reactors that could generate electricity and produce plutonium. It became operational in 1986 with a power output of only 5 MWe (megawatt electric), but that did not matter. The reactor was capable of producing enriched uranium and weapons-grade plutonium. In 1985, North Korea joined the Nuclear Non-Proliferation Treaty as a non-nuclear state, but it refused to sign a safeguards agreement with the International Atomic Energy Agency. The regime finally agreed to do so in 1991. But in 1993, unwilling to accept IAEA inspections, North Korea left the agency. There was now little doubt that the North Koreans were working on their own atomic bomb.[13]

The end of the Cold War resulted in the withdrawal of American nuclear weapons from South Korea—the original trigger for North Korea's extreme militarization and interest in nuclear weapons. But now the American weapons were only one of the many concerns of North Korea's aging leader, Kim Il-sung. With the collapse of the Soviet Union, Soviet assistance to North Korea came to an end. The fall of the former USSR buried most of the remaining communist world along with its satellite regimes in Eastern Europe. The reunification of Germany in 1989–90 showed what happens when the capitalist and socialist parts of a nation unite. With the Cold War over, the Soviet Union gone, and China continuing its economic liberalization, the North Korean leadership felt that it was losing allies committed to its protection and survival.[14]

The death of Kim Il-sung in July 1994 offered an opportunity to change course, and the new North Korean leader, Kim Il-sung's son and successor Kim Jong-il, seemed ready to do so. In October 1994 he signed an Agreed Framework with the United States. According to the new deal with the Americans, the country's archenemy during the Korean War and afterward, Kim Jong-il promised to shut down his dual-purpose reactor and stop the construction of two new ones, rejoin the NPT framework, and give inspectors access to North Korean nuclear sites. In exchange, he was supposed to get two high-powered light water reactors, a large quantity of oil, and security assurances from the United States. The normalization of relations with Washington portended the lifting of American sanctions, which had crippled the country's economy since the 1950s.

For President Bill Clinton, whose administration negotiated the deal, the Agreed Framework seemed to be one more success story on the road to nonproliferation. The Budapest Memorandum, signed with Ukraine, Belarus, and Kazakhstan in December 1994, was another important landmark on the same path. But unlike the denuclearization of Ukraine, that of North Korea, which was not yet in possession of the bomb, did not take place. In retrospect, there are serious doubts that the North Koreans were prepared to end their nuclear arms program completely, but there is no doubt that Congress, under Republican leadership, did not trust Kim Jong-il and was certainly not willing to support President Clinton's foreign policy initiatives. All these considerations doomed the North Korean denuclearization.

The promised money for oil was coming late if at all, and no funds were allocated for the construction of new reactors in North Korea. Meanwhile the country, suffering from economic mismanagement and lack of outside support, underwent an economic downturn and was hit by a famine that lasted from 1994 to 1997, costing the country of 22 million people somewhere between 500,000 and 3.5 million deaths from starvation. The North Koreans were desperate, as there seemed little prospect that they would obtain any of the promised American funds. In 1998 they threatened to resume their nuclear program, and by the following year, with the famine behind them, they began shopping on the international black market (formed largely through the efforts of Pakistan) for the equipment needed to produce weapons-grade uranium and, eventually, the bomb.

In June 2003 Pyongyang admitted its nuclear ambitions, arguing that the country needed an atomic deterrent as a cheaper alternative to conventional weapons. The money thus saved would be used for economic development. The North Korean leaders decided to withdraw into their orthodox communist shell, and as foolproof protection for themselves they chose nuclear weapons.[15]

On October 9, 2006, the North Korean Central News Agency shocked the world with an unexpected announcement: the government's mouthpiece declared that the country's scientists had "successfully conducted an underground nuclear test under secure conditions." According to the statement, "the nuclear test was conducted with indigenous wisdom and technology 100%" and would "contribute to defending peace and stability on the Korean peninsula and the area around it." While questions would linger for years about the success of the test, and the claim that it had been the product of purely North Korean "wisdom" was questionable at best, the reference to "peace and stability" indicated the true reason for the test and the nuclear program: the regime feared foreign invasion.[16]

Kim Jong-il completed the project launched by his father, Kim Il-sung. North Korea became a nuclear-armed state before South Korea. Admittedly, by that point South Korea was also in violation of IAEA regulations, refusing to disclose its experimental nuclear program completely. It never made the final leap to produce sufficient fissile material to arm a bomb but is now considered a nuclear-capable state. If South

Korean nuclear development had been North Korea's only reason for building a bomb, one would have expected the North's anxieties to disappear. They did not. The leadership continued to feel insecure and pushed ahead with nuclear and missile programs to achieve both military power and economic self-sufficiency. The result was increased isolation of the regime as economic sanctions were imposed on it.[17]

In 2007, with one nuclear test behind him, Kim Jong-il decided to reengage with the world, offering once again to shut down his nuclear weapons program in exchange for economic aid and the lifting of sanctions. This time the deal was offered not just to the United States but to a forum of five countries that had been involved in talks with North Korea since 2003: China, Russia, Japan, South Korea, and the United States. The North Koreans proceeded to shut down their plutonium-producing reactor and began receiving assistance, but the arrangement soon collapsed. The launch of a North Korean satellite in 2009 suggested that Pyongyang was now working on a ballistic missile program. By using the plutonium produced before 2007, it could readily become capable of sending its atomic bombs far beyond the Korean peninsula. The deal was off. North Korea bade farewell to it in May 2009 by conducting another nuclear test, this time with a yield of up to 7 kilotons.[18]

Thus North Korea became a full-fledged nuclear state, subscribing neither to the Non-Proliferation Treaty nor to the Comprehensive Nuclear Test Ban Treaty. For decades, it was on the radar of the UN Security Council and the International Atomic Energy Agency. Its nuclear weapons program became the most watched, discussed, and spectacular failure of efforts to enforce the nonproliferation regime. The main reason for the failure was the reclusive communist regime's dogged determination to ensure its survival no matter the cost. Conceived originally as a defense against the American nuclear presence in South Korea, the North Korean nuclear program allowed the regime to feel more secure not only abroad but also at home. First, through its nuclear ambitions, and then capabilities, it managed to gain the world attention it would otherwise have lacked and trumpet it through government media to strengthen the leadership's legitimacy at home.

In January 2016, the North Koreans announced that they had exploded a hydrogen bomb. While there are serious doubts about the veracity of the claim, it is generally accepted that North Korea has a growing arsenal of atomic bombs and improved missile capabilities that can allow it not only to threaten South Korea and Japan but also to reach Hawaii and other US territories in the Pacific. In 2017, Pyongyang successfully tested its intercontinental ballistic missiles and once again announced that it had tested a hydrogen weapon.[19]

American efforts to rein in the reclusive regime produced little result. The launch of the first North Korean ICBM was accompanied by an exchange of insults between President Trump and the leader of North Korea, Kim Jong-un. As Trump threatened to destroy North Korea and called Kim a "rocket man" on a "suicidal mission," the North Korean leader shot back, calling the American president a "mentally deranged old lunatic." In June 2018, Trump and Kim met for the first ever summit in US-North Korean history. Kim pledged to work toward the denuclearization of the North Korean peninsula. While the summit relieved international tensions, it turned out to be little more than a photo op for the two leaders.

In May 2019, less than a year after the summit, North Korea resumed its ballistic missile tests, limiting them for the time being to short-range ones. President Biden's efforts to resume the talks produced little result as North Korea launched its first solid-fuel ICBM, walked out of the talks with South Korea, and established close relations with Russia.[20]

The collapse of sanctions against India and Pakistan at the beginning of the new millennium and the inability of the US-led international community to stop the development of the North Korean nuclear program were major blows to nonproliferation. The reliance on nuclear weapons to ensure national security, improve international standing, and promote domestic legitimacy was a trend that had survived the Cold War era.

Chapter 27

PREEMPTIVE WAR

President George W. Bush, son of George H. W. Bush, who oversaw the end of the Cold War, began his own presidency in 2001 under very different circumstances. Less than eight months into his term, the Al-Qaeda attack of September 11, 2001, on New York and Washington, DC, shook America to its core and gave the president carte blanche to do whatever it took to protect the United States. In October of that year the United States, backed by the United Kingdom, launched a bombing campaign against the Taliban and Al-Qaeda bases in Afghanistan, helping a loose coalition of local warlords to topple the Taliban government in Kabul. An interim administration led by a Pashtun politician named Hamid Karzai established itself in the Afghan capital with the support of the United Nations before the end of the year.[1]

In January 2003, when George W. Bush took the Senate floor to deliver his first State of the Union address, he proposed a new strategy for the United States in a rapidly changing world. Denying to the present and potential adversaries access to the weapons of mass destruction, including nuclear devices, was an important part of that strategy. Summarized in the phrase "war on terror," the new strategy went far beyond the fight against terrorist organizations. The new enemies included not only organizations and networks such as Al-Qaeda but also states that Bush called the "axis of evil."

At the top of Bush's axis of evil was North Korea. He characterized it as "a regime arming with missiles and weapons of mass destruction, while starving its citizens." Then came Iran, which, Bush noted, "aggressively pursues these weapons and exports terror, while an unelected few repress the Iranian people's hope for freedom." Finally, there was Iraq, whose regime, according to Bush, continued "to flaunt its hostility toward America and to support terror." The president went on: "The Iraqi regime has plotted to develop anthrax and nerve gas and nuclear weapons for over a decade."

Bush promised to work with US allies to deny those states expertise, technology, and materials to achieve their goals. But that was not all. "We will develop and deploy effective missile defenses to protect America and our allies from sudden attack," declared Bush. That was a major departure from American policy of the previous thirty years. To build missile defenses one had first to do away with the Anti-Ballistic Missile Treaty signed by Richard Nixon and Leonid Brezhnev back in 1972—the mother of all nuclear arms control treaties. Bush was threatening to trigger a new arms race and meant what he was saying: a note informing Moscow about America's withdrawal from the treaty had been already sent to the Kremlin.[2]

By going after the "axis of evil," Bush was fighting what was known to the experts as the "horizontal proliferation" of nuclear weapons—the ability of states and non-state actors to acquire such weapons. By pledging to build anti-missile defenses, he was contributing to "vertical" proliferation, the nuclear buildup or improvement of missiles and other technologies related to nuclear weapons by states that already possessed them.

In June 2002, in defiance of an earlier United Nations resolution that called on Washington to abandon plans for an anti-missile system, the United States officially withdrew from the 1972 Anti-Ballistic Missile Treaty, opening the door to the building of an anti-missile system that most of the world opposed. In response to the Bush administration's denunciation of the treaty, Russia withdrew from the START II treaty on strategic nuclear weapons that had been signed by George H. W. Bush with President Boris Yeltsin in January 1993. This amounted to little more than the issuance of an official death certificate for a treaty that had never been implemented.[3]

The simultaneous official death of the 1972 Anti-Ballistic Missile Treaty and the 1993 Start II Treaty sent a potent signal throughout the world: the era of nuclear weapons control and reduction initiated in the early 1970s was now officially over. George W. Bush turned the clock back to the early 1980s, the first years of Ronald Reagan's presidency. Bush picked up where Reagan had left off with his Strategic Defense Initiative, initiating the creation of a more modest National Missile Defense. The main goal was not to allow "rogue" states to develop nuclear arms and delivery systems with which they could threaten the United States.[4]

Gone were the days when the American president had had to deal with one "evil empire." Bush confronted a number of states that his speechwriters defined for him and the rest of the world as an "axis of evil." Iraq became the first member of the "axis" to feel the brunt of the new American policy.

Iraq's nuclear program began in 1956 as part of the Atoms for Peace initiative with the encouragement of the United States but gained serious momentum in 1968, when the Soviets provided Iraq with a research reactor.

That year the Ba'ath Party, whose ideology combined socialism and nationalism, came to power in Iraq as the result of a coup. Saddam Hussein, age thirty-one, assumed the office of vice president in the new government and soon became the regime's most influential figure. In the early 1970s, it was on his initiative that Iraq secretly launched a nuclear weapons program. To develop it, he needed a bigger reactor capable of producing plutonium. The Iraqis approached the French, who agreed to build a 40 MWe light water reactor, which became known as Osirak in French and Tammuz in Iraqi parlance. Other countries were glad to help with dual-purpose equipment, as Iraq had petrodollars to spend. Its hard-currency reserves grew quickly during the oil crisis of the 1970s created by the Arab oil embargo, which in turn had been triggered by the Yom Kippur War.[5]

While Iraq sent its troops to fight on the Syrian side in the Yom Kippur War and was one of the initiators of the oil embargo, considerations involving the United States and Israel were only a partial incentive for launching its nuclear arms program. Saddam Hussein had a much more immediate concern on his mind—Iraq's next-door neighbor, Iran. Iraq is predominantly Arab, while Iran is mostly Persian. Their relations were further complicated by matters of religion. Shia Iran claimed the loyalty of most of Iraq's population, which is Shia, while Saddam and the country's ruling elite were predominantly Sunni.

There were also territorial disputes. Iraq claimed the Iranian province of Khuzestan, which had a large Arab minority and was located on the border between the two countries. Relations became particularly tense in the late 1960s because of a dispute over control of the waterways in Shatt al-Arab, the confluence of the Euphrates and Tigris rivers before they empty into the Persian Gulf. By 1971, when the two governments broke diplomatic relations with each other, the border dispute also involved a number of islands in the Persian Gulf.

Saddam had shielded Iranian opposition leaders on Iraq's territory, including the future ruler of Iran, Ayatollah Khomeini. Claiming the mantle of leadership in the Arab world after the death of Gamal Abdel Nasser in 1970, and faced with a challenge from US-backed Iran under the rule of Shah Mohammad Reza Pahlavi, Saddam believed that the future of his regime as a regional leader depended on acquiring the bomb. Nor did he neglect chemical and biological weapons, which were much cheaper to produce.

The time to strike came in September 1980, when Iraqi troops invaded Iran with the goal of seizing the disputed province of Khuzestan and the eastern bank of Shatt al-Arab. By that time Saddam had taken complete control of Iraq's government, sending his elderly predecessor as president of the country into retirement in 1979. There was also a change of government in Teheran. The Iranian revolution sent the shah into exile and brought to power Ayatollah Khomeini, a former refugee under Saddam's protection in Iraq.

Saddam did not count on Khomeini's gratitude—he had expelled him from Iraq a few years before the Iranian revolution during a temporary rapprochement with the shah. Instead, Saddam relied on weapons of mass destruction including chemical weapons, used numerous times during the war, which lasted until 1988. There are indications that biological weapons were used as well. They did not help Saddam to win the war but alienated world public opinion. The war ended with half a million people dead and no changes to the international borders.[6]

Neither before nor during the war did Saddam manage to advance his nuclear program sufficiently to produce an atomic bomb. In fact, the program suffered a major setback early in the war when the Osirak, a French-supplied Osiris class 40 MWe light water reactor still under construction at the time, was attacked from the air and completely destroyed. Surprisingly, the jets that destroyed the reactor did not come from Iran. They were fourteen American-built F-16 Fighting Falcon supersonic jets, part of the Israeli air force.

Since Iraq was a close ally of Syria and a participant in the Yom Kippur War, the Israelis had kept a close watch on the development of the Iraqi nuclear program. As the Israelis themselves were using a French reactor to produce plutonium and were experienced in concealing nuclear research behind claims of peaceful purposes, they believed that Saddam was attempting to build a bomb and found it imperative to stop his effort. They got in touch with the Iranian authorities, Iraq's sworn enemies in the ongoing war, asking them to bomb the site. Iranian planes tried to do so but failed. Tel Aviv then decided to send its own jets in an operation code-named "Opera." The attack took place on June 7, 1981.

This was one of the preemptive strikes that would become part of Israeli military policy and the most aggressive attempt to stop nuclear proliferation at the time. While President Kennedy had considered a commando operation against Chinese nuclear facilities and allegedly contemplated one against Israel as well, the United States did not take any such measure. The Israelis, with their long-term security on the line, were much more decisive if not reckless. Knowing that the United States would never approve such an initiative over concerns that it could ignite a much larger conflict in the region, Tel Aviv did not inform Washington about the operation until it was over. Once the mission was under way, the Israeli jets flew at the lowest possible altitude to avoid detection by American radar.[7]

In the White House, President Reagan was terrified by what had happened. "I swear I believe Armageddon is near," he noted in his diary. The United States canceled the agreed sale of F-16s to Israel and backed the UN resolution condemning the raid. But the Israelis achieved their goal: the Iraqi reactor was all but gone, and with it the hopes or fears that Iraq might acquire a bomb anytime soon. The Americans did not remain angry for long, and in 1991, with President George H. W. Bush launching a military invasion of Iraq in response to Saddam's takeover of oil-rich Kuwait, they even appreciated the Israeli action. Secretary of Defense Dick Cheney privately praised the Israelis for making the American task in Operation Desert Storm, the invasion of Iraq, much easier.[8]

Whether Saddam would have had the bomb by early 1991 if the Osirak reactor had still been in operation is doubtful, given that during the previous decade the country's resources had been consumed by the war with Iran, but the attack ensured that Iraq's road to the bomb would be much longer and more difficult than it would have been otherwise. While Iraqi scientists and engineers made significant progress on the design of the atomic bomb, the question of where and how to obtain fuel for it was never resolved. Iraqi scientists planned a new reactor capable of producing enriched uranium rather than plutonium, but the project was abandoned for lack of resources and much-needed international expertise, which no one was rushing to offer.

Desperate times call for desperate measures. Finding himself a pariah after his invasion of Kuwait in August 1990, Saddam launched a crash program to build an atomic bomb using enriched uranium from the fuel prepared for the Soviet-built research reactor and the French-built Osirak reactor almost destroyed by the Israeli raid. Whether the plan was technically feasible or not—some estimates suggest the possibility of building a bomb before the end of 1991—Saddam had simply run out of time. American airstrikes on Iraq's nuclear facilities in January and February 1991, during the first weeks of Operation Desert Storm, buried his hopes for the bomb by leveling the Iraqi nuclear complex. Gone were the fuel-processing plants and other facilities. The Americans had finished the job started ten years earlier by the Israelis.[9]

The United States and its allies in the Gulf War of 1991 wanted to make sure that Iraq would not restart its nuclear weapons program or its production of chemical and biological weapons. Iraq became the subject of numerous inspections mandated by the United Nations. The inspectors' task was to discover and destroy all weapons of mass destruction and facilities that were being used or could be used to produce them. The International Atomic Energy Agency became an important participant in that effort. The Iraqi government, still run by Saddam Hussein, who remained in power in Baghdad, did its best to obstruct the inspectors' efforts. Saddam refused to disclose the full scope of his nuclear arms program, which included a center for the development of centrifuges for uranium enrichment never detected by the Americans before the war and thus spared destruction.[10]

By 1998, the work of the inspectors had become extremely difficult because of continuing delays, denials, and occasional harassment on the part of Saddam's government. The greatest progress was reported by IAEA inspectors, who had managed to remove all fissionable material from the country, but chemical and biological weapons inspectors had not yet completed their work. There was no longer a consensus on whether they should continue their investigations, and under what conditions. Russia, China, and France believed that Saddam had complied with UN demands to destroy his weapons of mass destruction and that the sanctions imposed on the regime should be lifted. Washington disagreed.[11]

The inspection regime was crumbling. With no consensus in the Security Council, the United States decided to act alone. In October 1998 Congress passed the Iraq Liberation Act, its main objective regime change. In December of that year the United States and the United Kingdom began a four-day bombing campaign, its official purpose being to "degrade" sites used to produce weapons of mass destruction. Some suspected that the actual goal was to divert attention from the Clinton sex scandal then rocking Washington, which produced an impeachment process. Whatever the purpose, the bombings failed to force the Iraqi government to resume cooperation with the inspecting agencies.[12]

When in the fall of 2002, in response to a UN resolution, Saddam allowed the inspectors back, it was too late. The Americans had already made up their mind. On March 17, 2003, President Bush went on national television to deliver an ultimatum to Saddam Hussein, whom President George H. W. Bush had defeated in the Gulf War of 1990–91 and driven out of occupied Kuwait but not out of Iraq. The younger Bush was determined to complete the job not finished by his father. He was ready for war, making the case that Iraq possessed or could obtain nuclear weapons and other weapons of mass destruction that might eventually be passed on to terrorists.

"Intelligence gathered by this and other governments leaves no doubt that the Iraq regime continues to possess and conceal some of the most lethal weapons ever devised," declared Bush, making a case for preemptive war. "The danger is clear," he continued, "using chemical, biological or, one day, nuclear weapons, obtained with the help of Iraq, the terrorists could fulfill their stated ambitions and kill thousands or hundreds of thousands of innocent people in our country, or any other." Bush demanded that Saddam and his two sons leave the country; otherwise, the United States would begin a bombing campaign. Saddam was given two days.[13]

As expected, Saddam and his family refused to leave Iraq. The bombing began on March 19, followed by a land invasion. Baghdad fell to US troops in mid-April. Special units moved into the conquered country, looking for weapons of mass destruction. That turned out to be no easy task. In October 2004, the CIA Iraq Survey Group reported on the results of the search by more than 1,600 US and UN inspectors at 1,700 sites. It found what many had suspected from the outset: US intelligence and Bush's assumptions were wrong. Saddam had no programs to produce weapons of mass destruction, particularly nuclear weapons, and his capacity to develop such weapons had been diminishing, not increasing, at the time of the attack.[14]

Saddam Hussein, captured by American troops in December 2003, told his FBI interrogators why he had refused full cooperation with the inspectors. He did not want Iraq to look weak in the eyes of its enemies, especially Iran. The inspectors would have told the world that he had no such weapons, leaving Iraq without a deterrent. Saddam refused to come clean on the issue of weapons of mass destruction until it was too late, choosing the policy of strategic ambiguity. It clearly did not serve him well. Saddam Hussein was hanged on December 30, 2006.[15]

President Bush reluctantly admitted the mistake a year and half later. "The main reason we went into Iraq at the time was we thought he had weapons of mass destruction," he told a reporter in August 2006. "It turns out he didn't." In the same breath, Bush added: "but he had the capacity to make weapons of mass destruction." His other justification was the Al-Qaeda terrorist attack of 9/11, in which Saddam had not been involved. "Imagine," continued Bush in the same interview, "a world in which Saddam Hussein was there, stirring up even more trouble in a part of the world that had so much resentment and so much hatred that people came and killed 3,000 of our citizens."[16]

The collapse of the casus belli with the failure to find weapons of mass destruction in Iraq did not mean the end of the war. It would continue until 2011, costing the United States and its allies 4,700 dead and approximately 32,000 wounded. The war would kill more than 100,000 Iraqis, destabilizing the country and region for generations to come. The Iraq War, envisioned as part of the American "war on terror," encouraged terrorism in both the short and the long run. It destabilized the region and strengthened America's longtime nemesis, Iran, which emerged as a powerful force in the new Iraq, exercising enormous influence on the country's Shia majority and its government. Together with Syria and Hezbollah, Iran formed an "axis of resistance" that challenged the United States, Israel, and their allies in the Middle East.[17]

The Iraq War improved Iran's position significantly in more ways than one. Among the benefits of the war for Teheran was the opportunity to continue its nuclear program with little interference from the West. The United States was busy with Iraq. Disagreements between the permanent members of the UN Security Council on the use of force in Iraq made it difficult to chart a common course on Iran. The failure to find weapons of mass destruction in Iraq increased doubts about the Security Council's efforts to keep Iran's nuclear program under control. Meanwhile, that program had a longer history and was better developed than that of Iraq.

If Iraq's nuclear program had begun with a Soviet reactor, Iran's was initiated by a reactor from the United States. In 1957, under the auspices of Eisenhower's Atoms for Peace program, Washington and Teheran signed an agreement on cooperation in the nuclear sphere. Two years later a center for nuclear research was established at Teheran University. A decade later, in 1967, Iran had an American-supplied research reactor in service. By the 1970s Iran had become a party to the Non-Proliferation Treaty and was eager to jump on the nuclear energy bandwagon.

The shah of Iran, Mohammad Reza Pahlavi, was making plans to build almost two dozen reactors before the end of the century. His enthusiasm for nuclear power rivaled that of President Nixon, who called for the construction of 1,000 nuclear reactors in the United States by the year 2000. In 1975, documents were signed for the construction of the first Iranian nuclear power station in the city of Bushehr. The Germans, Swedes, and French were all eager to participate, and the Americans offered reprocessing facilities for the extraction of plutonium. Nonproliferation did not appear to be a significant concern with regard to friendly Iran. By 1978, Bushehr's first reactor was 80 percent complete. But the completion of the remaining 20 percent would take an unprecedented forty-three years.[18]

The reason for the "delay" was the Iranian Revolution. In 1979 the shah, a major proponent of nuclear energy, was ousted from Iran, and so were the Americans. The country's new ruler, Ayatollah Khomeini, did not believe in nuclear power. But by 1989, the year of his death, Teheran's nuclear ambitions had been rekindled. With the Americans still a sworn enemy, Iran signed a deal on nuclear cooperation with the Soviet Union, to be replaced after 1991 by Russia. The Russians agreed to pick up where the Germans and others had left off at Bushehr in 1978, and plans were made to complete the construction of the reactor there in four years. Owing to a host of financial and technical problems exacerbated by nonproliferation concerns, the plant was not connected to the country's electrical grid until September 2011, making the Bushehr construction project the longest in the history of nuclear energy.

This huge delay aside, the Bushehr reactor became the first nuclear power plant ever to have been built in the Middle East. Nuclear-armed Israel had never built one. Now Israel, the United States, and, to different degrees, other nuclear powers were concerned that Iran might become the first Islamic state in the Middle East to acquire nuclear weapons. In the early 2000s inspectors delegated by the IAEA began to report problems regarding Iran's compliance with the Non-Proliferation Treaty. The core issue was that of Iranian facilities for the enrichment of uranium. The United States pushed for stricter controls and advocated sanctions, but after the major failure to uncover weapons of mass destruction in Iraq, Washington's voice had lost its former authority.

The five permanent members of the UN Security Council split over the issue of how severely to treat Iran. Russia and China in particular refused to support the resolutions proposed by the majority of Security Council members and backed by Germany, whose government took part in the EU team negotiating with the Iranians. The deadlock was not resolved until the end of 2006, when everyone was finally on board. The Security Council voted to impose sanctions on Iran, encouraging its government to clean up its act and disclose all aspects of its nuclear program. Although the sanctions were enacted, they did not have the desired effect: Iran refused to suspend the enrichment of uranium.[19]

As negotiations kept going on and off, Iran played the game perfected by North Korea, promising to end parts of its nuclear program and then relaunching them. Experts and politicians alike found it impossible to decide whether this behavior would lead to the building of a bomb or to enrichment required for the production of nuclear energy. Mohamed Mustafa ElBaradei, the director general of the IAEA from 1997 to 2009, declared more than once that he and his inspectors had no proof that the Iranians were indeed developing a nuclear weapons program. The Americans and the Israelis disagreed, with some Israeli officials calling for ElBaradei's resignation. He eventually stepped down in November 2009 after three terms in office.[20]

June 2010 marked the beginning of a new stage in the history of nuclear nonproliferation, and possibly in the future of nuclear terrorism as well. That month computer scientists in the post-Soviet republic of Belarus, who had been asked to examine the Iranian computer systems that regulated the country's uranium-enriching centrifuges, discovered a 500-kilobyte computer worm. The malicious virus, which became known as "Stuxnet," had multiplied in the Windows computer systems before taking control of the gas centrifuges busily enriching uranium at the Natanz nuclear facility in central Iran. By making the centrifuges run at excessive speed and then slowing them down, Stuxnet destroyed as many as 1,000 centrifuges, or 10 percent of Iran's capacity. The actual damage was less important than the fact that cyberware had now effectively entered the sphere of nuclear energy, potentially changing it forever.

Although the origins of the virus are still unknown, there is a consensus among IT experts that while the Stuxnet worm can be used by anyone with a memory stick, its development requires the expertise, resources, and capacity of a state, or perhaps two states. Fingers were pointed at the United States and Israel. If the suspicions were accurate, then this was the second case in which Israel had used a preemptive strike to derail the nuclear program of a potential adversary: the first case was the bombing of the Iraqi Osirak reactor in June 1981. The Osirak attack had been executed with the early cooperation of Iran and against the wishes of the United States, but now the target was Iran, and the United States could be an ally.

The main target of the Stuxnet cyberattack was Iran's capacity to enrich uranium, as it was doing with the help of gas centrifuges at the Fordo and Natanz uranium enrichment plants. The attack raised the stakes in the global debate over the Iranian nuclear program but did little to halt the enrichment: the damaged centrifuges were soon replaced, and the process went on. American experts concluded that even a conventional attack on Iran's nuclear facilities by Israel would be unable to stop the nuclear weapons program if Iran decided to continue it.[21]

Iran's desire to acquire the bomb is often attributed to its regional ambitions—hostile relations with Israel, the desire to stop the American-inspired transformation of the Middle East, and rivalry with Saudi Arabia for leadership in the Muslim world. The insecurity of the Iranian regime, which relied on the radical Revolutionary Guards, and the presidency of the religious hardliner Mahmoud Ahmadinejad, who was in office from 2005 to 2013, strengthened the country's nuclear ambitions and sent more worrisome signals to the outside world.[22]

The North Korean nuclear program was a proof, if one were needed, that no sanctions could stop a reclusive regime from developing nuclear weapons if it was determined to do so. Iran had to be dissuaded by different means. That became the strategy behind the agreement signed by the five nuclear powers on the Security Council, China, France, Russia, UK, and US, plus Germany, and Iran in July 2015. Known as the Joint Comprehensive Plan of Action, the agreement required that Iran terminate the program potentially leading to its acquisition of weapons-grade uranium and plutonium. In exchange, Iran would gain access to $100 billion of its assets frozen after the introduction of nuclear-related sanctions in 2006 and would be allowed to export and sell oil on international markets. Sanctions in that sphere had cost Iran about $160 billion in the course of four years. Teheran committed itself to freezing its nuclear enrichment program for at least ten years.

The technical solution for the Iranian nuclear program was to allow the enrichment of uranium for nuclear energy production with a concentration of uranium-235 at the level of 3 to 4 percent. The deal included keeping Natanz as Iran's only enrichment facility and limiting the number and type of centrifuges that could be installed there. The maximum enrichment level allowed under the agreement was 3.67 percent. Iran also obliged itself to reduce its stockpile of low- and medium-enriched uranium by no less than 98 percent and redesign its planned heavy water reactor capable of producing plutonium. The agreement went into effect on January 16, 2016. By that day Iran had sold most of its low-enriched uranium to Russia, made the core of its heavy water reactor inoperable, and removed thousands of gaseous centrifuges from its nuclear enrichment plants.[23]

The "Iran deal" crowned President Obama's achievements in the nonproliferation sphere but encountered trouble immediately after his departure. The new president, Donald Trump, shared the conviction of the Israeli prime minister, Benjamin Netanyahu, that the agreement was doing little to stop the Iranian weapons program while giving the country a financial windfall from oil sales that it could exploit to enhance its standing in the region, fund terrorism, and eventually build a bomb.[24]

In May 2018 the Trump administration withdrew from the agreement negotiated with great difficulty by its predecessor. American sanctions were reimposed on Iran. The European signatories to the agreement were critical of the American position and introduced a barter system of trade exchanges with Iran. As a result, Iran continued to observe the conditions of the deal despite the American withdrawal.

The Trump administration pushed on. In July 2019 the United States introduced measures that completely halted the export of Iranian oil to international markets. In response, the Iranian government declared that it was beginning to enrich uranium beyond the agreed limit. The increase in enrichment was quite modest at the time, and Iran was indicating its readiness to return to the negotiating table, but there were no takers in Washington. That left Iran free to proceed with its enrichment program.[25]

The American war in Iraq, a classic example of preemptive war in the age of weapons of mass destruction, along with Israeli efforts to stop the Iraqi program with airstrikes and alleged Israeli-American attempts to stop the Iranian program with a cyberattack, raised the nonproliferation efforts of the nuclear powers to a new and extremely dangerous level. While no war has yet resulted from any country's acquisition of nuclear weapons, attempts to prevent Iraq from obtaining weapons of mass destruction have already contributed to an outbreak of a major war.[26]

Will similar wars follow? Chances are that they will. The Hamas attack on Israel in October 2023 and the war that followed destabilized the Middle East. Some members of the Israeli government considered a nuclear strike on the part of Gaza held by Hamas. In November 2024, after Iran launched a massive ballistic missile attack on Israel, Israel responded with an attack on the Iranian missile and nuclear infrastructure, destroying one of Iran's nuclear research facilities. Israel continued to rely on preemptive military action, if not a full-blown war, as the most effective way of enforcing the nonproliferation regime.[27]

As I write these words in the last days of 2024, it is difficult to predict what impact the exchange of the missile attacks between Iran and Israel will have on the Iranian nuclear program. Preemptive war is hardly a solution to the current failure of nonproliferation. If anything, it worsens the problem, driving non-nuclear countries to acquire atomic weapons out of fear. The Iraq War has shown that preemptive war only heightens regional and international tensions. If the existing nonproliferation regime is to be improved, then the nuclear-armed states must revive the solidarity that produced it in the first place.

The renewed warfare in the Middle East and the outbreak of the war between Russia and Ukraine put the solidarity of the nuclear-armed states in question and produced new threats not only to the nonproliferation regime but also to the taboo on the use of the nuclear weapons that have survived since the end of World War II. The world once again found itself on the brink of nuclear confrontation.

EPILOGUE

Early in the morning of February 24, 2022, Russian troops crossed the borders of Ukraine, turning the eight-year-old Russo-Ukrainian war, which began in the spring of 2014 with the Russian annexation of the Crimea, into the largest European conflict since 1945. As troops entered Ukraine from Russian territory in the east, Belarusian territory in the north, and the annexed Crimea in the south, Vladimir Putin released a prerecorded video address blaming the war on Western actions. He threatened "those who may be tempted to interfere in these developments from the outside" with consequences "such as you have never seen in your entire history." He added: "No matter how the events unfold, we are ready. All the necessary decisions in this regard have been taken. I hope that my words will be heard."[1]

The post–Cold War world order came to an end in the flames of the Russo-Ukrainian war, and a new international order began to emerge, with nuclear threats becoming part of its formation. Russia's war on Ukraine brought to the fore four key factors that characterize the new stage of the Nuclear Age and present a major threat to the survival of the world. Some of them are old. Others are new. The first one is the nuclear brinkmanship characterized by the nuclear threats against non-nuclear countries, the second is the renewed nuclear arms race between the key members of the nuclear club, the third is the challenges to the nonproliferation regime, brought by the war unleashed by a nuclear state on the state that gave up the nuclear weapons in its possession to comply with the nonproliferation regime, and the fourth is the coming of warfare on the nuclear energy sites, turning what became known as the Atoms for Peace into atoms for war.

Nuclear threats, as demonstrated above, became part of Russia's war strategy from the very first hours of the all-out aggression against Ukraine. Three days into the war, Putin ordered what he called "Russia's deterrence forces" to go on high alert, a "special regime of combat duty." Nuclear threats continued, primarily communicated by Putin's proxy, former Russian president Dmitry Medvedev, who warned Ukraine of nuclear Armageddon if it dared to attack Russian targets in Crimea or if the West continued to support Ukraine. In October 2022 after the successful Ukrainian counteroffensive in the Kharkiv region of Ukraine, Putin and his entourage considered using tactical nuclear weapons against their non-nuclear adversary. Washington believed that the chances of using tactical nuclear weapons were high enough for President Biden to send the CIA director, Bill Burns, to meet with one of Putin's aides and warn him about possible consequences of such actions. Secretary of State Antony Blinken later commented that Washington had reasons to believe "that China engaged Russia and said, 'don't go there'."[2]

The nuclear weapons were not used. But the nuclear threats have been the most successful Russian psychological operation (PSYOP) that has influenced U.S. policymakers and public attitudes toward Russian aggression against Ukraine. In response to the threats coming from Moscow, Washington and its allies adopted a strategy of "escalation management," providing Ukraine with older types of weapons and in insufficient numbers to dramatically change the situation on the battlefield. As a result, delayed supplies of HIMARS, Leopards, ATACMS, and F-16 fighters, along with restrictions on Ukraine's use of some of these weapons, have led to a longer and bloodier war and an increasing need for more armament.

In November 2024 Russia added a new gruesome "first" to the history of its aggression against Ukraine. For the first time in history an intermediate-range ballistic missile was used in battle, attacking an industrial complex in the city of Dnipro, allegedly the same factory that produced the Soviet missiles delivered to Cuba in 1962. A week later Putin himself threatened the use of similar missiles, called *Oreshnik* (hazel tree), against the centers of decision making and other targets in Ukraine. He claimed that "the power of the strike would be comparable to the use of nuclear weapons." Oreshnik indeed is capable of carrying multiple nuclear warheads and its use from the start has been nothing else but a nuclear threat aimed not only against Ukraine but also against the world. The launch of Oreshnik took place one day after Putin revised Russia's nuclear doctrine, permitting the use of nuclear weapons against a conventional attack on Russia from a non-nuclear state if it is supported by a nuclear one.[3]

Ukraine and the West refused to blink in the face of Putin's nuclear blackmail. Putin's next step in raising the nuclear stakes was the decision to deploy Oreshnik missiles in Belarus. Putin's client there, Aliaksandr Lukashenka, declared that his country was ready to host the Oreshnik missile on its territory—an addition to the Russian tactical nuclear weapons moved there earlier. In June 2024 Russia and Belarus conducted joint nuclear drills and announced the preparation of new nuclear doctrines. They included the use of nuclear weapons by Belarus and a lowered threshold for the use of nuclear weapons by Russia. On Russian television, propagandists called for nuclear strikes against the West.[4]

Russia's progression from nuclear threats to the use of new missile technology against Ukraine brings into focus the dangers presented by the renewed nuclear arms race between the key members of the nuclear club. Russia is not the Soviet Union of the past, and its present-day economy does not allow it to engage in an arms race on the scale that characterized the Cold War. Nevertheless, the Russian nuclear arsenal, estimated at 5,889 nuclear warheads in 2023, is the largest in the world. Until recently, nuclear weapons were more important to Moscow politically than militarily, as they allowed it to claim great-power status, even though Russia's economy is smaller than that of Italy. Since the start of the Russo-Ukrainian War, such weapons have become an instrument of nuclear blackmail of the West in an effort to force it to abandon its support for Ukraine. The vast majority of Russia's nuclear capabilities are of post-Soviet vintage, and its nuclear modernization program is underway despite war in Ukraine.[5] There is no international regime in place to stop Russia from engaging in the nuclear buildup. The Cold War–era treaties have been unraveling one by one ever since George W. Bush withdrew from the Anti-Ballistic Missile Treaty of 1972. The official reason for the withdrawal was the revival of plans to build the anti-missile system envisioned by President Reagan, this time against North Korea and Iran. Russia and China suspected that the anti-ballistic missile systems were really aimed at them.[6]

President Obama tried to reverse the trend when in 2010 he signed with President Dmitrii Medvedev of Russia the New START Treaty. Donald Trump opposed a five-year extension of the treaty; it was extended by Joe Biden in February 2021.[7] One year earlier Moscow and Washington had abandoned the Intermediate-Range Nuclear Forces Treaty, signed by Ronald Reagan and Mikhail Gorbachev in 1987, which imposed limitations on the number of missiles with a range of less than 5,500 kilometers (3,418 miles). Both countries are now free to develop their intermediate-range missile capabilities as they wish.[8]

The United States has been especially concerned about Russia's efforts to develop a satellite that could bring nuclear weapons into space and produce an explosion that would destroy satellites which provide not only information but also communication services most of the world depends on. Meanwhile, the United States is reviewing its own nuclear arsenal, both hardware and software. When it comes to hardware, the Sentinel ICBM program has been launched to upgrade and replace Minuteman III missiles. Work is underway to produce B61–13, a new model of thermonuclear gravity bomb, and a nuclear-armed sea-launched cruise missile. The F-35 Stealth Strike Fighters and B61–12 thermonuclear bombs are deployed to Europe to modernize NATO nuclear capabilities. "[Nuclear] intermission is over, and we are clearly in the next act," Dr. Vipin Narang, Acting Assistant Secretary of Defense for Space Policy, stated in August 2024. He added: "We have an obligation to continually assess our policies and capabilities and consider whether we are doing enough to protect the United States and our allies and partners."[9]

Another active participant in the new arms race is China. In the fall of 2020, the Pentagon released a report stating that China intended to double the number of its nuclear warheads within the next ten years. The increased arsenal, currently estimated at fewer than 410 warheads, would still be a poor match for Russia's stockpiles numbering about 5,890 warheads and an American arsenal estimated at 5,225 warheads. Nevertheless, Beijing's decision to double its stockpiles sent chills through Washington. Some members of the US Senate are concerned about the possibility of a tripling or quadrupling of Chinese warheads, which would give China a "nuclear overmatch" vis-à-vis the United States. This concern has been heightened by indications that the Chinese have been preparing to resume nuclear testing and turn the Lop Nur nuclear firing range into a year-round operating facility.[10]

In March 2021 Britain jumped on the nuclear rearmament bandwagon, releasing its "Global Britain" program, which increases the cap on the number of warheads for the Trident nuclear-armed submarines from 180 to 260. Britain's uncertain international standing after Brexit and the possible future independence of Scotland, which serves as the United Kingdom's base for the Trident submarines, contributed to the decision. Britain was also responding to the changing international environment. France, another historical great power with some 300 warheads, announced plans to modernize its nuclear submarine fleet.

Given uncertainity about the United States's continuing membership in NATO, Paris is also trying to involve other European countries in improving the French nuclear shield, presented as an all-European one.[11]

The third major nuclear threat produced by the Russo-Ukrainian War is the increased chance of unraveling the nonproliferation regime.

Russia's annexation of the Crimea in 2014 and the start of the all-out war on Ukraine in February 2022 have been major blows to already troubled nonproliferation efforts. Ukraine, which gave up nuclear weapons and received security assurances in exchange under the terms of the Budapest Memorandum of 1994, remains under attack by one of the guarantors of that agreement, which has occupied part of its territory. The other guarantors, including the United States, the United Kingdom, France, to say nothing about China, did not do enough to help non-nuclear Ukraine stop the aggression of nuclear-armed Russia, thereby sending an unmistakable message around the world: nuclear weapons are the best available guarantors of sovereignty. Unless Ukraine repels Russian aggression and regains its territorial integrity, nonproliferation is likely to be a dead letter.

In 2022, Vipin Narang defined four different strategies used by states to acquire nuclear weapons. In the first category are "sprinters" like the USSR, France, and China, which tried to acquire such weapons as soon as they could. Next are countries that employed a "sheltered pursuit" strategy. They include Israel, Pakistan, and North Korea, which benefited from their great-power patrons' benevolent neglect of their nuclear proliferation efforts. "Hiders" include such countries as South Africa, whose government developed its nuclear program in secret from the rest of the world. But the strategy to which Narang pays special attention is "hedging." "Hedgers" develop key elements of their nuclear weapons program, but stop short of weaponizing them for strategic, geopolitical, or domestic reasons. India and West Germany resorted to that strategy in the past, and Australia, Japan, Germany,

Sweden, Switzerland, Brazil, and Argentina make use of it today. The danger is that some of them may turn into "sprinters" overnight if they decide to go nuclear.[12]

If the international system loses its current balance and the global and regional rivalries that have fueled the nuclear proliferation of recent decades veer out of control, we may soon see as many as forty additional nuclear-armed countries in the world, about five times as many as there are today. If a country as poor as North Korea can build a bomb and produce missiles to deliver it, then most countries can do likewise if they are sufficiently threatened by their neighbors or regional powers. Should proliferation continue, the threat of intentional or accidental use of nuclear weapons will increase dramatically.[13]

Finally, the Russian war on Ukraine produced a new risk, unknown to the old nuclear age: the warfare on the nuclear sites.

In an important respect, the Russo-Ukrainian war went nuclear on the very first day of the all-out invasion. Before the end of February 24, Russian troops advancing from Moscow's client state of Belarus captured the Chernobyl nuclear power plant. A few days later, an astonished world watched in real time as Russian forces attacked the Zaporizhia nuclear power plant—the largest nuclear power station in Europe. Tanks and missiles were used against the Ukrainian defenders of the plant, and one of the shots produced a fire at the plant's training facility. Fortunately, it was extinguished by Ukrainian firefighters before it could spread to the rest of the nuclear complex.[14]

Relentless bombing of Ukrainian energy infrastructure put Ukraine's three remining nuclear power stations in danger of a nuclear catastrophe. To cause a nuclear incident one does not need to attack the plants directly but rather to continue bombing the electricity distribution infrastructure, cutting the nuclear power plants off from the electricity grid and creating conditions for a disaster akin to the one at the Fukushima nuclear power plant in Japan in 2011. Back then, a tsunami destroyed the power lines necessary to cool the overheated reactor cores. Without electricity, the pumps failed, leading to a meltdown. There have been several "Fukushima moments" involving the cutting of external electricity supply lines at the Russia-occupied Chernobyl nuclear power plant and the Zaporizhia nuclear power plant, the largest in Europe still under Russia occupation.[15]

The Russian seizure of Ukraine's nuclear plants in Chernobyl and Zaporizhia and bombing the country's power grid put in question the future of the nuclear energy viewed as an alternative to the nuclear weapons project and a hope to stop global climate change. By blurring the line between atoms for peace and atoms for war, it suggested the possibility that any of the world's 440 existing nuclear reactors might become a dirty bomb. That line was further blurred by statements in the Russian media that a perceived threat to a nuclear plant on Russian or Russian-occupied territory posed by a non-nuclear state might serve as a basis for nuclear "retaliation" on the part of Russia.[16]

The development of nuclear energy, which American leaders of the 1950s encouraged as a way to stop nuclear arms proliferation, produced opposite results, turning into a back door through which nations can join the nuclear club without the approval of its original members. What happened in Ukraine in 2022, with the Russian takeover of the nuclear power plants, demonstrates that nuclear industry facilities can be used as weapons of war. Yet the world has not fully recognized this new threat and remains committed to Cold War–era thinking, denying any commonality between atoms for war and atoms for peace. Consequently, we are on the brink of a renaissance of the nuclear industry, with new modular reactors under construction, increasing the potential for future conventional wars to go nuclear.

What, if anything, can be done to bring nuclear brinkmanship to an end, put the new arms race of the old nuclear powers under control, and make the nonproliferation regime work? And can the history of the first decades of the Nuclear Age help us in this process?

The history of the Nuclear Age has been very much a chain reaction produced by fear of the Nazi nuclear bomb, with the United States developing such a bomb in the course of World War II. Other countries tried to build their own bombs often in response either to a perceived American threat or to threats from other countries with nuclear weapons. That chain reaction was curbed only by the solidarity of the nuclear powers that first acquired the bomb. Feeling threatened by one another, the superpowers designed a system of nuclear arms control that made the world a safer place. We see little sign of such nuclear-club solidarity today, suggesting that in historical terms we are back to the nuclear arms race of the 1950s and early 1960s, which was not regulated by international agreements.

As the Cold War crises demonstrated, those susceptible to nuclear blackmail face real danger, as occurred during the Cuban missile crisis when Kennedy's public concerns about nuclear war encouraged Khrushchev to position missiles in Cuba. The crisis was resolved first and foremost because Kennedy refused to be provoked or blackmailed by Khruschev's actions and threats. The lessons that one can draw from the Cuban missile crisis are probably more important today than ever before. After all, one of the reasons why Russia decided not to act on its nuclear threats in the fall of 2022 was the potential for Western retaliation in response to the use of nuclear weapons.

History rarely repeats itself to the letter, and the realities that we observe today point toward the further development of nuclear multipolarity, in which nuclear threats are numerous and not limited to the great powers. Still, all changes aside, one factor remains the same. It is the fear of nuclear annihilation. While states acquire and give up nuclear weapons on the basis of national interest as they individually understand it, there is general agreement that it is in no one's interest to die in a global nuclear war. We must enhance the instinct of self-preservation shared by friends and foes alike to save the world once again.

ACKNOWLEDGMENTS

This book began with the question of what makes countries acquire and give up their nuclear weapons, which was very much informed by the Budapest Memorandum of 1994 and the nuclear disarmament of Ukraine, Belarus, and Kazakhstan. When I began researching that question, opinions were divided on whether the deal negotiated between the former nuclear states and the two nuclear superpowers, the United States and Russia, with the former giving up their weapons in exchange for the latter's security assurances, was a model for the rest of the world to achieve nuclear disarmament and nonproliferation. By the time I finished my research and writing, the answer—a resounding "no"—came with the outbreak of the Russo-Ukrainian War, the largest military conflict in Europe since 1945. The world is riper for nuclear proliferation now than at any time since the end of the Cold War, and this book seems timelier than it did when I began to write it.

I would like to use this opportunity to thank those who helped me most with researching and writing this book. I am grateful to Yuri Kostenko, a key participant in the process of Ukrainian nuclear disarmament and the author of a study on that subject, for his insights. I owe a debt of gratitude to Professor David Wolff and his colleagues at Hokkaido University for providing me with a home for researching the history of nuclear disasters in the summer of 2023, and for helping me to gain access to key figures in the Japanese government and diplomatic service concerned with issues of nuclear nonproliferation. At Harvard, my thanks go to my colleague Mariana Budjeryn, the author of the definitive study of Ukrainian nuclear disarmament, and her colleagues at the Kennedy School Belfer Center's Managing the Atom Project for welcoming me to their seminars and events. Graham Allison and his Applied History Project at the same Center gave me an opportunity to meet key scholars and practitioners of international politics who assessed Russian nuclear threats in the course of the Russo-Ukrainian War.

I had the pleasure working once again with Myroslav Yurkevich, who helped me with editing the manuscript and provided helpful advice on its structure. John Glusman at W. W. Norton and Casiana Ionita at Allen Lane did an excellent job of editing the manuscript and guiding it through the publication process. Sarah Chalfant and her colleagues at the Wiley Agency, Emma Smith and Rebecca Nagel, were most supportive and helped with this project in numerous ways. As always, it was my wife, Olena, who provided the inspiration and support that made this book possible. I am grateful to everyone mentioned above and many others who helped me indirectly while I was working on the history of the Nuclear Age, which refused to be just a history.

NOTES

Preface

1. J. Robert Oppenheimer, "Now I am become death...," Atomic Archive.
2. John Pickrell, "Introduction: The Nuclear Age," *New Scientist*, September 4, 2006.
3. Matthew Bunn, "Reducing Nuclear Dangers," *Science* 384, no. 6702 (June 20, 2024): 1277.
4. On the reasons for nations to acquire or give up nuclear weapons, see Scott D. Sagan, "Why Do States Build Nuclear Weapons? Three Models in Search of a Bomb," *International Security* 21, no. 3 (Winter 1996–1997): 54–86; Sagan, "The Causes of Nuclear Weapons Proliferation," *Annual Review of Political Science* 14, no. 1 (June 2011): 225–44; Francis J. Gavin, "History and the Unanswered Questions of the Nuclear Age," in *The Age of Hiroshima*, ed. Michael G. Gordin and G. John Ikenberry (Princeton, NJ, 2020), 295–312; Sico van der Meer, "States' Motivations to Acquire or Forgo Nuclear Weapons: Four Factors of Influence," *Journal of Military and Strategic Studies* 17, no.1 (2016); Kenneth Waltz, "The Spread of Nuclear Weapons: More May Be Better," *Adelphi Papers*, 171 (London, 1981); Jacques E. C. Hymans, *The Psychology of Nuclear Proliferation: Identity, Emotions, and Foreign Policy* (Cambridge, UK, 2006).
5. On the category of fear in international relations, see Arash Heydarian Pashakhanlou, *Realism and Fear in International Relations: Morgenthau, Waltz and Mearsheimer Reconsidered* (Berlin, 2017); Ioannis Evrigenis, *Fear of Enemies and Collective Action* (Cambridge, UK, 2007); Corey Robin, *Fear: The History of a Political Idea* (Oxford, 2004). Cf. Serhii Plokhy, *Nuclear Folly: A History of the Cuban Missile Crisis* (New York, 2021).

Chapter 1: Prophecy

1. Herbert George Wells, *The World Set Free: A Story of Mankind* (New York, 1914); W. Warren Wagar, *H. G. Wells: Traversing Time* (Middletown, CT, 2004), 54–58, 139–41.
2. Wells, *The World Set Free*, 34.

3. Wells, *The World Set Free*, dedication page; Frederick Soddy, *The Interpretation of Radium, Being the Substance of Three Popular Experimental Lectures Delivered at the University of Glasgow*, 3rd ed., revised and enlarged (New York, 1912), 4, 167.
4. Bruce Cameron Reed, *The History and Science of the Manhattan Project* (Berlin and Heidelberg, 2014), 26; Richard Reeves, *A Force of Nature: The Frontier Genius of Ernest Rutherford* (New York and London, 2008), 55–56.
5. Reeves, *A Force of Nature*, 56–57; Alex Keller, *The Infancy of Atomic Physics: Hercules in His Cradle* (Oxford, 1983), 98–114.
6. Wells, *The World Set Free*, 113.
7. Wells, *The World Set Free*, 117.
8. Wells, *The World Set Free*, 152.
9. Peter Herrlich, "The responsibility of the scientist. What can history teach us about how scientists should handle research that has the potential to create harm?" *EMBO Report* 14 (9) (September 2013): 759–64; Igor Novak, *Science: A Many-Splendored Thing* (Singapore, 2011), 306–16; Margit Szöllösi-Janze, *Fritz Haber, 1868–1934: Eine Biographie* (Munich, 1998).
10. Derek Hodson, "Victor Grignard (1871–1935)," *Chemistry in Britain* 23 (February 1987): 141–42; Oliver Lepick, *La Grande Guerre chimique. 1914–1918* (Paris, 1998).
11. "Marie Curie—War Duty (1914–1919), Parts 1–2," American Institute of Physics; Susan Quinn, *Marie Curie: A Life* (New York, 1996), 353–75.
12. "To the Civilized World," *The North American Review* 210, no. 765 (August 1919): 284–87; Jeffrey Johnson, "Science and Technology," in *International Encyclopedia of the First World War, 1914–1918*.
13. "All Nobel Prizes in Physics," The Nobel Prize; "All Nobel Prizes in Chemistry," The Nobel Prize.
14. John Campbell, *Rutherford: Scientist Supreme* (Christchurch, 1999), 361–64.
15. Reeves, *A Force of Nature*, 92–93.
16. Reeves, *A Force of Nature*, 102–3.
17. Wells, *The World Set Free* (1921) (Redditch, Worcestershire, 2016), Preface; Wells, *The World Set Free* (1914), 40.
18. Gennadi Sardanashvily, "Dmitri Ivanenko (1904–1994). In Honor of the 110th Year Anniversary," *Science Newsletter* 1 (2014): 16; Joan Bromberg, "The Impact of the Neutron: Bohr and Heisenberg," *Historical Studies in the Physical Sciences* 3 (1971): 307–41, here 332.
19. William Lanouette, *Genius in the Shadows: A Biography of Leo Szilard, the Man Behind the Bomb* (New York, 2013), 135–42.

Chapter 2: Fright

1. Chuck Rothman, "Albert Einstein and My Grandfather," Chuck Rothman Google Site, https://sites.google.com/site/chuckrothmansf/einstein; Richard Rhodes, *The Making of the Atomic Bomb* (New York, 1988), 304–5; Ferdinand Kuhn Jr., "Chamberlain Bars Any Coup in Danzig," *New York Times*, July 11, 1939, 1.
2. Walter Isaacson, *Einstein: His Life and Universe* (New York, 2008), 394–447; William Lanouette, *Genius in the Shadows: A Biography of Leo Szilard, the Man Behind the Bomb* (New York, 2013), 204–5.
3. Lanouette, *Genius in the Shadows*, 86–89, 204–5.

4. Tom Zoellner, *Uranium: War, Energy, and the Rock That Shaped the World* (New York, 2010), 1–14, 77–78; Lanouette, *Genius in the Shadows*, 205.
5. Rhodes, *The Making of the Atomic Bomb*, 251–53.
6. Rhodes, *The Making of the Atomic Bomb*, 249–62; Ruth Lewin Sime, *Lise Meitner: A Life in Physics* (Berkeley and Los Angeles, 1997), 184–230.
7. "December 1938: Discovery of Nuclear Fission," This Month in Physics History, *APS News* 16, no. 11 (December 2007); Leona Marshall Libby, *The Uranium People* (New York, 1979), 48–50.
8. Edward Teller, *Memoirs: A Twentieth Century Journey in Science and Politics* (New York, 2002), 139–41; Rhodes, *The Making of the Atomic Bomb*, 268–71.
9. Lanouette, *Genius in the Shadows*, 205–6; Rhodes, *The Making of the Atomic Bomb*, 280–81, 305.
10. Lanouette, *Genius in the Shadows*, 194–95, 204–5; Rhodes, *The Making of the Atomic Bomb*, 291–92, 302–4.
11. Lanouette, *Genius in the Shadows*, 205.
12. Lanouette, *Genius in the Shadows*, 205–6; Rhodes, *The Making of the Atomic Bomb*, 305.
13. "Papers of Alexander Sachs," in Franklin D. Roosevelt Presidential Library and Museum; Lanouette, *Genius in the Shadows*, 206–7; Rhodes, *The Making of the Atomic Bomb*, 305–6.
14. Lanouette, *Genius in the Shadows*, 207–9; Rhodes, *The Making of the Atomic Bomb*, 307–8.
15. "Albert Einstein to F. D. Roosevelt," in *The Manhattan Project: The Birth of the Atomic Bomb in the Words of Its Creators, Eyewitnesses and Historians*, ed. Cynthia C. Kelly (New York, 2007), 43–44.
16. John U. Bacon, *The Great Halifax Explosion: A World War I Story of Treachery, Tragedy, and Extraordinary Heroism* (New York, 2017), 181–224.
17. "Albert Einstein to F. D. Roosevelt," 43–44.
18. Rhodes, *The Making of the Atomic Bomb*, 308; Gerard J. DeGroot, *The Bomb: A Life* (Cambridge, MA, 2005), 20.
19. Henry C. Lee and Frank Tirnady, *Blood Evidence: How DNA Is Revolutionizing the Way We Solve Crimes* (New York, 2003), 139; World War II Timeline: August 24, 1939–August 31, 1939.
20. Robert Jungk, *Brighter than a Thousand Suns: The Story of the Men Who Made the Bomb* (New York, 1958), 109–11; Rhodes, *The Making of the Atomic Bomb*, 312–14; DeGroot, *The Bomb*, 21–23.

Chapter 3: Nazis and Their Friends

1. David C. Cassidy, *Uncertainty: The Life and Science of Werner Heisenberg* (New York, 1993).
2. Philip Ball, *Serving the Reich: The Struggle for the Soul of Physics under Hitler* (Chicago, 2014); Heinrich Himmler to Werner Heisenberg, July 21, 1938, in *Physics and National Socialism: An Anthology of Primary Sources*, ed. and trans. Klaus Hentschel and Ann M. Hentschel (Birkhäuser, 1996), 176.
3. Werner Heisenberg, "Research in Germany on the Technical Application of Atomic Energy [August 16, 1947]," in *Physics and National Socialism*, 363.

4. "Vision Earth Rocked by Isotope Blast: Scientists Say Bit of Uranium Could Wreck New York," *New York Times*, April 30, 1939.
5. Richard Rhodes, *The Making of the Atomic Bomb* (New York, 1988), 291–97; Katja Grace, *Leó Szilárd and the Danger of Nuclear Weapons: A Case Study in Risk Mitigation* (Berkeley, CA, 2015), 5–6.
6. Heisenberg, "Research in Germany on the Technical Application of Atomic Energy," 362; Rhodes, *The Making of the Atomic Bomb*, 296; Gerard J. DeGroot, *The Bomb: A Life* (Cambridge, MA, 2005), 15–16, 19.
7. Rhodes, *The Making of the Atomic Bomb*, 311; Heisenberg, "Research in Germany on the Technical Application of Atomic Energy," 364; Klaus Hentschel and Ann M. Hentschel, "Introduction," in *Physics and National Socialism*, lxxxii.
8. Klaus Hentschel and Ann M. Hentschel, "Introduction," lxviii; Heisenberg, "Research in Germany on the Technical Application of Atomic Energy," 364.
9. Richard Dean Burns and Joseph M. Siracusa, *A Global History of the Nuclear Arms Race: Weapons, Strategy and Politics*, vol. 1 (Santa Barbara, CA, 2013), 3; Klaus Hentschel, and Ann M. Hentschel, "Introduction," lxxxii; Heisenberg, "Research in Germany on the Technical Application of Atomic Energy," 365–66.
10. Walter E. Grunden, Mark Walker, and Masakatsu Yamazaki, "Wartime Nuclear Weapons Research in Germany and Japan," *Osiris* 20, no. 1 (2005): 10–30, here 115–16; "Japanese Atomic Project," Atomic Heritage Foundation.
11. William L. Laurence, "Vast Power Source in Atomic Energy Opened by Science: Report on New Source Power," *New York Times*, May 5, 1940, 1, 51.
12. Alan D. Ferguson, "George Vernadsky, 1887–1973," *Russian Review* 32, no. 4 (October 1973): 456–58; K. E. Bailes, *Science and Russian Culture in an Age of Revolutions: V. I. Vernadsky and His Scientific School, 1863–1945* (Bloomington and Indianapolis, 1990).
13. David Holloway, *Stalin and the Bomb: The Soviet Union and Atomic Energy, 1939–1956* (New Haven and London, 1996), 29–34, 59–60; "Zapiska V. I. Vernadskogo i V. G. Khlopina P. I. Stepanovu," June 25, 1940, in *Atomnyi proekt SSSR*, ed. L. D. Riabev et al., vol. 1, pt. 1 (Moscow, 1998), 113–14; "Iz zapiski V. I. Vernadskogo i V. G. Khlopina v prezidium AN SSSR," July 12, 1940, ibid., 123–24.
14. Holloway, *Stalin and the Bomb*, 9–28, 34–36; Monis Sominskii, *Abram Fedorovich Ioffe (1880–1960)* (Moscow and Leningrad, 1964); Zhores Alferov, "Papa Ioffe i ego detskii sad," in Alferov, *Nauka i kul'tura. Izbrannye lektsii* (St. Petersburg, 2009), 127–67.
15. Holloway, *Stalin and the Bomb*, 41–44; Iu. V. Pavlenko, Iu. N. Raniuk, and Iu. A. Khramov, *"Delo" UFTI, 1935–1938* (Kyiv, 1998); V. V. Vlasov and V. D. Khodusov, "K 40-letiiu fiziko-tekhnicheskogo fakul'teta Khar'kovskogo natsional'nogo universiteta imeni V. N. Karazina," *Vestnik Khar'kovskogo natsional'nogo universiteta*, Seriia fizicheskaia 114, no. 4 (2002).
16. "Zaiavka na izobretenie V. A. Maslova i V. S. Shpinelia 'Ob ispol'zovanii urana v kachestve vzryvchatogo i otravliaiushchego veshchestva,'" October 17, 1940, in *Atomnyi proekt SSSR*, vol. 1, pt. 1, 193–95.
17. "Zakliuchenie . . . na zaiavku na izobreteniia sotrudnikov UFTI," after January 24, 1941, in *Atomnyi proekt SSSR*, vol. 1, pt. 1, 220–21; "Zakliuchenie . . . na zaiavki sotrudnikov UFTI," April 17, 1941, ibid., 228–29.
18. Norman Davies, *No Simple Victory: World War II in Europe, 1939–1945* (New York, 2007), 94–98.

19. Heisenberg, "Research in Germany on the Technical Application of Atomic Energy," 371.
20. Rhodes, *The Making of the Atomic Bomb*, 383–86; Jim Baggott, *The First War of Physics: The Secret History of the Atom Bomb, 1939–1949* (New York, 2010), 75–92.

Chapter 4: Transatlantic Alliance

1. Franklin Delano Roosevelt, Message to the Congress on the Arming of Merchant Ships, October 9, 1941, in The American Presidency Project; John N. Petrie, *American Neutrality in the 20th Century: The Impossible Dream* (Washington, DC, 1995), 79–80.
2. Franklin D. Roosevelt, Day by Day, October 9, 1941, A Project of the Pare Lorentz Center at the FDR Presidential Library; John C. Culver and John Hyde, *American Dreamer: A Life of Henry A. Wallace* (New York and London, 2000), 266–68.
3. G. Pascal Zachary, *Endless Frontier: Vannevar Bush, Engineer of the American Century* (New York, 1997), 81–146.
4. "1938, Welles Scares Nation," This Day In History, History.com.
5. Richard Rhodes, *The Making of the Atomic Bomb* (New York, 1988), 338; Zachary, *Endless Frontier*, 194.
6. Rhodes, *The Making of the Atomic Bomb*, 373–74.
7. Margaret Gowing, *Britain and Atomic Energy 1939–1945* (London, 1964), 34–36.
8. Gowing, *Britain and Atomic Energy 1939–1945*, 36–40; Richard Dean Burns and Joseph M. Siracusa, *A Global History of the Nuclear Arms Race: Weapons, Strategy and Politics*, vol. 1 (Santa Barbara, CA, 2013), 6.
9. Gowing, *Britain and Atomic Energy 1939–1945*, 41–42; Gerard J. DeGroot, *The Bomb: A Life* (Cambridge, MA, 2005), 24.
10. Otto R. Frisch and Rudolf Peierls, "From Memorandum on the Properties of a Radioactive Super-Bomb," in *The Manhattan Project: The Birth of the Atomic Bomb in the Words of Its Creators, Eyewitnesses and Historians*, ed. Cynthia C. Kelly (New York, 2007), 45–48.
11. Gowing, *Britain and Atomic Energy 1939–1945*, 45–64.
12. "From Report on the Use of Uranium in a Bomb, Outline of Present Knowledge, MAUD Committee, March 1941," in *The Manhattan Project*, 51–55; Gowing, *Britain and Atomic Energy 1939–1945*, 67–68, 76–78.
13. Gowing, *Britain and Atomic Energy 1939–1945*, 80, 91–106.
14. Rhodes, *The Making of the Atomic Bomb*, 374.
15. Samuel K. Allison, "Arthur Holly Compton 1892–1962," *Biographical Memoirs*, National Academy of Sciences 38 (1965): 81–110; Martin D. Saltzman, "James Bryant Conant: The Making of an Iconoclastic Chemist," *Bulletin for the History of Chemistry* 28, no. 2 (2003): 84–94.
16. Rhodes, *The Making of the Atomic Bomb*, 375–76.
17. Rhodes, *The Making of the Atomic Bomb*, 376–77; Frederick Dainton, "George Bogdan Kistiakowsky. 18 November 1900–7 December 1982," Biographical Memoirs of Fellows of the Royal Society 31 (1985): 376–408; Susan Eva Heuman, *Kistiakovsky: The Struggle for National and Constitutional Rights in the Last Years of Tsarism* (Cambridge, MA, 1998).
18. Rhodes, *The Making of the Atomic Bomb*, 377–79.

19. "German Declaration of War with the United States: December 11, 1941," *The Avalon Project. Lillian Goldman Law Library*; "Franklin Delano Roosevelt's Note to Vannevar Bush, January 19, 1942," in Manhattan Project. An Interactive History.

Chapter 5: Manhattan Project

1. Andrew Brown, *The Neutron and the Bomb: A Biography of Sir James Chadwick* (New York, 1997), 217.
2. James G. Hershberg, *James B. Conant: Harvard to Hiroshima and the Making of the Nuclear Age*, vol. 1 (New York, 1993), 160.
3. Hershberg, *James B. Conant*, 1: 160–61.
4. "Appointment of the Military Policy Committee, September 23, 1942," The George C. Marshall Foundation, Leslie L. Groves Collection.
5. James Kunetka, *The General and the Genius: Groves and Oppenheimer: The Unlikely Partnership That Built the Atom Bomb* (Washington, DC, 2015), 25–33; Richard Rhodes, *The Making of the Atomic Bomb* (New York, 1988), 424–25.
6. Rhodes, *The Making of the Atomic Bomb*, 436–42; "Chicago Pile I," Atomic Heritage Foundation.
7. Charles Johnson and Charles Jackson, *City Behind a Fence: Oak Ridge, Tennessee, 1942–1946* (Knoxville, TN, 1981), 3–64; Denise Kiernan, *The Girls of Atomic City: The Untold Story of the Women Who Helped Win World War II* (New York, 2013), 3–34.
8. Vincent C. Jones, *Manhattan: The Army and the Atomic Bomb* (Washington, DC, 1985), 117–83; Johnson and Jackson, *City Behind a Fence*, 65–166; Kiernan, *The Girls of Atomic City*, 35–108; "Historical Gold Prices 1833 to Present," National Mining Association, https://nma.org/wp-content/uploads/2016/09/historic_gold_prices_1833_pres.pdf.
9. Jones, *Manhattan: The Army and the Atomic Bomb*, 108–11, 184–226; Kate Brown, *Plutopia: Nuclear Families, Atomic Cities, and the Great Soviet and American Plutonium Disasters* (New York, 2013), 15–74.
10. Leslie R. Groves, *Now It Can Be Told* (New York, 1962), 38–40.
11. Kai Bird and Martin J. Sherwin, *American Prometheus: The Triumph and Tragedy of J. Robert Oppenheimer* (New York, 2005), 111–65.
12. Rhodes, *The Making of the Atomic Bomb*, 381, 445–48.
13. Bird and Sherwin, *American Prometheus*, 179–222.
14. *In the Matter of J. Robert Oppenheimer. Transcript of Hearing Before Personnel Security Board*, United States Atomic Energy Commission (Washington, DC, 2005), 12–13.
15. Glenmore S. Trenear-Harvey, *Historical Dictionary of Atomic Espionage* (Lanham, MD, 2011), 81.
16. Bird and Sherwin, *American Prometheus*, 228–29.
17. Jones, *Manhattan: The Army and the Atomic Bomb*, 506–11.
18. "George Kistiakowsky's Interview," January 15, 1982, Voices of the Manhattan Project; Rhodes, *The Making of the Atomic Bomb*, 542–43, 574–78.
19. Jones, *Manhattan: The Army and the Atomic Bomb*, 507; 610–12, 655.

Chapter 6: Unequal Partners

1. *Churchill and Roosevelt: The Complete Correspondence*, ed. Warren F. Kimball (Princeton, NJ, 1984), 1: 249–50.
2. *Churchill and Roosevelt: The Complete Correspondence*, 1: 279; cf. Prime Minister

Churchill to the President's Special Assistant (Hopkins), London, February 27, 1943, in *Foreign Relations of the United States*, Conferences at Washington and Quebec, 1943, no. 4.
3. Andrew Brown, *The Neutron and the Bomb: A Biography of Sir James Chadwick* (New York, 1997), 224–26.
4. Barton J. Bernstein, "The Uneasy Alliance: Roosevelt, Churchill, and the Atomic Bomb, 1940–1945," *Western Political Quarterly* (University of Utah) 29, no. 2 (June 1976): 202–30, here 206, 208–9.
5. Richard G. Hewlett and Oscar E. Anderson, *The New World, 1939–1946* (University Park, PA, 1962), 108.
6. Bernstein, "The Uneasy Alliance," 209–10.
7. Bernstein, "The Uneasy Alliance," 211–19.
8. Bernstein, "The Uneasy Alliance," 220–23.
9. Kai Bird and Martin J. Sherwin, *American Prometheus: The Triumph and Tragedy of J. Robert Oppenheimer* (New York, 2005), 268.
10. Jeffrey Richelson, *Spying on the Bomb: American Nuclear Intelligence from Nazi Germany to Iran and North Korea* (New York, 2007), 24–26; Klaus Hentschel, "Introduction," in *Physics and National Socialism: An Anthology of Primary Sources*, ed. Klaus Hentschel, trans. Ann M. Hentschel (Birkhäuser, 1996), xlviii.
11. Hentschel, "Introduction," lxxxii.
12. Hentschel, "Introduction," lxix.
13. Albert Speer, *Inside the Third Reich: Memoirs* (New York, 1970), 269–72; Richard Rhodes, *The Making of the Atomic Bomb* (New York, 1988), 404–5.
14. Hentschel, "Introduction," lxviii–lxix.
15. Churchill's Copy of Hyde Park Aide-Memoire, September 19, 1944, in *The Manhattan Project: The Birth of the Atomic Bomb in the Words of Its Creators, Eyewitnesses and Historians*, ed. Cynthia C. Kelly, intro. Richard Rhodes (New York, 2007), 104–5.
16. Thomas Powers, *Heisenberg's War: The Secret History of the German Bomb* (New York, 2000), 384–405.
17. Niels Bohr to Winston Churchill, May 22, 1941, in *The Manhattan Project*, 101–3.
18. Niels Bohr's Memorandum to President Roosevelt, July 1944, Atomic Archive.
19. Churchill's Copy of Hyde Park Aide-Memoire, September 19, 1944, in *The Manhattan Project*, 105.
20. Margaret Gowing, *Britain and Atomic Energy 1939–1945* (London, 1964), 358; Rhodes, *The Making of the Atomic Bomb*, 538.
21. Joseph Rotblat, "Leaving the Bomb Project," *Bulletin of the Atomic Scientists* 41, no. 7 (August 1985): 16–19.
22. Bird and Sherwin, *American Prometheus*, 287–89.

Chapter 7: American Bomb

1. Herbert George Wells, *The World Set Free: A Story of Mankind* (New York, 1914), 117; Kevin Ruane, *Churchill and the Bomb in War and Cold War* (London, 2018), 12.
2. "Albert Einstein to F. D. Roosevelt," in *The Manhattan Project. The Birth of the Atomic Bomb in the Words of Its Creators, Eyewitnesses and Historians*, ed. Cynthia C. Kelly (New York, 2007), 43–44.
3. "'From Report on the Use of Uranium in a Bomb, Outline of Present Knowledge,' MAUD Committee, March 1941," in *The Manhattan Project*, 53–54.

4. Don Hornig, interview on "The Story with Dick Gordon," WUNC, in *The Manhattan Project*, 298–99.
5. William Lanouette, *Genius in the Shadows: A Biography of Leo Szilard, the Man Behind the Bomb* (New York, 2013), 275.
6. "Report of the Committee on Political and Social Problems, Manhattan Project 'Metallurgical Laboratory,'" University of Chicago, June 11, 1945, in *The Manhattan Project*, 288–89.
7. "Science Panel's Report to the Interim Committee," June 16, 1945, in *The Manhattan Project*, 290–91; Richard Rhodes, *The Making of the Atomic Bomb* (New York, 1988), 696–97.
8. The Commanding Officer of the Manhattan Project (Groves) to the Secretary of War (Stimson), Washington, July 18, 1945, in *Foreign Relations of the United States. The Conference of Berlin (The Potsdam Conference) 1945* (Washington, DC, 1960), 2: 1366–68.
9. Brigadier General Thomas F. Farrell, in The Commanding Officer of the Manhattan Project (Groves) to the Secretary of War (Stimson), Washington, July 18, 1945, 364–66; Thomas O. Jones, "Eye Witness Accounts of the Trinity test," Memoranda to Leslie R. Groves, July 23, 1945 and July 30, 1945, in *The Manhattan Project*, 310.
10. "Szilard Petition," Atomic Heritage Foundation.
11. Michael Neiberg, *Potsdam: The End of World War II and the Remaking of Europe* (New York, 2015).
12. *Harry S. Truman and the Bomb: A Documentary History*, ed. Robert H. Ferrell (Glendo, WY, 1996), 30. Cf. "Hiroshima: Harry Truman's Diaries and Papers," www.doug-long.com.
13. *Harry S. Truman and the Bomb*, 30. Cf. "Hiroshima: Harry Truman's Diaries and Papers," www.doug-long.com.
14. S. M. Plokhy, *Yalta: The Price of Peace* (New York, 2010), 216–28, 285–92.
15. Wilson D. Miscamble, *From Roosevelt to Truman: Potsdam, Hiroshima, and the Cold War* (New York, 2007), 200–202.
16. "Science Panel's Report to the Interim Committee," June 16, 1945, in *The Manhattan Project*, 290–91.
17. Harry S. Truman, *Year of Decisions* (New York, 1955), 416; Winston Churchill, *Triumph and Tragedy: The Second World War* (New York, 1986), 381. Cf. "Truman-Stalin Conversation, Tuesday, July 24, 1945, 7:30 P.M.," in *Foreign Relations of the United States. The Conference of Berlin (The Potsdam Conference) 1945* (Washington, DC, 1960), 2: 378–79.
18. "Proclamation Defining Terms for Japanese Surrender Issued, at Potsdam," July 26, 1945, Atomic Archive.
19. Walter E. Grunden, Mark Walker, and Masakatsu Yamazaki, "Wartime Nuclear Weapons Research in Germany and Japan," *Osiris* 20, no. 1 (2005): 107–30; "Japanese Atomic Project," Atomic Heritage Foundation.
20. "Press Conference Statement by Prime Minister Suzuki," July 24, 1945, in *Foreign Relations of the United States. The Conference of Berlin (The Potsdam Conference) 1945* (Washington, DC, 1960), 2: 1293.
21. *Harry S. Truman and the Bomb: A Documentary History*, 31.
22. Emily Strasser, "The weight of a butterfly," *Bulletin of the Atomic Scientists*, February 25, 2015.

23. "Notes of Meeting of the Interim Committee, June 1, 1945," 8–9, 14, The Harry S Truman Library and Museum.
24. Rhodes, *The Making of the Atomic Bomb*, 710–11; "Hiroshima and Nagasaki Bombing Timeline," Atomic Heritage Foundation.
25. Meilan Solly, "Nine Eyewitness Accounts of the Bombings of Hiroshima and Nagasaki," *Smithsonian Magazine*, August 5, 2020.
26. "The Atomic Bombings of Hiroshima and Nagasaki. General Description of Damage Caused by the Atomic Explosions" and "The Atomic Bombings of Hiroshima and Nagasaki. Total Casualties," Atomic Archive.
27. William M. Rigdon, *Log of the President's Trip to the Berlin Conference, July 6, 1945 to August 7, 1945* (Washington, DC, 1945), 49–50.
28. "Truman Statement on Hiroshima," Atomic Heritage Foundation.
29. "The Atomic Bombings of Hiroshima and Nagasaki. Total Casualties," Atomic Archive.
30. *Reports of General MacArthur* (Washington, DC, 1966), vol. 2, pt. 2, 715–26; Rhodes, *The Making of the Atomic Bomb*, 742–43.
31. Emperor Hirohito, Radio Broadcast, August 14, 1945, in *Reports of General MacArthur*, vol. 2, pt. 2, 727–28; Rhodes, *The Making of the Atomic Bomb*, 744–46; Tsuyoshi Hasegawa, *Racing the Enemy: Stalin, Truman and the Surrender of Japan* (Cambridge, MA, 2005), 215–51; Tsuyoshi Hasegawa, "The Atomic Bombs and the Soviet Invasion: What Drove Japan's Decision to Surrender?," *Asia Pacific Journal* 5, no. 8 (August 2007); Michael Kort, "Racing the Enemy: A Critical Look," *Historically Speaking* 7, no. 3 (January/February 2006): 22–24.

Chapter 8: Stolen Secret

1. Albert Speer, *Inside the Third Reich: Memoirs* (New York, 1970), 271; Vladislav Zubok and Constantine Pleshakov, *Inside the Kremlin's Cold War: From Stalin to Khrushchev* (Cambridge, MA, 1996), 41.
2. V. N. Pavlov, "Avtobiograficheskie zametki," *Novaia i noveishaia istoriia*, 2000, no. 4: 94–111, here 109; Michael D. Gordin, "How Much Did Stalin Know?" *The History Reader*, June 16, 2011; David Holloway, *Stalin and the Bomb: The Soviet Union and Atomic Energy, 1939–1956* (New Haven and London, 1996), 116–33.
3. Georgii Zhukov, *Vospominaniia i razmyshleniia* (Moscow, 1971), 685; David Holloway, "Entering the Nuclear Arms Race: The Soviet Decision to Build the Atomic Bomb, 1939–45," *Social Studies of Science*, 1981-05-01, vol. 11 (2): 159–97, here 180.
4. "Conversation, August 8, 1945. Present: W. A. Harriman, George F. Kennan, Generalissimus Stalin, M. V. Molotov, Mr. Pavlov"; William Averell Harriman and Elie Abel, *Special Envoy to Churchill and Stalin, 1941–1946* (New York, 1975), 491; Tsuyoshi Hasegawa, *Racing the Enemy: Stalin, Truman, and the Surrender of Japan* (Cambridge, MA, 2006), 192–94.
5. Simon Sebag Montefiore, *Stalin: The Court of the Red Tsar* (New York, 2004), 202; Richard Rhodes, *The Dark Sun: The Making of the Hydrogen Bomb* (New York, 2012), 177.
6. "Conversation, August 8, 1945. Present: W. A. Harriman, George F. Kennan, Generalissimus Stalin, M. V. Molotov, Mr. Pavlov."

7. Serhii Hrabovs'kyi, "Kharkiv-1940: atomna preliudiia," *Den'*, September 2, 2009; Holloway, *Stalin and the Bomb*, 68–71; L. D. Blokhintsev and A. N. Grum-Grzhimailo, "Kratkaia biografiia Vladimira Semenovicha Shpinelia," in *Osnovopolozhniki. Oni sozdali nash institut. NII iadernoi fiziki im. D. B. Skobel'tseva MGU im. M. V. Lomonosova* (Moscow, 2016), 15.
8. "Spravka 1-go upravleniia NKVD SSSR o soderzhanii poluchennoi iz Londona agenturnoi informatsii o 'soveshachanii komiteta po uranu,'" *Atomnyi proekt SSSR. Dokumenty i materialy*, ed. L. D. Riabev et al. (Moscow, 1998), vol. 1 *(1938–1945)*, pt. 1, 239–40.
9. Kate Brown, *Plutopia: Nuclear Families, Atomic Cities and the Great Soviet and American Plutonium Disasters* (New York, 2013), 78–79; Holloway, *Stalin and the Bomb*, 85–86; "Rasporiazhenie GKO 'Ob organizatsii rabot po uranu,' September 28, 1942, in *Atomnyi proekt SSSR. Dokumenty i materialy*, vol. 1, pt. 1, 269–71; Campbell Craig and Sergey S. Radchenko, *The Atomic Bomb and the Origins of the Cold War* (New Haven and London, 2008), 48–49.
10. Holloway, "Entering the Nuclear Arms Race: The Soviet Decision to Build the Atomic Bomb, 1939–45," 165–70.
11. Brown, *Plutopia*, 79–80.
12. "Zapiska I. V. Kurchatova, 'Sostoianie rabot po uranu na 1.VII.1943," *Dokumenty i materialy*, vol. 1, pt. 1, 348–54.
13. Holloway, *Stalin and the Bomb*, 100–115.
14. "Zapiska I. V. Kurchatova L. P. Berii o neudovletvoritel'nom proizvodstvennom sostoianii rabot po probleme," September 29, 1944, in *Atomnyi proekt SSSR. Dokumenty i materialy*, ed. L. D. Riabev et al. (Moscow, 2002), vol. 1, pt. 2, 127.
15. Holloway, "Entering the Nuclear Arms Race: The Soviet Decision to Build the Atomic Bomb, 1939–45," 183.
16. S. G. Kochariants and N. N. Gorin, *Stranitsy istorii iadernogo tsentra Arzamas-16* (Arzamas-16, 1993), 13–14; Brown, *Plutopia*, 83–85.
17. "Postanovlenie GOKO 'O Spetsial'nom komittete GOKO,'" in *Atomnyi proekt SSSR. Dokumenty i materialy*, ed. L. D. Riabev et al. (Moscow, 1999), vol. 2 *(Atomnaia bomba, 1945–1954)*, pt. 1, 11–14; Holloway, *Stalin and the Bomb*, 129.
18. "Memorandum by the Secretary of War (Stimson) to President Truman," September 11, 1945, in *Foreign Relations of the United States: Diplomatic Papers, 1945, General: Political and Economic Matters*, vol. 2, no. 13.
19. Holloway, *Stalin and the Bomb*, 155.
20. John Lewis Gaddis, *The United States and the Origins of the Cold War, 1941–1947* (New York, 2000), 263–67.
21. Holloway, *Stalin and the Bomb*, 156.
22. *Molotov Remembers: Inside Kremlin Politics*, ed. Albert Resis (Chicago, 1993), 58.
23. "28-ia godovshchina Velikoi Oktiabr'skoi sotsialisticheskoi revoliutsii. Doklad V. M. Molotova na torzhestvennom zasedanii Moskovskogo soveta, 6 noiabria, 1945 g.," *Pravda*, November 7, 1945.

Chapter 9: United Nations

1. W. H. Laurence, "Nagasaki Flames Rage for Hours," *New York Times*, August 10, 1945, 1, 5; David W. Moore, "Majority Supports Use of Atomic Bomb on Japan in WWII. Say bombing saved American lives by shortening the war, but divided on whether it saved Japanese lives," *Gallup*, August 5, 2005.

2. "Atomic Culture," Atomic Age Foundation; "When the Atom Bomb Fell," performed by Karl and Harty, recorded December 1945, written by Davis and Taylor, WWII in American Music: Axis and Allies.
3. Norman Cousins, "Modern Man Is Obsolete," in Cousins, *Present Tense: An American Editor's Odyssey* (New York, 1967), 120–30, here 120–21.
4. Allen Pietrobon, "Peacemaker in the Cold War: Norman Cousins and the Making of a Citizen Diplomat in the Atomic Age," PhD Thesis, American University, Washington, DC, 2016, 14–35; Cousins, "Modern Man Is Obsolete," 14–15, 123.
5. Pietrobon, "Peacemaker in the Cold War," 27–28; Milton S. Katz, *Ban the Bomb: A History of SANE, the Committee for a Sane Nuclear Policy, 1957–1985* (New York, 1986), 7.
6. Katz, *Ban the Bomb*, 2–3.
7. The British Prime Minister (Attlee) to President Truman, October 16, 1945, *Foreign Relations of the United States, Diplomatic Papers, 1945*, General: political and economic matters, 2, 58–59.
8. Kevin Ruane, *Churchill and the Bomb in War and Cold War* (London, 2016), 144–47; John Baylis and Kristen Stoddart, *The British Nuclear Experience: The Roles of Beliefs, Culture and Identity* (Oxford, 2014), 19–21.
9. President Harry S. Truman, "Message to Congress on the Atomic Bomb," Washington, DC, October 3, 1945, Atomic Archive; Harry S. Truman, Public Papers of the Presidents of the United States: Harry S. Truman, 1945, 382.
10. Felix Belair Jr., "3 Nations Offer Atomic Bomb to UNO on Reciprocal Basis, Ban on War Is Aim," *New York Times*, November 16, 1945, 1; Harry S. Truman, Clement Attlee, and William Lyon Mackenzie King, "Text of the Agreed Declaration on Atomic Energy," November 15, 1945, in US Senate, Atomic Energy Act of 1946 (Washington, DC, 1946), 33–35, here 33–34.
11. Truman, Attlee, and King, "Text of the Agreed Declaration on Atomic Energy," 34. Memorandum by the Director of the Office of Scientific Research and Development (Bush) to the Secretary of State, November 5, 1945. Subject: Coming conference with Mr. Attlee, in *Foreign Relations of the United States, Diplomatic Papers, 1945*, General: political and economic matters, 2.
12. Wilson D. Miscamble, *From Roosevelt to Truman: Potsdam, Hiroshima, and the Cold War* (New York, 2007), 268–69; John Lewis Gaddis, *The United States and the Origins of the Cold War, 1941–1947* (New York, 2000), 277–78.
13. James B. Conant, *My Several Lives: Memoirs of a Social Inventor* (New York, 1970), 480–81; Record of Conversation, Prepared by the United Kingdom Delegation at the Moscow Conference of Foreign Ministers, Moscow, December 17, 1945, *Foreign Relations of the United States, Diplomatic Papers, 1945*, General: political and economic matters, 2; United States Delegation Minutes, Sixth Formal Session, Conference of Foreign Ministers, Spiridonovka, Moscow, December 22, 1945, 5:10 p.m., *Foreign Relations of the United States, Diplomatic Papers, 1945*, General: political and economic matters, 2.
14. Record of Conversation, Prepared by the United Kingdom Delegation at the Moscow Conference of Foreign Ministers, Moscow, December 17, 1945.
15. United States Delegation Minutes of an Informal Meeting, Conference of Foreign Ministers, Moscow, Spiridonovka, December 24, 1945, 3:15 p.m., in *Foreign Relations of the United States, Diplomatic Papers, 1945*, General: political and economic matters, 2.

16. George F. Kennan, *Memoirs, 1925–1950* (Boston, 1967), 287–88; Miscamble, *From Roosevelt to Truman*, 270–72; Gaddis, *The United States and the Origins of the Cold War*, 279–81.
17. Miscamble, *From Roosevelt to Truman*, 273–76.
18. Establishment of a Commission to deal with the problems raised by the discovery of atomic energy. UN General Assembly (1st sess.: 1946), London and Flushing Meadows, NY).
19. Robert L. Beisner, *Dean Acheson: A Life in the Cold War* (New York, 2006), 33–35; Steven Neuse, *David E. Lilienthal: The Journey of an American Liberal* (Knoxville, TN, 1996), 167–77.
20. *A Report on the International Control of Atomic Energy* (Washington, DC, March 16, 1946), 21.
21. *A Report on the International Control of Atomic Energy*, 51–61.
22. Gaddis, *The United States and the Origins of the Cold War*, 333–34.
23. "The Baruch Plan, presented to the United Nations Atomic Energy Commission, June 14, 1946," in Sokolski, *Best of Intentions*, 115–22.
24. Sokolski, *Best of Intentions*, 19–20.
25. Gromyko's statement in *Bulletin of the Atomic Scientists* 2, nos. 5–6 (September 1946): 11–18; Sokolski, *Best of Intentions*, 14–24.
26. Acting Secretary of State [Acheson] to Certain Diplomatic Representatives, May 4, 1946, in *Foreign Relations of the United States, 1946*, vol. 6: Eastern Europe, the Soviet Union (Washington, DC, 1969), 751–52; "Operation Crossroads: The Effect of the Atomic Bomb on Naval Power," *Bulletin of the Atomic Scientists* 1, no. 5 (February 1946): 1, 11; "Bikini A-Bomb tests, July 1946," National Security Archive.
27. "Molotov at the UN, October 29, 1946," Past Daily; David Holloway, "The Soviet Union and the Baruch Plan," Wilson Center, History and Public Policy Program, June 11, 2020.

Chapter 10: Union Jack

1. The British Prime Minister (Attlee) to President Truman, August 8, 1945, *Foreign Relations of the United States*, Diplomatic Papers, 1945, General: Political and Economic Matters, 2: 37.
2. "The Atomic Bomb. Memorandum by the Prime Minister," in *Cabinets and the Bomb*, ed. Peter Hennessy (Oxford, 2007), 36–38.
3. The British Prime Minister (Attlee) to President Truman, August 8, 1945; Harry S. Truman, Clement Attlee, and William Lyon Mackenzie King, "Text of the Agreed Declaration on Atomic Energy," November 15, 1945, in US Senate, *Atomic Energy Act of 1946* (Washington, DC, 1946), 33–35; Septimus H. Paul, *Nuclear Rivals: Anglo-American Nuclear Relations* (Columbus, OH, 2000), 76–77; Margaret Gowing with the assistance of Lorna Arnold, *Independence and Deterrence: Britain and Atomic Energy, 1945–1952*, vol. 1, *Policy Making* (London, 1974), Appendix 4, 82–84; John Baylis and Kristan Stoddart, *The British Nuclear Experience: The Roles of Beliefs, Culture and Identity* (Oxford, 2014), 21–22.
4. Paul, *Nuclear Rivals*, 78–82; Baylis and Stoddart, *The British Nuclear Experience*, 18–24.

5. Paul, *Nuclear Rivals*, 94–95.
6. Paul, *Nuclear Rivals*, 88–89, 99–101; Gowing and Arnold, *Independence and Deterrence*, 1: 92–112, 124–30.
7. Graham Farmelo, *Churchill's Bomb: A Hidden History of Science, War and Politics* (London, 2013), 321.
8. James Gill, *Britain and the Bomb: Nuclear Diplomacy* (Stanford, CA, 2014), 11–15.
9. Gowing and Arnold, *Independence and Deterrence*, 1: 163–64, 174.
10. Guy Hartcup and T. E. Allibone, *Cockcroft and the Atom* (Bristol, 1984), 133–46; Gowing and Arnold, *Independence and Deterrence*, 1: 161–83.
11. "Cabinet. Atomic Energy. Note of a Meeting of Ministers held at No. 10 Downing Street, S.W.1., on Tuesday, 18th December, 1945, at 10.45 a.m.," in *Cabinets and the Bomb*, 40–42.
12. Gowing and Arnold, *Independence and Deterrence*, 1: 165–68, 172.
13. Gowing and Arnold, *Independence and Deterrence*, 1: 39–40, 175–78.
14. Gowing and Arnold, *Independence and Deterrence*, 1: 178–79; David Kynaston, *Austerity Britain, 1945–1951* (London, 2008), 93–184; Gill, *Britain and the Bomb: Nuclear Diplomacy*, 16–19.
15. "Cabinet. Atomic Energy. Note of a Meeting of Ministers held at No. 10 Downing Street, S.W.1., on Friday, 26th October, 1946, at 2.15 p.m.," in *Cabinets and the Bomb*, 44–47.
16. Baylis and Stoddart, *The British Nuclear Experience*, 32; Gowing and Arnold, *Independence and Deterrence*, 1: 179; "Cabinet. Atomic Energy. Note of a Meeting of Ministers held at No. 10 Downing Street, S.W.1., on Friday, 26th October, 1946, at 2.15 p.m.," in *Cabinets and the Bomb*, 45–46.
17. *Cabinets and the Bomb*, 48.
18. "Memorandum by the Minister of Supply. Note by the Controller of Production of Atomic Energy," December 31, 1946, in *Cabinets and the Bomb*, 50–51; ibid., 48.
19. "Research on Atomic Weapons. Note for the Meeting of Ministers," in *Cabinets and the Bomb*, 53–54.
20. "Cabinet. Atomic Energy. Note of a Meeting of Ministers held at no. 10 Downing Street, S.W.1., on Wednesday, 8th January, 1947 at 3:00 p.m.," in *Cabinets and the Bomb*, 55–56; "Meeting of Ministers. Confidential Annex, Minute 1 (8th January, 1947—3:00 p.m.). Research in Atomic Weapons," ibid., 57–58; Gowing and Arnold, *Independence and Deterrence*, 1: 183.
21. Gill, *Britain and the Bomb: Nuclear Diplomacy*, 14–15.
22. Gowing and Arnold, *Independence and Deterrence*, 1: 180–83; Farmelo, *Churchill's Bomb*, 368.
23. Richard Rhodes, *The Making of the Atomic Bomb* (New York, 1988), 497–500, 547–48, 557–60; Gowing and Arnold, *Independence and Deterrence*, 1: 189–93, 241–65; Arnold, *Windscale 1957* (London, 2007), 9–11; Farmelo, *Churchill's Bomb*, 366–75.
24. Arnold, *Windscale, 1957*, 13–18.
25. Gill, *Britain and the Bomb*, 13–15; John Baylis and Kristan Stoddart, *The British Nuclear Experience: The Roles of Beliefs, Culture and Identity* (Oxford, 2014), 30–39.
26. Paul, *Nuclear Rivals*, 158–59; Nancy Greenspan, *Atomic Spy: The Dark Lives of Klaus Fuchs* (New York, 2020), 175–94.

Chapter 11: Stalin's Bomb

1. "V. N. Merkulov L. P. Berii o napravlenii razvedmaterialov," August 17, 1945, in *Atomnyi proekt SSSR. Dokumenty i materialy*, ed. L. D. Riabev et al. (Moscow, 2002), vol. 1, *(1938–1945)*, pt. 2, 353; "Spravka I. V. Kurchatova i. K. Kikoina o sostoianii i rezul'tatakh nauchno-issledovatel'skikh rabot," August 1945, in *Atomnyi proekt SSSR. Dokumenty i materialy*, ed. L. D. Riabev et al. (Moscow, 2000), vol. 2 *(Atomnaia bomba, 1945–1954)*, pt. 2, 307–12; "Tezisy soobshcheniia I. V. Kurchatova na pervom zasedanii Spetsial'nogo komiteta o sostoianii rabot po uranu v SSSR," August 24, 1945, in *Atomnyi proekt SSSR. Dokumenty i materialy*, ed. L. D. Riabev et al. (Moscow, 1999), vol. 2 *(Atomnaia bomba, 1945–1954)*, pt. 1, 612–13.
2. Pavel Rubinin, "Svobodnyi chelovek v nesvobodnoi strane," *Vestnik Rossiiskoi akademii nauk* 64, no. 6 (1994): 497–510.
3. "Pis'mo P. L. Kapitsy I. V. Stalinu ob organizatsii rabot po probleme atomnoi bomby i svoem osvobozhdenii ot raboty v Spetsial'nom komitete i Tekhnicheskom sovete," November 25, 1945, in *Atomnyi proekt SSSR. Dokumenty i materialy*, ed. L. D. Riabev et al. (Moscow, 1999), vol. 2 *(Atomnaia bomba, 1945–1954)*, pt. 1, 613–19; Yu. N Smirnpv, "Stalin i atomanai bomba," *Vestnik instituta istorii estestvoznaniia i tekhniki*, no. 2 (1994): 125–30, here 127.
4. Loren R. Graham, "Peter Kapitsa, a Man of Many Parts," in Graham, *Moscow Stories* (Bloomington, IN, 2006), 128–38, here 133, 138; P. E. Rubinin, "Kapitsa, Beriia i bomba," in *Nauka i obshchestvo. Istoriia Sovetskogo atomnogo proekta (40-e-50-e gody)* (Moscow, 1999), 2: 260–79.
5. "Zapis' besedy I. V. Kurchatova s I. V. Stalinym," in Iu. N. Smirnov, "Stalin i atomnaia bomba," in *Kurchatovskii institut. Istoriia atomnogo proekta* (Moscow, 1998), vyp. 13, 146–56, here 153–54.
6. John Lewis Gaddis, *George F. Kennan: An American Life* (New York, 2011), 201–24.
7. "Zapis' besedy I. V. Kurchatova s I. V. Stalinym," 153.
8. "Zapis' besedy I. V. Kurchatova s I. V. Stalinym," 153.
9. Mariia Zezina, "Material'noe stimulirovanie nauchnogo truda v SSSR, 1945–1985," *Vestnik Rossiiskoi Akademii nauk* 67, no. 1 (1997): 20–27; Sergei I. Zhuk, *Soviet Americana: The Cultural History of Russian and Ukrainian Americanists* (London and New York, 2018), 38–39.
10. I. A. Andriushin, A. K. Chernyshev, and Iu. A. Iudin, *Ukroshchenie iadra. Stranitsy istorii iadernogo oruzhiia i iadernoi infrastruktury SSSR* (Sarov, Saransk, 2003), 35–66.
11. "Spravka 1-go upravleniia NKVD SSSR o soderzhanii poluchennoi iz Londona agenturnoi informatsii o 'soveshchanii komiteta po uranu,'" *Atomnyi proekt SSSR. Dokumenty i materialy*, ed. L. D. Riabev et al. (Moscow, 1998), vol. 1 *(1938–1945)*, pt. 1, 239–40; "John Cairncross," Atomic Heritage Foundation; "Dzhon Kernkross," Sluzhba vneshnei razvedki Rossii; Christopher Andrew and Vasili Mitrokhin, *The Sword and the Shield: The Mitrokhin Archive and the Secret History of the KGB* (New York, 1999), 57–67, 137–44, 154–60.
12. Robert Chadwell Williams, *Klaus Fuchs: Atom Spy* (Cambridge, MA, 1987); Nancy Thorndike Greenspan, *Atomic Spy: The Dark Lives of Klaus Fuchs* (New York, 2020).
13. Allen M. Hornblum, *The Invisible Harry Gold: The Man Who Gave the Soviets the Atom Bomb* (New Haven and London, 2010).

14. Ronald Radosh and Joyce Milton, *The Rosenberg File: A Search for the Truth* (New York, 1983); Sam Roberts, *The Brother: The Untold Story of the Rosenberg Case* (New York, 2014).
15. John Earl Haynes, Harvey Klehr, and Alexander Vassiliev, *Spies: The Rise and Fall of the KGB in America* (New Haven and London, 2009), 110–17; Alan S. Cowell, "Theodore Hall, Prodigy and Atomic Spy, Dies at 74," *New York Times,* November 10, 1999.
16. Sharon Weinberger, "Why did the atomic spy do it?" *Nature* (London), 2020–08, vol. 584 (7819): 34–35.
17. Kate Brown, *Plutopia: Nuclear Families, Atomic Cities and the Great Soviet and American Plutonium Disasters* (New York, 2013), 97–104.
18. David Holloway, *Stalin and the Bomb: The Soviet Union and Atomic Energy, 1939–1956* (New Haven and London, 1996), 180–84; "Dokladnaia zapiska L. P. Beriia, I. V. Kurchatova, B. L. Vannikova, M. G. Pervukhina na imia I. V. Stalina o puske 25 dekabria 1946 goda opytnogo uran-grafitovogo reaktora," in *Atomnyi proekt SSSR. Dokumenty i materialy,* ed. L. D. Riabev et al. (Moscow, 1999), vol. 2 *(Atomnaia bomba, 1945–1954),* pt. 1, 631–32.
19. Brown, *Plutopia,* 87–123; Holloway, *Stalin and the Bomb,* 184–89; "Dokladnaia zapiska I. V. Kurchatova, B. G. Muzrukova, E. P. Slavskogo na imia L P. Beriia ob osushchestvlenii tsepnoi reaktsii v pervom promyshlennom reaktore kombinata no. 817," June 11, 1948, ibid., 635–36; Anatoli Diakov, "The History of Plutonium Production in Russia," *Science & Global Security* 19 (2011): 28–45, 33.
20. Holloway, *Stalin and the Bomb,* 189–92.
21. Holloway, *Stalin and the Bomb,* 196–201, 213–19; Andriushin et al., *Ukroshchenie iadra. Stranitsy istorii iadernogo oruzhiia i iadernoi infrastruktury SSSR,* 69–78; "Soviet Closed Cities," Atomic Heritage Foundation; "Arzamas-16. Kak Sarov stal tsentrom Sovetskogo atomnogo proekta," *Argumenty i Fakty,* April 1, 2016.
22. "Ispytanie pervoi atomnoi bomby v SSSR," *RIA Novosti,* August 29, 2015; Andriushin et al., *Ukroshchenie iadra. Stranitsy istorii iadernogo oruzhiia i iadernoi infrastruktury SSSR,* 79–81.
23. "Doklad L. P. Beriia i I. V. Kurchatova I. V. Stalinu o predvaritel'nykh dannykh, poluchennykh pri ispytanii atomnoi bomby," August 30, 1949, in *Atomnyi proekt SSSR. Dokumenty i materialy,* ed. L. D. Riabev et al. (Moscow, 1999), vol. 2 *(Atomnaia bomba, 1945–1954),* pt. 1, 639–43.

Chapter 12: British Hurricane

1. William L. Laurence, "Soviet Achievement Ahead of Predictions by 3 Years: Soviet Achieved the Bomb Quickly," *New York Times,* September 24, 1949, 1; Donald P. Steury, "How the CIA Missed Stalin's Bomb," *Studies in Intelligence* 49 (2005); "Detection of the First Soviet Nuclear Test, September 1949," National Security Archive.
2. "Truman Statement on Atom," *New York Times,* September 24, 1949, 1; "Russian Development of the Atomic Bomb, 1949," *Los Angeles Times,* September 24, 1949.
3. John D. Morris, "Atom News Spurs European Arms Aid: Administration to Press Drive for Full $1,314,010,000—Foes Hold Land Force Now Futile," *New York Times,* September 25, 1949, 1.

4. "Soobshchenie TASS," *Pravda*, September 23, 1949; "Soviet Union Has 'Atomic Weapon,' Moscow Says as to U.S. Statement," *New York Times*, September 25, 1949, 1.
5. "Announcement by Three Powers," *New York Times*, September 24, 1949, 4.
6. Michael S. Goodman, *Spying on the Nuclear Bear: Anglo-American Intelligence and the Soviet Bomb* (Stanford, CA, 2007), 38–39.
7. Septimus H. Paul, *Nuclear Rivals: Anglo-American Atomic Relations, 1941–1952* (Columbus, OH, 2000), 158–59; *Cabinets and the Bomb*, ed. Peter Hennessy (Oxford, 2007), 64, 76.
8. Paul, *Nuclear Rivals*, 158–59.
9. Paul, *Nuclear Rivals*, 161–62.
10. Paul, *Nuclear Rivals*, 162–64.
11. Nancy Thorndike Greenspan, *Atomic Spy: The Dark Lives of Klaus Fuchs* (New York, 2020), 193–270; Christopher Andrew, *Defend the Realm: The Authorized History of MI5* (New York, 2010), 387–88; Robert Chadwell Williams, "Who Is Trying to Keep What Secret from Whom and Why? MI5-FBI Relations and the Klaus Fuchs Case," *Journal of Cold War Studies* 7, no. 3 (1987): 124–46.
12. Greenspan, *Atomic Spy*, 306–11; Christopher Andrew and Vasili Mitrokhin, *The Sword and the Shield: The Mitrokhin Archive and the Secret History of the KGB* (New York, 1999), 127–28, 131.
13. Frank Close, *Half-Life: The Divided Life of Bruno Pontecorvo, Physicist or Spy* (New York, 2015), 147–99.
14. Andrew and Mitrokhin, *The Sword and the Shield*, 157–61; Andrew Lownie, *Stalin's Englishman: The Lives of Guy Burgess* (London, 2016), 217–56.
15. Paul, *Nuclear Rivals*, 174–87.
16. "1951: Churchill back in power at last," Past Elections, *BBC News*; David Edgerton, *Warfare State: Britain, 1920–1970* (New York, 2006), 105; Margaret Gowing and Lorna Arnold, *Independence and Deterrence, 2: Policy Execution (Britain and Atomic Energy, 1945–1952)* (London, 1974), 56–57; Kevin Ruane, *Churchill and the Bomb in War and Cold War* (London and New York, 2016), 195.
17. Ruane, *Churchill and the Bomb*, 195.
18. Ruane, *Churchill and the Bomb*, 197; Graham Farmelo, *Churchill's Bomb: How the United States Overtook Britain in the First Nuclear Arms Race* (New York, 2013), 381–83.
19. Ruane, *Churchill and the Bomb*, 199–205.
20. J. L. Symonds, *A History of British Atomic Tests in Australia* (Canberra, 1982), 88–108; Lorna Arnold and Mark Smith, *Britain, Australia and the Bomb: The Nuclear Tests and Their Aftermath* (Basingstoke, 2006), 39–44.
21. Churchill's statement to Parliament, in "Atom Bomb Test, Australia, Volume 505": debated on Thursday, October 23, 1952.

Chapter 13: Managing Fear

1. Robert North Roberts, Scott John Hammond, and Valerie A. Sulfaro, *Presidential Campaigns, Slogans, Issues, and Platforms: The Complete Encyclopedia*, vol. 1 (Santa Barbara, CA, 2012), 255; William I. Hitchcock, *The Age of Eisenhower: America and the World in the 1950s* (New York, 2019), 66–86.
2. David Alan Rosenberg, "The Origins of Overkill: Nuclear Weapons and Ameri-

can Strategy, 1945–1960," in Steven E. Miller, *Strategy and Nuclear Deterrence* (Princeton, NJ, 1984), 113–82, here 137–38.
3. Rosenberg, "The Origins of Overkill," 136–37.
4. Memorandum of Discussion at a Special Meeting of the National Security Council on Tuesday, March 31, 1953, in *Foreign Relations of the United States, 1952–1954, Korea*, vol. 15, pt. 1, no. 427.
5. Rosenberg, "The Origins of Overkill," 136; Michio Kaku and Daniel Axelrod, *To Win a Nuclear War: The Pentagon's Secret War Plans* (Boston, 1987), 95–96.
6. Samuel F. Wells, Jr., "The Origins of Massive Retaliation," *Political Science Quarterly* 96, no. 1 (Spring 1981): 31–52, here 45–46; Report to the National Security Council by the Executive Secretary (Lay), NSC 162/2, Washington, October 30, 1953, in *Foreign Relations of the United States, 1952–1954*, National Security Affairs, vol. 2, pt. 1, no. 101.
7. Guy Oaks, *The Imaginary War: Civil Defense and American Cold War Culture* (New York and Oxford, 1994), 33–39.
8. Oaks, *The Imaginary War*, 64.
9. Oaks, *The Imaginary War*, 47–51, 63–71.
10. *Duck and Cover* (1951), Bert the Turtle, www.youtube.com/watch?v=IKqXu-5jw6o; Spencer R. Weart, *The Rise of Nuclear Fear* (Cambridge, MA, 2012), 70–75.
11. Report by the Panel of Consultants of the Department of State to the Secretary of State, January 1953, in *Foreign Relations of the United States, 1952–1954*, National Security Affairs, vol. 2, pt. 2, no. 67; Henry D. Sokolski, *Best of Intentions: America's Campaign against Strategic Weapons Proliferation* (Westport, CT, and London, 2001), 26–27.
12. "Project 'Candor,'" July 22, 1953, Eisenhower Presidential Library.
13. Susanna Schrafsetter and Stephen Twigge, *Avoiding Armageddon: Europe, the United States, and the Struggle for Nuclear Non-Proliferation, 1945–1970* (Westport, CT, 2004), 53; Ira Chernous, *Eisenhower's Atoms for Peace* (College Station, TX, 2002), 61; Chernous, "Operation Candor: Fear, Faith, and Flexibility," *Diplomatic History* 29, no. 5 (November 2005): 779–809; "Memorandum by the President to the Secretary of State, September 8, 1953," in *Foreign Relations of the United States, 1952–1954*, vol. 2, National Security Affairs, pt. 1 (Washington, DC, 1984), 461.
14. Vladislav Zubok and Konstantin Pleshakov, *Inside the Kremlin's Cold War: From Stalin to Khrushchev* (Cambridge, MA, 1996), 166.
15. Weart, *The Rise of Nuclear Fear*, 80.
16. "Atoms for Peace," *New York Times*, May 27, 1958, 30; "Address by Mr. Dwight D. Eisenhower, President of the United States of America, to the 470th Plenary Meeting of the United Nations General Assembly," Tuesday, December 8, 1953, International Atomic Energy Agency; "Draft of the Presidential Speech before the General Assembly of the United Nations, Draft 5, November 28, 1953, Eisenhower Presidential Library.
17. Sokolski, *Best of Intentions*, 28–29; Weart, *The Rise of Nuclear Fear*, 79–85; Schrafstetter and Twigge, *Avoiding Armageddon*, 52–55.
18. Chernous, *Eisenhower's Atoms for Peace*, xi–xix, 79–118.
19. Eisenhower, "Annual Message to the Congress on the State of the Union, January 07, 1954."

20. John Foster Dulles, "The Evolution of Foreign Policy," Before the Council on Foreign Relations, New York, N.Y., Department of State, Press Release No. 81 (January 12, 1954).
21. Wells, "The Origins of Massive Retaliation," 41–43; William W. Kaufmann, "The Requirements of Deterrence," in *US Nuclear Strategy: A Reader*, ed. Philip Bobbitt (Houndmills and London, 1989), 169–70; Marc Trachtenberg, *A Constructed Peace: The Making of the European Settlement, 1945–1963* (Princeton, NJ, 1999), 185–86.
22. Memorandum of Discussion at the 203rd Meeting of the National Security Council, Wednesday, June 23, 1954, in *Foreign Relations of the United States, 1952–1954*, National Security Affairs, vol. 2, pt. 2, no. 230.
23. "Report of the Committee on Political and Social Problems," Manhattan Project "Metallurgical Laboratory," University of Chicago, June 11, 1985, in *The Manhattan Project: The Birth of the Atomic Bomb in the Words of Its Creators, Eyewitnesses and Historians*, ed. Cynthia C. Kelly (New York, 2007), 288.
24. Robert Norris and Hans M. Kristensen, "Global nuclear weapons inventories, 1945–2010," *Bulletin of the Atomic Scientists* 66, no. 4 (July 1, 2010): 77–83.

Chapter 14: Super Bomb

1. Ralph E. Lapp, *The Voyage of the Lucky Dragon* (New York, 1958), 6–26; Matashichi Ōishi, *The Day the Sun Rose in the West: Bikini, the Lucky Dragon, and I* (Honolulu, 2011), 18–19; Mark Schreiber, "Lucky Dragon's Lethal Catch," *Japan Times*, March 18, 2012.
2. Lapp, *The Voyage of the Lucky Dragon*, 27–54; James R. Arnold, "Effects of Recent Bomb Tests on Human Beings," *Bulletin of the Atomic Scientists* 10, no. 9 (1954): 347–48; "Statement of Lewis Strauss," March 22, 1955, *AEC-FCDA Relationship: Hearings Before the Subcommittee on Security of the Joint Committee on Atomic Energy* (Washington, DC, 1955), 6–9.
3. Major General P. W. Clarkson, *History of Operation Castle*, Pacific Proving Ground Joint Task Force Seven (United States Army, 1954), 121; "Operation Castle, 1954—Pacific Proving Ground," The Nuclear Weapon Archive.
4. Richard Rhodes, *Dark Sun: The Making of the Hydrogen Bomb* (New York, 1996), 379–80.
5. Rhodes, *Dark Sun*, 248; Gregg Herken and Richard C. Leone, *Cardinal Choices: Presidential Science Advising from the Atomic Bomb to SDI* (Stanford, CA, 2000), 34–35.
6. Rhodes, *Dark Sun*, 207–9; Herken and Leone, *Cardinal Choices*, 36–38; David Holloway, *Stalin and the Bomb: The Soviet Union and Atomic Energy, 1939–1956* (New Haven and London, 1996), 299–300; Gregg Herken, *Brotherhood of the Bomb: The Tangled Lives and Loyalties of Robert Oppenheimer, Ernest Lawrence and Edward Teller* (New York, 2002), 127.
7. Holloway, *Stalin and the Bomb*, 301; Herken and Leone, *Cardinal Choices*, 40.
8. Kai Bird and Martin J. Sherwin, *American Prometheus: The Triumph and Tragedy of J. Robert Oppenheimer* (New York, 205), 421–23; "The H-Bomb Decision," GlobalSecurity.org.
9. Lewis Strauss, Chairman of the Atomic Energy Commission to President Harry S. Truman, November 25, 1949, AtomicArchive.com; John Lewis Gaddis, *The Cold War: A New History* (New York, 2005), 61–63.

10. "The President Orders Exploration of a Super Bomb," *Bulletin of the Atomic Scientists* (March 1960): 66; Herken and Leone, *Cardinal Choices*, 35–49; "The H-Bomb Decision," GlobalSecurity.org; Rhodes, *Dark Sun*, 382–408; Rhodes, "Stanislaus Ulam's Interview," (1983), Voices of the Manhattan Project.
11. "Vita—Excerpts from Adventures of a Mathematician by S. N. Ulam," *Los Alamos Science*, 15 (1987): 8–22; Holloway, *Stalin and the Bomb*, 302–3.
12. Holloway, *Stalin and the Bomb*, 307–14; Andrei Sakharov, *Memoirs* (New York, 1990), 101–5.
13. Rhodes, *Dark Sun*, 484–512.
14. Sakharov, *Memoirs*, 162–81; Holloway, *Stalin and the Bomb*, 303–9; German A. Goncharov, "American and Soviet H-bomb development programmes: historical background," *Physics-Uspekhi* 39, no. 10 (1996): 1033–44.
15. "Operation Castle, 1954—Pacific Proving Ground"; Lapp, *The Voyage of the Lucky Dragon*, 6–26; Mark Schreiber, "Lucky Dragon's Lethal Catch," *Japan Times*, March 18, 2012.
16. Guy Oaks, *The Imaginary War: Civil Defense and American Cold War Culture* (New York and Oxford, 1994), 149; Spencer R. Weart, *The Rise of Nuclear Fear* (Cambridge, MA, 2012), 96.
17. Holloway, *Stalin and the Bomb*, 312–17; Sakharov, *Memoirs*, 188–96.
18. Kevin Ruane, *Churchill and the Bomb in War and Cold War* (London, 2016), 245–81; Barbara Leaming, *Jack Kennedy: The Education of a Statesman* (New York, 2006), 235–57; Churchill's 1955 speech, "The Hydrogen Bomb: Churchill's last major speech in Parliament, UK Parliament," 2.
19. Weart, *The Rise of Nuclear Fear*, 102–3, 111.
20. Ralph E. Lapp, "Civil Defense Faces New Peril," *Bulletin of the Atomic Scientists* 9 (*November 1954*): 349–51; Lapp, "Radioactive Fallout," *Bulletin of the Atomic Scientists*, 1 (*February 1955*): 45–51; Milton S. Katz, *Ban the Bomb: A History of SANE, the Committee for a Sane Nuclear Policy* (Santa Barbara, CA, 1987), 14–15.
21. "The Russell-Einstein Manifesto, London, 9 July 1955," Atomic Heritage Foundation.
22. Oaks, *The Imaginary War*, 150–51.

Chapter 15: Missile Gap

1. Eugene Rabinowitch, "The Narrowing Way," *Bulletin of the Atomic Scientists* 9, no. 8 (1953): 294–95, 298.
2. Vladislav Zubok and Constantine Pleshakov, *Inside the Kremlin's Cold War: From Stalin to Khrushchev* (Cambridge, MA, 1996), 163–69.
3. Susanna Schrafstetter and Stephen Twigge, *Avoiding Armageddon: Europe, the United States, and the Struggle for Nuclear Non-Proliferation, 1945–1970* (Westport, CT, 2004), 56–57.
4. Zubok and Pleshakov, *Inside the Kremlin's Cold War*, 163–69; Schrafstetter and Twigge, *Avoiding Armageddon*, 57.
5. Zubok and Pleshakov, *Inside the Kremlin's Cold War*, 163–69, 188–89.
6. Nikita Khrushchev, *Khrushchev Remembers* (Boston, 1971), 265–362; William Taubman, *Khrushchev: The Man and His Era* (New York and London), 208–324.
7. Cited in Taubman, *Khrushchev*, 351–52; Khrushchev, *Khrushchev Remembers*, 432–34.

8. Taubman, *Khrushchev*, 351–52; Zubok and Pleshakov, *Inside the Kremlin's Cold War*, 163–69.
9. "Rocket R-7," *Energia*; Gaddis, *The Cold War: A New History* (New York, 2005), 71.
10. Austin Jersild, "Sharing the Bomb among Friends: The Dilemmas of Sino-Soviet Strategic Cooperation," Cold War International History Project, Woodrow Wilson Center; Mao cited in Benjamin Shobert, *Blaming China: It Might Feel Good But It Won't Fix America's Economy* (Lincoln, NE, 2018), 35.
11. Sergei Khrushchev, *Rozhdenie sverkhderzhavy. Kniga ob ottse* (Moscow, 2003), 241; Vladimir Platonov, "R-12. 'Gadkii utenok' s beregov Dnepra," *Dzerkalo tyzhnia*, June 22, 2007.
12. Christopher A. Preble, "'Who Ever Believed in the 'Missile Gap?': John F. Kennedy and the Politics of National Security," *Presidential Studies Quarterly* 33, no. 4 (December 2003): 801–26; W. J. Rorabaugh, *The Real Making of the President: Kennedy, Nixon, and the 1960 Election* (Lawrence, KS, 2009); Gary Donaldson, *The First Modern Campaign: Kennedy, Nixon, and the Election of 1960* (Lanham, MD, 2007), 127–48; Barbara Leaming, *Jack Kennedy: The Education of a Statesman* (New York, 2006), 230.
13. Donaldson, *The First Modern Campaign*, 61–126; "Remarks of Senator John F. Kennedy in the United States Senate, National Defense, Monday, February 29, 1960," Archives, John F. Kennedy Presidential Library and Archive.
14. Spencer R. Weart, *The Rise of Nuclear Fear* (Cambridge, MA, 2012), 124; Nevil Shute, *On the Beach* (New York, 1957).
15. Nelson W. Polsby, *Political Innovation in America: The Politics of Policy Initiation* (New Haven and London, 1984), 57–58; Richard Dean Burns and Joseph Siracusa, *A Global History of the Nuclear Arms Race: Weapons, Strategy, and Politics*, 2 vols. (Santa Barbara, CA, 2013), 1: 236–37, 247–48; Glenn T. Seaborg with the assistance of Benjamin S. Loeb, *Kennedy, Khrushchev and the Test Ban* (Berkeley, Los Angeles, and London, 1981), 8.
16. Polsby, *Political Innovation in America*, 59; Gerard J. DeGroot, *The Bomb: A Life* (Cambridge, MA, 2005), 255; Allen Pietrobon, "Peacemaker in the Cold War: Norman Cousins and the Making of the Citizen Diplomat in the Atomic Age," PhD diss., American University, 2016, 99–199, 201; Martin S. Katz, *Ban the Bomb: A History of SANE, the Committee for Sane Nuclear Policy* (Westport, CT, 1987), 15–16.
17. Katz, *Ban the Bomb*, 17–25.
18. Russell Baker, "Eisenhower Bars 'Total' Outlawing of Nuclear Tests," *New York Times*, June 6, 1957, 1, 15.
19. Katherine Magraw, "Teller and the 'Clean Bomb' Episode," *Bulletin of the Atomic Scientists* (May 1988): 32–37, here 32–34; Pietrobon, "Peacemaker in the Cold War," 210–11.
20. Frederick Kempe, *Berlin 1961: Kennedy, Khrushchev, and the Most Dangerous Place on Earth* (New York, 2011), 29–30.
21. Zubok and Pleshakov, *Inside the Kremlin's Cold War*, 200–201; "Khrushchev Arrives in Washington," *This Day in History*, September 15, 1959; "Khrushchev Ends Trip to the United States," *This Day in History*, September 27, 1959.
22. Aleksandr Fursenko and Timothy Naftali, *One Hell of a Gamble: Khrushchev, Castro, and Kennedy, 1958–1964: The Secret History of the Cuban Missile Crisis* (New York, 1997), 80.

23. Dino A. Brugioni and Doris G. Taylor, *Eyes in the Sky: Eisenhower, the CIA, and Cold War Aerial Espionage* (Annapolis, MD, 2010), 343–46.
24. Zubok and Pleshakov, *Inside the Kremlin's Cold War*, 189; Michael Beschloss, *Mayday: Eisenhower, Khrushchev, and the U-2 Affair* (New York, 1986); Bruce Geelhoed and Anthony O. Edmond, *Eisenhower, Macmillan, and Allied Unity, 1957–1961* (New York, 2003), 101.
25. "Remarks of Senator John F. Kennedy in the Senate, Washington, DC, June 14, 1960," Archives, John F. Kennedy Presidential Library and Archive.
26. Zubok and Pleshakov, *Inside the Kremlin's Cold War*, 202–5.

Chapter 16: Bombe Atomique

1. William Burr, "The U.S. Nuclear Presence in Western Europe, 1954–1962, Part I," National Security Archive.
2. Henry D. Sokolski, *Best of Intentions: America's Campaign against Strategic Weapons Proliferation* (Westport, CT, and London, 2001), 40–41.
3. "Nuclear Threats. Second Berlin Crisis, 1960," GlobalSecurity.org; Benjamin Varat, "Point of Departure: A Reassessment of Charles de Gaulle and the Paris Summit of May 1960," *Diplomacy & Statecraft* 19, no. 1 (2008): 96–124.
4. "Frédéric Joliot-Curie (1900–1958)," Atomic Archive.
5. Wilfred L. Kohl, *French Nuclear Diplomacy* (Princeton, NJ, 2015), 16–17; Vincent Bugeja, "Joliot-Curie Rips America for Atomic Energy Report," *New York Herald Tribune*, European edition, June 15, 1947.
6. Kohl, *French Nuclear Diplomacy*, 16.
7. "Frédéric Joliot-Curie (1900–1958)," Atomic Archive; "France's Nuclear Weapons. Origin of the Force de Frappe," Nuclear Weapons Archive.
8. Stephanie Cooke, *In Mortal Hands: A Cautionary History of the Nuclear Age* (New York, 2009), 133–34; Wolf Mendl, "The Background of French Nuclear Policy," *International Affairs* 41, no. 1 (January 1965): 22–36, here 27–28.
9. Cooke, *In Mortal Hands*, 145–46.
10. Cooke, *In Mortal Hands*, 138–39; "Marcoule: G1, G2 and G3 reactors for plutonium production," fissilematerials.org.
11. Derek Varble, *The Suez Crisis, 1956* (Oxford, 2003), 28–90; Odd Arne Westad, *The Cold War: A World History* (New York, 2017), 272–75.
12. Kohl, *French Nuclear Diplomacy*, 18–22, 29–31, 35–36.
13. Kohl, *French Nuclear Diplomacy*, 26–29; Bruno Barrillot, "French Nuclear Tests in the Sahara: Open the Files," Science for Democratic Action, Institute for Energy and Environmental Research, April 2008.
14. Kohl, *French Nuclear Diplomacy*, 26–29; Barrillot, "French Nuclear Tests in the Sahara."
15. Kohl, *French Nuclear Diplomacy*, 88–90.
16. "Memorandum of Conversation, Paris, July 5, 1958. The Secretary's talks with General de Gaulle in Paris, July 5, 1958," in *Foreign Relations of the United States, 1958–1960*, Western Europe, vol. 7, pt. 2, no. 34.
17. Kohl, *French Nuclear Diplomacy*, 62–64; Jean-Marc Regnault, "France's Search for Nuclear Test Sites, 1957–1963," *Journal of Military History* 67, no. 4 (2003): 1223–48, here 1227; Richard Dean Burns and Joseph M. Siracusa, *A Global History of the Nuclear Race: Weapons, Strategy and Politics* (Santa Barbara, CA, 2013), 2: 330–31.

18. Kohl, *French Nuclear Diplomacy*, 91–93, 106–7; Varat, "Point of Departure," 99.
19. "Discontinuance of Nuclear Tests," United States Congress. *The Congressional Record: Proceedings and Debates of the 86th Congress, Second Session*, vol. 106, pt. 2 (January 27–February 16, 1960), 2079; Cooke, *In Mortal Hands*, 145–46.
20. Kohl, *French Nuclear Diplomacy*, 6–7; Cooke, *In Mortal Hands*, 147; Burns and Siracusa, *A Global History of the Nuclear Race*, 2: 328–29.
21. Barrillot, "French Nuclear Tests in the Sahara"; Jean Lacouture, *De Gaulle, the Ruler, 1945–1970* (New York, 1993), 423.

Chapter 17: China Syndrome

1. "V Sovete ministrov SSSR," *Pravda*, January 18, 1955.
2. John Lewis Gaddis, *The Cold War: A New History* (New York, 2005), 48–50, 58–60; Carl A. Posey, "How the Korean War Almost Went Nuclear," *Air & Space Magazine*, July 2015.
3. John Wilson Lewis and Litai Xue, *China Builds the Bomb* (Stanford, CA, 1991), 13–14.
4. Lewis and Xue, *China Builds the Bomb*, 12; Victor Gobarev, "Soviet Policy toward China: Developing Nuclear Weapons, 1949–1969," *Journal of Slavic Military History* 12, no. 4 (December 1999): 1–53, here pp. 3–4.
5. Gobarev, "Soviet Policy toward China," 10–11.
6. Henrietta Harrison, "Popular Responses to the Atomic Bomb in China 1945–1955," *Past & Present* 218 (suppl. 8): 98–116, here pp. 106–7.
7. Lewis and Xue, *China Builds the Bomb*, 22–29, 32–35; Austin Jersild, *The Sino-Soviet Alliance: An International History* (Chapel Hill, NC, 2016), 145; H. W. Brands, "Testing Massive Retaliation: Credibility and Crisis Management in the Taiwan Strait," *International Security* 12, no. 4 (1988): 124–51; Gordon H. Chang, "To the Nuclear Brink: Eisenhower, Dulles, and the Quemoy-Matsu Crisis," *International Security* 12, no. 4 (1988): 96–123.
8. Lewis and Xue, *China Builds the Bomb*, 36–46.
9. "Address by Zhou Enlai at the Plenary Session of the Fourth Meeting of the State Council (Excerpt)," January 31, 1955, History and Public Policy Program Digital Archive, Dang de wenxian (Party Historical Documents), no. 3 (1994): 16–19, trans. Neil Silver.
10. "Mao Zedong, 'The Chinese People Cannot Be Cowed by the Atom Bomb,'" January 28, 1955, History and Public Policy Program Digital Archive, Selected Works of Mao Tse-Tung, vol. 5, 152–53; "At the Plenary Session of the Fourth Meeting of the State Council (Excerpt)," January 31, 1955, History and Public Policy Program Digital Archive, Dang de wenxian (Party Historical Documents), no. 3 (1994): 16–19, trans. Neil Silver.
11. "Address by Zhou Enlai."
12. "Talk by Mao Zedong at an Enlarged Meeting of the Chinese Communist Party Central Committee Politburo (Excerpts)," April 25, 1956, History and Public Policy Program Digital Archive, Mao Zedong wenji (Selected Writings of Mao Zedong), vol. 7 (Beijing: Renmin chubanshe, 1999), 27, trans. Neil Silver.
13. Jersild, *The Sino-Soviet Alliance*, 146–47; Lorenz Lüthi, *The Sino-Soviet Split: Cold War in the Communist World* (Princeton, NJ, 2008), 1–17; Sergey Radchenko, *Two Suns in the Heavens: The Sino-Soviet Struggle for Supremacy, 1962–1967* (Washington, DC, and Stanford, CA, 2009), 1–22; Jersild, "Sharing the Bomb among

Friends: The Dilemmas of Sino-Soviet Strategic Cooperation," *Cold War International History Project, Nuclear Proliferation International History Project*; "Request by the Chinese Leadership to the Soviet Leadership for Help in Establishing a Chinese Nuclear Program," January 15, 1956, History and Public Policy Program Digital Archive, TsKhSD (Center for the Storage of Contemporary Documentation), f. 5, op. 30, d. 164, ll. 7a, 48–9. Obtained by Tatiana Zazerskaia and translated by David Wolff.

14. Jersild, *The Sino-Soviet Alliance*, 146; Khrushchev, *Khrushchev Remembers*, trans. Strobe Talbott (Boston, 1971), 517–37.
15. "Excerpt from the Unedited Translation of Mao Zedong's Speech at the Moscow Conference of Communist and Workers' Parties," November 18, 1957, History and Public Policy Program Digital Archive, RGANI: fond 10, opis' 1, delo 32, listy 1-91. Reproduced in *Nasledniki Kominterna: Mezhdunarodnye soveshchaniia predstavitelei kommunisticheskikh i rabochikh partii v Moskve*, ed. N. G. Tomilina (Moscow, 2013), trans. Sergey Radchenko. [Translation edited for style.]
16. Khrushchev, *Khrushchev Remembers*, 519–20; Jersild, *The Sino-Soviet Alliance*, 147; "Letter from the Communist Party of the Soviet Union Central Committee to the Chinese Communist Party Central Committee on the Temporary Halt in Nuclear Assistance," June 20, 1959, History and Public Policy Program Digital Archive, PRC FMA 109-02563-01, 1–3, trans. Neil Silver.
17. Jersild, *The Sino-Soviet Alliance*, 147; William Burr and Jeffrey T. Richelson, "Whether to 'Strangle the Baby in the Cradle': The United States and the Chinese Nuclear Program, 1960–64," *International Security* 25, no. 3 (Winter 2000/01): 54–99, here 57–59; Odd Arne Westad, *Brothers in Arms: The Rise and Fall of the Sino-Soviet Alliance, 1945–63* (Stanford, CA, 1998), 157–59, 206–7; Victor Gobarev, "Soviet Policy toward China: Developing Nuclear Weapons, 1949–1969," *Journal of Slavic Military History* 12, no. 4 (December 1999), 17–31.
18. "Mao Zedong's Talk at the Beidaihe Central Committee Work Conference (Excerpt)," July 18, 1960, History and Public Policy Program Digital Archive, Jiangguo yilai Mao Zedong junshi wengao (Mao Zedong's Manuscripts on Military Affairs since the Founding of the PRC), vol. 3 (Beijing, 2010), 100, trans. Neil Silver; Burr and Richelson, "Whether to 'Strangle the Baby in the Cradle,'" 58; Lewis and Xue, *China Builds the Bomb*, 177–78.
19. "Zhou Enlai's Discussion with a Kenyan African National Federation Delegation (Excerpt)," September 5, 1963, History and Public Policy Program Digital Archive, Dang de wenxian (Party Historical Documents), no. 3 (1994): 15–16, trans. Neil Silver.
20. "JFK on Nuclear Weapons and Non-Proliferation," Carnegie Endowment for International Peace, November 17, 2003.
21. Burr and Richelson, "Whether to 'Strangle the Baby in the Cradle,'" 60–63.
22. Wilfred L. Kohl, *French Nuclear Diplomacy* (Princeton, NJ, 2015), 130.

Chapter 18: Cuban Gamble

1. "JFK on Nuclear Weapons and Non-Proliferation," Carnegie Endowment for International Peace, November 17, 2003.
2. Memorandum of Conversation, Vienna, June 4, 1961, in *FRUS*, 1961–1963, vol. 5: Soviet Union, no. 87; Richard Reeves, *President Kennedy: Profile in Power* (New

York, 1993), 175; Michael Dobbs, *One Minute to Midnight: Kennedy, Khrushchev and Castro on the Brink of Nuclear War* (New York, 2008), 229.
3. "Radio and Television Report to the American People on the Berlin Crisis, July 25, 1961," JFK Presidential Library and Museum.
4. Aviva Chomsky, *A History of the Cuban Revolution* (UK, 2015), 28–44; Jim Rasenberger, *The Brilliant Disaster: JFK, Castro, and America's Doomed Invasion of Cuba's Bay of Pigs* (New York, 2011).
5. Aleksandr Fursenko and Timothy Naftali, *"One Hell of a Gamble": Khrushchev, Castro and Kennedy, 1958–1964* (New York and London, 1997), 158–65; Serhii Plokhy, *Nuclear Folly: A History of the Cuban Missile Crisis* (New York, 2021), 37–57.
6. Fursenko and Naftali, *"One Hell of a Gamble,"* 166–97, 158–72; Plokhy, *Nuclear Folly,* 58–96.
7. *The Kennedy Tapes: Inside the White House during the Cuban Missile Crisis,* ed. Ernest R. May and Philip D. Zelikow, concise ed. (New York and London, 2002), 3–137; Plokhy, *Nuclear Folly,* 132–57.
8. "Central Committee of the Communist Party of the Soviet Union Presidium Protocol 60," October 23, 1962; cf. *Prezidium TsK KPSS, 1954–1964,* ed. Aleksandr Fursenko (Moscow, 2003), vol. 1, protocol no. 60, 617; Anastas Mikoian, "Diktovka o poezdke na Kubu," January 19, 1963, in Aleksandr Lukashin and Mariia Aleksashina, "My voobshche ne khotim nikuda brosat' rakety, my za mir," *Rodina,* January 1, 2017; Sergo Mikoyan, *The Soviet Cuban Missile Crisis: Castro, Mikoyan, Kennedy, Khrushchev, and the Missiles of November* (Stanford, CA, 2014), 148, 157.
9. "Radio and Television Report to the American People on the Berlin Crisis, July 25, 1961," JFK Presidential Library and Museum; "Letter from President Kennedy to Chairman Khrushchev," October 22, 1962, in *FRUS, 1961–1963,* vol. 6, Kennedy-Khrushchev Exchanges, no. 60; "Telegram from the Embassy in the Soviet Union to the Department of State," Moscow, October 23, 1962, 5 p.m., *FRUS, 1961–1963,* vol. 6, Kennedy-Khrushchev Exchanges, no. 61.
10. "Central Committee of the Communist Party of the Soviet Union Presidium Protocol 60," October 23, 1962.
11. Anatoly Dobrynin, *In Confidence: Moscow's Ambassador to America's Six Cold War Presidents (1962–1986)* (New York, 1995), 82–83.
12. Robert F. Kennedy, *Thirteen Days: A Memoir of the Cuban Missile Crisis,* with a new foreword by Arthur Schlesinger Jr. (New York, 1999), 53–54.
13. Executive Committee Meeting of the National Security Council, Wednesday, October 24, 1962, 10:00 a.m.; *The Kennedy Tapes,* 231–33.
14. G. M. Kornienko, *Kholodnaia voina. Svidetel'stvo ee uchastnika* (Moscow, 2001), 124; Dobrynin, *In Confidence,* 83, 129; Scott D. Sagan, *The Limits of Safety: Organizations, Accidents, and Nuclear Weapons* (Princeton, NJ, 1993), 68–69.
15. "Central Committee of the Communist Party of the Soviet Union Presidium Protocol 61," October 25, 1962, History and Public Policy Program Digital Archive, RGANI, f. 3, op. 16, d. 947, l. 42-42 ob., trans. and ed. Mark Kramer, with assistance from Timothy Naftali.
16. *The Diary of Anatoly S. Chernyaev, 1976,* trans. Anna Melyakova, 2, National Security Archive.
17. "Central Committee of the Communist Party of the Soviet Union Presidium Pro-

tocol 62," October 27, 1962, History and Public Policy Program Digital Archive, RGANI, f. 3, op. 16, d. 947, l. 43–44, trans. and ed. Mark Kramer, with assistance from Timothy Naftali; Letter from Chairman Khrushchev to President Kennedy, Moscow, October 27, 1962, *FRUS, 1961–1963*, vol. 6, Kennedy-Khrushchev Exchanges, no. 66.
18. Plokhy, *Nuclear Folly*, 234–45.
19. Dobbs, *One Minute to Midnight*, 297–334; Plokhy, *Nuclear Folly*, 257–88.
20. "Poslanie Pervogo sekretaria TsK KPSS Nikity Sergeevicha Khrushcheva, prezidentu Soedinennykh Shtatov Ameriki, Dzhonu F. Kennedi," *Pravda*, October 29, 1962, 1; cf. "Letter from Chairman Khrushchev to President Kennedy," Moscow, October 28, 1962, *FRUS, 1961–1963*, vol. 6, Kennedy-Khrushchev Exchanges, no. 68.
21. Fursenko and Naftali, *"One Hell of a Gamble,"* 290–318; Mikoyan, *The Soviet Cuban Missile Crisis*, 195–234; Plokhy, *Nuclear Folly*, 319–35.

Chapter 19: Banning the Bomb

1. "Telegram from the Department of State to the Embassy in the Soviet Union," Washington, October 28, 1962, in *Foreign Relations of the United States, 1961–1963*, vol. 6, Kennedy-Khrushchev Exchanges, no. 69.
2. Glenn T. Seaborg with Benjamin S. Loeb, foreword by W. Averell Harriman, *Kennedy, Khrushchev and the Test Ban* (Berkeley, Los Angeles, and London, 1981), 21–66.
3. Nikita Khrushchev, *Khrushchev Remembers* (Boston, 1971), 536.
4. "Letter from Chairman Khrushchev to President Kennedy," Moscow, October 30, 1962, *Foreign Relations of the United States, 1961–1963*, vol. 6, Kennedy-Khrushchev Exchanges, no. 71.
5. "Letter from Chairman Khrushchev to President Kennedy," Moscow, October 30, 1962.
6. "Letter from Chairman Khrushchev to President Kennedy," Moscow, December 19, 1962, *Foreign Relations of the United States, 1961–1963*, vol. 6, Kennedy-Khrushchev Exchanges, no. 85.
7. "Message from President Kennedy to Chairman Khrushchev," Washington, December 28, 1962, in *Foreign Relations of the United States, 1961–1963*, vol. 6, Kennedy-Khrushchev Exchanges, no. 87; "Letter from Chairman Khrushchev to President Kennedy," Moscow, January 7, 1963, ibid., no. 92; Vojtech Mastny, "The 1963 Nuclear Test Ban Treaty: A Missed Opportunity for Détente?" *Journal of Cold War Studies* 10, no. 1 (2008): 3–25, here 8–9; Seaborg, *Kennedy, Khrushchev and the Test Ban*, 181–85.
8. Norman Cousins, *The Improbable Triumvirate: John F. Kennedy, Pope John, Nikita Khrushchev* (New York, 1972), 47–51; Allen Pietrobon, "Peacemaker in the Cold War: Norman Cousins and the Making of the Citizen Diplomat in the Atomic Age," PhD diss., American University, 2016, 326–27.
9. Mastny, "The 1963 Nuclear Test Ban Treaty," 12.
10. William Krasner, "Baby Tooth Survey—First Results," *Environment: Science and Policy for Sustainable Development* 55, no. 2 (2013): 18–24; "St. Louis Baby Tooth Survey, 1959–1970," Washington University School of Dental Medicine.
11. Cousins, *The Improbable Triumvirate*, 85–86; Seaborg, *Kennedy, Khrushchev and the Test Ban*, 191, 209–10.

12. Mastny, "The 1963 Nuclear Test Ban Treaty," 14–15.
13. "Telegram from the Department of State to the Embassy in the Soviet Union," Washington, May 30, 1963, *Foreign Relations of the United States, 1961–1963*, vol. 6, Kennedy-Khrushchev Exchanges, no. 105; "Letter from Chairman Khrushchev to President Kennedy," Moscow, June 8, 1963, ibid., no. 106; Seaborg, *Kennedy, Khrushchev and the Test Ban*, 199.
14. John F. Kennedy, "Commencement Address at American University," Washington, DC, June 10, 1963, John F. Kennedy Library and Archive; Cousins, *The Improbable Triumvirate*, 122–26.
15. "Rech tovarishcha N. S. Khrushcheva na mitinge v Berline 2 iiulia 1963 goda," *Pravda*, July 3, 1963, 1–2; Mastny, "The 1963 Nuclear Test Ban Treaty," 17–18.
16. Mastny, "The 1963 Nuclear Test Ban Treaty," 19–21.
17. Cousins, *The Improbable Triumvirate*, 127–48; Martin S. Katz, *Ban the Bomb: A History of SANE, the Committee for Sane Nuclear Policy* (Westport, CT, 1987), 84–67; Dennis Hevesi, "Dr. Louise Reiss, Who Helped Ban Atomic Testing, Dies at 90," *New York Times*, January 10, 2011.
18. "Letter from President Kennedy to Chairman Khrushchev," Washington, December 14, 1962, in *Foreign Relations of the United States, 1961–1963*, vol. 6, Kennedy-Khrushchev Exchanges, no. 84; "Letter from Chairman Khrushchev to President Kennedy, Moscow, December 19, 1962," ibid., no. 85.
19. Seaborg, *Kennedy, Khrushchev and the Test Ban*, 23, 181–82; Jeffrey T. Richelson and William Burr, "Whether to 'Strangle the Baby in the Cradle': The United States and the Chinese Nuclear Program, 1960–64," *International Security* 25, no. 3 (2000): 54–99, here 67–72.
20. Burr and Richelson, "Whether to 'Strangle the Baby in the Cradle,'" 72–73.
21. Burr and Richelson, "Whether to 'Strangle the Baby in the Cradle,'" 86–96.
22. William Taubman, *Khrushchev: The Man and His Era* (New York, 2003), 614–22.
23. "Statement of the Government of the People's Republic of China," October 16, 1964, History and Public Policy Program Digital Archive, PRC FMA 105-01262-01, 22–26. Obtained by Nicola Leveringhaus.
24. "Statement of the Government of the People's Republic of China," October 16, 1964.
25. Michael S. Gerson, *The Sino-Soviet Border Conflict: Deterrence, Escalation, and the Threat of Nuclear War in 1969*, Defense Threat Reduction Agency Advanced Systems and Concepts Office (2010), 46–52; Yang Kuisong, "The Sino-Soviet Border Clash of 1969: From Zhenbao Island to Sino-American Rapprochement," *Cold War History* 1 (2000): 21–52; David Reynolds, *Summits: Six Meetings that Shaped the Twentieth Century* (New York, 2007), 226.
26. Henry D. Sokolski, *Best of Intentions: America's Campaign against Strategic Weapons Proliferation* (Westport, CT, and London, 2001), 41–45.
27. Sokolski, *Best of Intentions*, 46–52.
28. Joseph Cirincione, *Bomb Scare: The History and Future of Nuclear Weapons* (New York, 2007), 29–31; *Treaty on the Non-Proliferation of Nuclear Weapons*, International Atomic Energy Agency, Information Circular, April 22, 1970.
29. Cirincione, *Bomb Scare*, 32–34; David Fischer, *History of the International Atomic Energy Agency* (Vienna, 1997), 27–70.

Chapter 20: Star of David

1. Benny Morris, *1948: The First Arab-Israeli War* (New Haven, 2008).
2. Avner Cohen, *Israel and the Bomb* (New York, 1998), 10–12, 66.
3. Avner Cohen, "Before the Beginning: The Early History of Israel's Nuclear Project (1948–1954)," *Israel Studies* 3, no. 1 (1998): 112–39; Cohen, *Israel and the Bomb*, 55; Stephanie Cooke, *In Mortal Hands: A Cautionary History of the Nuclear Age* (New York, 2009), 150–54.
4. Cohen, *Israel and the Bomb*, 53–54; S. Ilan Troen, "The Protocol of Sèvres: British/French/Israeli Collusion against Egypt, 1956," *Israel Studies* 1, no. 2 (Fall 1996): 124–39; Avi Shlaim, "The Protocol of Sèvres, 1956: Anatomy of a War Plot," *International Affairs* 73, no. 3 (1997): 509–30.
5. Cohen, *Israel and the Bomb*, 56–65, 70–73; Cooke, *In Mortal Hands*, 148–50; Memorandum of Conversation, August 6, 1959, between Gunnar Randers and Richard J. Kerry, National Security Archive.
6. Cohen, *Israel and the Bomb*, 66–68, 75–78; Cooke, *In Mortal Hands*, 152–53; Peter Pry, *Israel's Nuclear Arsenal* (London and New York, 1984), 12.
7. Cohen, *Israel and the Bomb*, 73–76; Julian Schofield, *Strategic Nuclear Sharing* (New York, 2014), 50–58; Avner Cohen and William Burr, "Israel's Quest for Yellowcake: The Secret Argentina-Israel Connection, 1963–1966," Wilson Center.
8. Memorandum of Conversation. US Secretary of State Christian Herter and UK Ambassador to the US Sir Harold Caccia, Subject: Safeguards for Reactors, November 25, 1960, National Security Archive; Cohen, *Israel and the Bomb*, 81–87.
9. "Memorandum of Discussion at the 470th Meeting of the National Security Council," December 8, 1960, in *Foreign Relations of the United States, 1958–1960*, vol. 13, Arab-Israeli Dispute; United Arab Republic; North Africa, no. 177; Avner Cohen and William Burr, "The Eisenhower Administration and the Discovery of Dimona: March 1958–January 1961," National Security Archive.
10. Cohen, *Israel and the Bomb*, 88.
11. Cohen, *Israel and the Bomb*, 88–93; Cohen and Burr, "The Eisenhower Administration and the Discovery of Dimona."
12. *Special National Intelligence Estimate: Implication of the Acquisition by Israel of a Nuclear Weapons Capability*, December 8, 1960, National Security Archive; Avner Cohen and William Burr, "Kennedy, Dimona and the Nuclear Proliferation Problem: 1961–1962," National Security Archive.
13. Cohen and Burr, "Kennedy, Dimona and the Nuclear Proliferation Problem: 1961–1962"; "Memorandum of Conversation, President Kennedy and Prime Minister Ben Gurion," New York, May 30, 1961, in *Foreign Relations of the United States, 1961–1963*, vol. 17, Near East, 1961–1962, no. 57.
14. Cohen and Burr, "Kennedy, Dimona and the Nuclear Proliferation Problem: 1961–1962"; *National Intelligence Estimate: Outlook for Israel*, October 5, 1961, National Security Archive; Howard Furnas, Office of Special Assistant to Secretary of State for Atomic Energy and Outer Space, to Dwight Ink, Atomic Energy Commission, November 15, 1961, National Security Archive.
15. "Circular Airgram from the Department of State to Certain Posts, Washington, October 31, 1962, Subject: Israel's Dimona Reactor," in *Foreign Relations of the United States, 1961–1963*, vol. 18, Near East, 1962–1963, no. 87; Cohen and Burr,

"Kennedy, Dimona and the Nuclear Proliferation Problem: 1961–1962"; Pry, *Israel's Nuclear Arsenal*, 7–9.
16. Avner Cohen and William Burr, "The Battle of the Letters, 1963: John F. Kennedy, David Ben-Gurion, Levi Eshkol, and the U.S. Inspections of Dimona," National Security Archive; "State Department Telegram 938 to U.S. Embassy Israel," June 15, 1963, National Security Archive.
17. Cohen and Burr, "The Battle of the Letters, 1963"; "State Department Telegram 193 to U.S. Embassy Israel," July 4, 1963, National Security Archive.
18. U.S. Inspection Team Visit to Israeli Atomic Energy Installation, January 16–20, 1964, 1–2, National Security Archive; "Israel's Nuclear Weapon Capability: An Overview," Wisconsin Project on Nuclear Arms Control, *The Risk Report* 2, no. 4 (July-August 1996).
19. Cohen, *Israel and the Bomb*, 237; Richard Dean Burns and Joseph M. Siracusa, *A Global History of the Nuclear Race: Weapons, Strategy and Politics* (Santa Barbara, CA, 2013), 2: 348–49.
20. Pry, *Israel's Nuclear Arsenal*, 31–32.
21. Seymour M. Hersh, *The Samson Option: Israel's Nuclear Arsenal and American Foreign Policy* (New York, 1991), 217, 222–29; Richard Sale, "Yom Kippur: Israel's 1973 Nuclear Alert," UPI, September 16, 2002; Elbridge Colby, Avner Cohen, William McCants, Bradley Morris, and William Rosenau, *The Israeli "Nuclear Alert" of 1973: Deterrence and Signaling in Crisis*, CAN Analysis and Solutions, April 2013; Abraham Rabinovich, *The Yom Kippur War: The Epic Encounter That Transformed the Middle East* (New York, 2017), 534–56.
22. William A. Schwartz and Charles Derber, *The Nuclear Seduction: Why the Arms Race Doesn't Matter—and What Does* (Berkeley, Los Angeles. and Oxford, 1989), 109–17.
23. Pry, *Israel's Nuclear Arsenal*, 1–2, 30–33; "Israel's Nuclear Weapon Capability: An Overview."

Chapter 21: MAD Men

1. "Dr. Strangelove Or: How I Learned To Stop Worrying And Love The Bomb (1964)," *Filmsite Movie Review*; "Doomsday Machine," in Ace G. Pilkington, *Science Fiction and Futurism: Their Terms and Ideas* (Jefferson, NC, 2017), 66–70.
2. Louis Menand, "Fat Man. Herman Kahn and the Nuclear Age," *New Yorker*, June 19, 2005; Daniel Deudney, *Whole Earth Security: A Geopolitics of Peace*, Worldwatch Paper 55 (Washington, DC, 1983), 80.
3. James Nissenbaum, "John von Neumann: The Mathematician Who You May Not Realize Changed Your World," *Thomas Insights*, March 15, 2019.
4. Gerard J. DeGroot, *The Bomb: A Life* (Cambridge, MA, 2005), 270, 297; cf. McNamara interview in "Cold War MAD, 1960–1972," CNN Documentary Series.
5. Joseph Cirincione, *Bomb Scare: The History and Future of Nuclear Weapons* (New York, 2008), 36; Warner R. Schilling, "US Strategic Nuclear Concepts in the 1970s: The Search for Sufficiently Equivalent Countervailing Parity," in *Strategy and Nuclear Deterrence*, ed. Steven E. Miller (Princeton, NJ, 1984), 183–214, here 184.
6. DeGroot, *The Bomb*, 270.
7. John Lewis Gaddis, *The Cold War: A New History* (New York, 2007), 156–88; Jonathan Haslam, *Russia's Cold War: From the October Revolution to the Fall of the Wall* (New Haven and London. 2011), 214–94.

8. "Paper prepared in the Department of Defense," undated, Washington, in *Foreign Relations of the United States, 1969–1976*, vol. 32, SALT I, 1969–1972 (Washington, DC, 2010), 3; "Memorandum from K. Wayne Smith of the National Security Council Staff to the President's Assistant for National Security Affairs (Kissinger)," Washington, October 6, 1971, ibid., 618; DeGroot, *The Bomb*, 271.
9. "A Brief History of Anti-Ballistic Missile Systems," *Smithsonian National Air and Space Museum*.
10. David Reynolds, *Summits: Six Meetings that Shaped the Twentieth Century* (New York, 2007), 233–36, 246–50.
11. Memorandum of Conversation. Participants: Leonid I. Brezhnev, General-Secretary of Central Committee of CPSU . . . and Mr. Henry A. Kissinger . . . , Moscow, April 22, 1972, in *Foreign Relations of the United States, 1969–1976*, vol. 32, SALT I, 1969–1972, 775–77.
12. Reynolds, *Summits*, 247–61; Iu. Latysh, "Petro Shelest, Volodymyr Shcherbyts'kyi i Richard Nikson: Amerykans'kyi slid u vidstavtsi pershoho sekretaria TsK Kompartiï Ukraïny," *Visnyk Kyïvs'koho natsional'noho universytetu im. Tarasa Shevchenka. Istoriia*, 1 (144)/2020, 34–36.
13. Anatoly Dobrynin, *In Confidence: Moscow's Ambassador to America's Six Cold War Presidents* (New York, 1995), 253–54; Memorandum of Conversation. Participants: Leonid I. Brezhnev, General Secretary of the Central Committee of the CPSU . . . and the President, Dr. Henry A. Kissinger, Moscow, May 23, 1972, *Foreign Relations of the United States, 1969–1976*, vol. 32, SALT I, 1969–1972, 851.
14. Dobrynin, *In Confidence*, 251–52; Reynolds, *Summits*, 264–72; Gaddis, *Cold War*, 203–4.
15. Reynolds, *Summits*, 274; Gaddis, *Cold War*, 199–200.
16. Reynolds, *Summits*, 276–81.
17. *Military Implications of the Treaty on the Limitations of the Anti-Ballistic Missile Systems and the Interim Agreement on Limitation of Strategic Offensive Arms*. Hearing before the Committee on Armed Services, United States Senate, 92nd Congress, 1st Session (Washington, DC, 1972).
18. "Vstrecha Leonida Brezhneva i prezidenta SShA Dzheral'da Forda (1974)," *RIA Novosti*; Dobrynin, *In Confidence*, 327–34.
19. Richard Rhodes, *Arsenals of Folly: The Making of the Nuclear Arms Race* (New York, 2007), 118–37.
20. The Diary of Anatoly S. Chernyaev 1976.
21. *The SALT II Treaty*. Hearings before the Committee on Foreign Relations, United States Senate, 96th Congress, 1st Session (Washington, DC, 1979).
22. Dobrynin, *In Confidence*, 422–27; Gaddis, *Cold War*, 200–203.

Chapter 22: Smiling Buddha

1. "Richard Nixon Discusses Energy Policy," September 26, 1971, Richard Nixon Foundation, www.youtube.com/watch?v=uRCib2l6TuY.
2. Richard Nixon, "Special Message to the Congress on Energy Policy, April 18, 1973," in *Public Papers of the Presidents of the United States* (Washington, DC, 1975), 310; Rod Adams, "Why did Richard Nixon so strongly endorse nuclear energy in April 1973?" *Atomic Insights*, September 22, 2015.
3. "Environment: Project Dubious," *Time*, April 9, 1973.

4. Michael Corbett, "Oil Shock of 1973–74 (October 1973–January 1974)," Federal Reserve History; David Hammes and Douglas T. Wills, "Black Gold: The End of Bretton Woods and the Oil Price Shocks of the 1970s," *Independent Review* 9, no. 4 (Spring 2005): 501–11; Luke Phillips, "Nixon's Nuclear Energy Vision," October 20, 2016, Richard Nixon Foundation.
5. James Mahaffey, *Nuclear Awakening: A New Look at the History and Future of Nuclear Power* (New York, 2010), 300–326; John Abbotts, "The Long, Slow Death of the Fast Flux Facility," *Bulletin of the Atomic Scientists*, September 1, 2004.
6. Mahaffey, *Atomic Awakening*, 232–37; Scott Kirsch, *Proving Grounds: Project Plowshare and the Unrealized Dream of Nuclear Earthmoving* (New Brunswick, NJ, and London, 2005); Milo D. Nordyke, "The Soviet program for peaceful uses of nuclear explosions," Lawrence Livermore National Laboratory, July 24, 1996; Valentina Semiashkina, "Atomnyi kotlovan," *Ėkologiia i pravo*, December 24, 2002.
7. William Burr, "The Nixon Administration and the Indian Nuclear Program, 1972–1974," Nuclear Proliferation International History Project, Woodrow Wilson International Center for Scholars.
8. George Perkovich, *India's Nuclear Bomb: The Impact on Global Proliferation* (Berkeley and Los Angeles, 2001), 14; Raj Chengappa, *Weapons of Peace: The Secret Story of India's Quest To Be a Nuclear Power* (New Delhi, 2000), 66–71.
9. William George Penney, "Homi Jehangir Bhabha, 1909–1966," *Biographical Memoirs of Fellows of the Royal Society*, 13 (1967): 35–55; Perkovich, *India's Nuclear Bomb*, 16–21.
10. Perkovich, *India's Nuclear Bomb*, 22–26; "Indian Nuclear Program," Atomic Heritage Foundation.
11. Perkovich, *India's Nuclear Bomb*, 27–28; "India's Nuclear Weapons Program. The Beginning: 1944–1960," Nuclear Weapon Archive; Manpreet Sethi, "The Indo-Canadian Nuclear Relationship: Possibilities and Challenges," *International Journal* 69, no. 1 (March 2014): 35–47, here 36–37.
12. Perkovich, *India's Nuclear Bomb*, 28–40; "Plutonium Processing Plant," India Facilities, Nuclear Threat Initiative.
13. Perkovich, *India's Nuclear Bomb*, 44–47.
14. Perkovich, *India's Nuclear Bomb*, 64–85.
15. Chengappa, *Weapons of Peace*, 103–13.
16. Chengappa, *Weapons of Peace*, 111–12, 116–30; "India's Nuclear Weapons Program. Smiling Buddha: 1974," Nuclear Weapon Archive.
17. "US Embassy Airgram A-20 to State Department, 'India's Nuclear Intentions,'" January 21, 1972, History and Public Policy Program Digital Archive, National Archives, Record Group 59, SN 70-73, Def 18-8 India. Obtained and contributed by William Burr and included in NPIHP Research Update #4.
18. Chengappa, *Weapons of Peace*, 114–15; "India's Nuclear Weapons Program. Smiling Buddha: 1974."
19. "Intelligence Community Staff, Post Mortem Report, 'An Examination of the Intelligence Community's Performance Before the Indian Nuclear Test of May 1974,'" July 1974, History and Public Policy Program Digital Archive, Mandatory declassification review request; release by the Interagency Security Classification Appeals Panel. Obtained and contributed by William Burr and included in NPIHP Research Update #4.

20. "US Embassy India Cable 6598 to State Department, 'India's Nuclear Explosion: Why Now?,'" May 18, 1974, History and Public Policy Program Digital Archive, Access to Archival Databases (AAD), National Archives and Records Administration, Record Group 59, Central Foreign Policy File, document number 1974NEWDE06598. Obtained and contributed by William Burr and included in NPIHP Research Update #4; "India's Nuclear Weapons Program. Smiling Buddha: 1974."
21. "Intelligence Community Staff, Post Mortem Report, 'An Examination of the Intelligence Community's Performance Before the Indian Nuclear Test of May 1974.'"
22. "National Security Study Memorandum (NSSM) 202 on Nuclear Proliferation," May 23, 1974, History and Public Policy Program Digital Archive, Nixon Presidential Library, National Security Council Institutional Files, Study Memorandums (1969–1974), Box H-205. Obtained by Fundação Getúlio Vargas; Henry D. Sokolski, *Best of Intentions: America's Campaign against Strategic Weapons Proliferation* (Westport, CT, and London, 2001), 63–65.
23. David Fischer, *History of the International Atomic Energy Agency* (Vienna, 1997), 262–63.
24. Carey Sublette, "Pakistan's Nuclear Weapons Program. The Beginning," Nuclear Weapon Archive; Samina Ahmed and David Cortright, "Pakistani Public Opinion and Nuclear Weapons Policy," in Ahmed and Cortright, eds., *Pakistan and the Bomb: Public Opinion and Nuclear Options* (Notre Dame, IN, 1998), 3–46, here 9–10.
25. William Burr, "U.S. and British Combined to Delay Pakistani Nuclear Weapons Program in 1978–1981, Declassified Documents Show," Nuclear Proliferation International History Project, Woodrow Wilson International Center for Scholars; "US Embassy Paris cable 31540 to State Department, 'Elysée Views on Reprocessing Issues,'" September 23, 1978, History and Public Policy Program Digital Archive, Mandatory Declassification Review request. Obtained and contributed by William Burr and included in NPIHP Research Update #3; Shubhangi Pandey, *U.S. Sanctions on Pakistan and Their Failure as Strategic Deterrent*, Observer Research Foundation Issue Brief 251 (August 2018).

Chapter 23: Star Wars

1. Frank Warner, "New World Order Seventeen Years Ago," *The Morning Call*, March 5, 2000.
2. James Cant, "The SS-20 Missile—Why Were You Pointing at Me?" in *Russia: War, Peace and Diplomacy*, ed. Ljubica Erickson and Mark Erickson (London, 2007), 240–53; Taylor Downing, *1983: Reagan, Andropov and a World on the Brink* (New York, 2018), 78–79.
3. Christopher Paine, "Pershing II: The Army's Strategic Weapon," *Bulletin of the Atomic Scientists* 36, no. 8 (October 1980): 25–31.
4. Joshua Woodyatt, "War Scares and (Nearly) the End of the World: The Euromissiles Crisis of 1977–1987," *E-International Relations*, May 2, 2020; John T. Correll, "The Euromissile Showdown," *Air Force Magazine*, February 1, 2020; Peter E. Quint, "Civil Disobedience and the German Courts: The Pershing Missile Protests in Comparative Perspective," *Book Gallery*, 34 (2008).

5. Ronald Reagan, Speech on the Strategic Arms Reduction Talks, November 18, 1981, Miller Center; "Bonn Stands by 'Zero Option,'" *New York Times*, January 30, 1983, A 3; "Brezhnev's Condemnation of NATO's Plans," October 6, 1979, in *The Cold War Through Documents: A Global History*, ed. Edward H. Judge and John W. Langdon, 3rd ed. (Lanham, MD, Boulder, New York, and London, 2018), 287; Ronald Reagan, "Speech on the Strategic Arms Reduction Talks," November 18, 1981.
6. "Andropov's Speech on Reduction in Nuclear Missiles, 21 December 1982," in *The Cold War Through Documents*, 310–11.
7. Martin S. Katz, *Ban the Bomb: A History of SANE, the Committee for Sane Nuclear Policy* (Westport, CT, 1987), 149–52; Warner, "New World Order Seventeen Years Ago."
8. Ronald Reagan, "Evil Empire" Speech, March 8, 1983, Miller Center; "Evil Empire" Speech by President Reagan—Address to the National Association of Evangelicals, YouTube.
9. Warner, "New World Order Seventeen Years Ago."
10. Ronald Reagan, "Address to the Nation on National Security," March 23, 1983, Miller Center; Steven R. Weisman, "Reagan Proposes U.S. Seek New Way to Block Missiles," *New York Times*, March 24, 1983, A 1.
11. Reagan, "Address to the Nation on National Security," March 23, 1983.
12. Reagan, "Address to the Nation on National Security," March 23, 1983; Andrew Glass, "President Reagan calls for launching 'Star Wars' initiative, March 23, 1983," *Politico*, March 23, 2017; George P. Shultz, *Turmoil and Triumph: My Years as Secretary of State* (New York, 1993), 26; Donald R. Baucom, *The Origins of SDI, 1944–1983* (Lawrence, KS, 1992), 132–33, 153; Frances FitzGerald, *Way Out There in the Blue: Reagan, Star Wars and the End of the Cold War* (New York, 2000), 127–35, 145–46.
13. Edward Teller with Judith Shoolery, *Memoirs: A Twentieth-Century Journey in Science and Politics* (New York, 2001), 530–32; Reagan, "Address to the Nation on National Security," March 23, 1983; Taylor Downing, *1983: Reagan, Andropov, and a World on the Brink* (New York, 2018), 90–105.
14. Sharon Watkins Lang, "Where do we get 'Star Wars'?" *The Eagle*, March 2007; Anatoly Dobrynin, *In Confidence: Moscow's Ambassador to America's Six Cold War Presidents* (New York, 1995), 528.
15. "Otvety Iuriia Vladimirovicha Andropova na voprosy korrespondenta 'Pravdy,'" *Pravda*, March 27, 1983, 1.
16. Dobrynin, *In Confidence*, 532.
17. "KGB Chairman Andropov to KGB Members," March 25, 1981, National Security Archive; Chairman Yuri Andropov at the National Consultation Meetings of the Leadership and Members of the KGB, May 25, 1981, National Security Archive.
18. KGB Headquarters, Moscow to the London KGB Residency, February 17, 1983; Christopher Andrew and Vasily Mitrokhin, *The Sword and the Shield: The Mitrokhin Archive and the Secret History of the KGB* (New York, 2000), 392–93; Downing, *1983: Reagan, Andropov, and a World on the Brink*, 68–89.
19. Benjamin B. Fischer, *A Cold War Conundrum: The 1983 Soviet War Scare*; Downing, *1983: Reagan, Andropov and the World on the Brink*, 149–68.
20. Dobrynin, *In Confidence*, 537–38; Ronald Reagan, "Speech on the Soviet Attack

on a Korean Airliner," September 5, 1983; Artem Krechetnikov, "Tragediia koreiskogo Boinga: chto bylo na samom dele?," August 21, 2013, BBC Russian Service.
21. Pavel Aksenov, "Stanislav Petrov: The man who may have saved the world," BBC Russian Service, September 26, 2013; "Obituary: Stanislav Petrov was declared to have died on September 18th," *Economist*, September 30, 2017.
22. "Zaiavlenie general'nogo sekretaria TsK KPSS, Predsedatelia Verkhovnogo Soveta SSSR Iu. V. Andropova," *Pravda*, September 29, 1983, 1; Andrei Gromyko, *Memories* (London, 1989), 384–86; Anatoly Dobrynin, *In Confidence: Moscow's Ambassador to America's Six Cold War Presidents* (New York, 1995), 540.
23. "Segodnia den' raketnykh voisk i artillerii," *Pravda*, November 19, 1983, 1; Dmitrii Ustinov, "Borot'sia za mir, ukrepliat' oboronosposobnost'," *Pravda*, November 19, 1983, 4.

Chapter 24: The Fall of the Nuclear Colossus

1. Ronald Reagan, *White House Diaries*, October 10, 1983, in Ronald Reagan Presidential Foundation and Institute; Reagan, *White House Diaries*, November 18, 1983, ibid. Alistair Cooke, "The Day After—25 November 1983," BBC Radio 4.
2. Christopher Andrew and Oleg Gordievsky, *KGB: The Inside Story of Its Foreign Operations from Lenin to Gorbachev* (London, 1990), 502–4; Ronald Reagan, *An American Life: The Autobiography* (New York, 2011), 585, 588–89; Steven F. Hayward, *The Age of Reagan: The Conservative Counterrevolution, 1980–1989* (New York, 2010), 331–32; Downing, *1983: Reagan, Andropov and the World on the Brink* (New York, 2018), 118–127.
3. Ronald Reagan, "Address to the Nation and Other Countries on United States-Soviet Relations," January 16, 1984, Ronald Reagan Presidential Library and Museum.
4. Mikhail Gorbachev interviewed for the CNN Cold War Mini-Series, Episode 22: Star Wars, min. 27:29, YouTube; Ronald Reagan, *White House Diaries*, November 5, 1985, in Ronald Reagan Presidential Foundation and Institute.
5. David Reynolds, *Summits: Six Meetings That Shaped the Twentieth Century* (New York, 2007), 370–75.
6. Reynolds, *Summits*, 358, 379–80; An American participant in the Geneva summit interviewed for the CNN Cold War Mini-Series, Episode 22: Star Wars, min. 28:04.
7. *V Politbiuro TsK KPSS. Po zapisiam Anatoliia Cherniaeva, Vadima Medvedeva, Georgiia Shakhnazarova (1985–1991)* (Moscow, 2006), 43, 78.
8. Mark Harrison, "Soviet Economic Growth Since 1928: The Alternative Statistics of G. I. Khanin," *Europe-Asia Studies* 45, no. 1 (1993): 141–67; *V Politbiuro*, 102–3; Vladislav Zubok, *A Failed Empire: The Soviet Union in the Cold War from Stalin to Gorbachev* (Chapel Hill, NC, 2007), 291, 299.
9. *V Politbiuro*, 35, 39.
10. Serhii Plokhy, *Chernobyl: The History of a Nuclear Catastrophe* (New York, 2018).
11. *V Politbiuro*, 43; Zubok, *A Failed Empire*, 288.
12. Reynolds, *Summits*, 389–90.
13. *V Politbiuro*, 88–89.
14. Dobrynin, *In Confidence*, 623–24; *V Politbiuro*, 168.
15. *V Politbiuro*, 162, 182.

16. Mikhail Gorbachev, *Perestroika: New Thinking for Our Country and the World* (New York, 1987), 12, 234.
17. "Treaty Between the United States of America and the Union of Soviet Socialist Republics on the Elimination of Their Intermediate-Range and Shorter-Range Missiles (INF Treaty)," U.S. Department of State; *SIPRI Yearbook 2007: Armaments, Disarmament, and International Security*, Stockholm International Peace Research Institute (Stockholm, 2007), 683; Reynolds, *Summits*, 394–95.
18. Ronald Reagan, *White House Diaries*, December 8, 1987, in Ronald Reagan Presidential Foundation and Institute; Reagan, *White House Diaries*, December 9, 1987, ibid.; Reynolds, *Summits*, 396.
19. Odd Arne Westad, *The Cold War: A World History* (New York, 2017), 579–608.
20. START I at a Glance, Strategic Arms Association.
21. Plokhy, *The Last Empire: The Final Days of the Soviet Union* (New York, 2015), 15; Dobrynin, *In Confidence*, 627.
22. "Otstavka Gorbacheva. Zaiavlenie," December 25, 1991, YouTube; Plokhy, *The Last Empire*, 376–77.

Chapter 25: Giving Up the Bomb

1. "Strategic Arms Reduction Treaty (START II)," Federation of American Scientists; "Brief Chronology of START II," Arms Control Association.
2. On Brazil and Argentina giving up their nuclear programs, see Matias Spektor, "The Long View: How Argentina and Brazil Stepped Back from a Nuclear Race," *Americas Quarterly*, October 28, 2015.
3. Zondi Masiza, "A Chronology of South Africa's Nuclear Program," *The Nonproliferation Review* (Fall 1993): 35–55; "Nuclear Power in South Africa," World Nuclear Association; "South Africa. Nuclear," Nuclear Threat Initiative.
4. Roger Jardine W. De-Villiers and Mitchell Reiss, "Why South Africa Gave Up the Bomb," *Foreign Affairs* 72 (November/December 1993): 98–101.
5. Leonard Thompson, *A History of South Africa*, 3rd ed. (New Haven and London, 2001), 187–219; "Mandela and South African Communist Party," South African History Online.
6. "South Africa withdraws from the Commonwealth, 15 March 1961," South African History Online; Richard Dale, *The Namibian War of Independence, 1966–1989: Diplomatic, Economic and Military Campaigns* (Jefferson, NC, 2014), 74–77, 93–95.
7. "South Africa. Missile," Nuclear Threat Initiative.
8. De-Villiers and Reiss, "Why South Africa Gave Up the Bomb," 102; "South Africa. Nuclear," Nuclear Threat Initiative; Sasha Polakow-Suransky, *The Unspoken Alliance: Israel's Secret Relationship with Apartheid South Africa* (New York, 2010), 118–53.
9. "South Africa. Nuclear," Nuclear Threat Initiative; De-Villiers and Reiss, "Why South Africa Gave Up the Bomb," 103–4; Kathryn O'Neill and Barry Munslow, "Ending the Cold War in Southern Africa," *Third World Quarterly* 12, no. 3/4 (1990–1991): 81–96.
10. Serhii Plokhy, *The Last Empire: The Final Days of the Soviet Union* (New York, 2015), 199–200.
11. Telcon of President George H. W. Bush with President Yeltsin of the Republic of Russia, December 8, 1991, 1, George H. W. Bush Presidential Library and Museum.

12. Mariana Budjeryn, "Looking Back: Ukraine's Nuclear Predicament and the Nonproliferation Regime," Arms Control Association; Budjeryn, *Inheriting the Bomb: The Collapse of the USSR and the Nuclear Disarmament of Ukraine* (Baltimore, 2023), 62–83; "The Lisbon Protocol at a Glance," Arms Control Association.
13. Steven Pifer, *The Eagle and the Trident: U.S–Ukraine Relations in Turbulent Times* (Washington, DC, 2017), 4–5; Mariana Budjeryn, *Inheriting the Bomb*, 34–36; Yuri Kostenko, *Ukraine's Nuclear Disarmament: A History* (Cambridge, MA, 2020), 34; "End of the Soviet Union. Text of Bush's Address to Nation on Gorbachev's Resignation," *New York Times*, December 26, 1991, A 16; George H. W. Bush, Address on Gorbachev Resignation, December 25, 1991, C-SPAN; "Ukraine, Nuclear Weapons, and Security Assurances at a Glance," Arms Control Association.
14. "Deklaratsiia pro derzhavnyi suverenitet Ukraïny," Verkhovna Rada Ukraïny.
15. Budjeryn, "Looking Back: Ukraine's Nuclear Predicament and the Nonproliferation Regime."
16. "Zaiava pro bez'iadernyi status Ukraïny," Verkhovna Rada Ukraïny; Kostenko, *Ukraine's Nuclear Disarmament*, 41.
17. Sergei Shargorodsky, "Ukraine Suspends Removal of Tactical Nuclear Weapons With AM-Soviet-Unrest," *Associated Press*, March 12, 1992; Pifer, *The Eagle and the Trident*, 11; Kostenko, *Ukraine's Nuclear Disarmament*, 35, 39, 45–46; Serge Schmemann, "Ukrainian Uses Summit to Berate Russians and the Commonwealth: Animosity clouds the meeting of the ex-Soviet republics," *New York Times*, March 21, 1992, 1.
18. Kostenko, *Ukraine's Nuclear Disarmament*, 64–68.
19. "Postanova Verkhovnoï Rady Ukraïny Pro dodatkovi zakhody shchodo zabezpechennia nabuttia Ukraïnoiu bez'iadernoho statusu," Verkhovna Rada Ukraïny; Pifer, *The Eagle and the Trident*, 11.
20. Strobe Talbott, *The Russia Hand: A Memoir of Presidential Diplomacy* (New York, 2002), 79; John J. Mearsheimer, "The Case for a Ukrainian Nuclear Deterrent," *Foreign Affairs* 72, no. 3 (Summer, 1993): 50–66.
21. Kostenko, *Ukraine's Nuclear Disarmament*, 230–33.
22. Jane Perlez, "Economic Collapse Leaves Ukraine with Little to Trade but Its Weapons," *New York Times*, January 13, 1994; Serhii Plokhy, *The Gates of Europe: A History of Ukraine* (New York, 2016), 328–29; "Ukraine Inflation Rate 1993–2021," Macrotrends.
23. Serhii Plokhy, *Chernobyl: The History of a Nuclear Catastrophe* (New York, 2018), 329–30.
24. Budapest Memorandums on Security Assurances, 1994, Council on Foreign Relations.
25. Budjeryn, "Looking Back: Ukraine's Nuclear Predicament and the Nonproliferation Regime"; "Nuclear Disarmament. Ukraine," Nuclear Threat Initiative; "Tretia pislia Rosiï ta SShA. Iak vyhliadav iadernyi potentsial Ukraïny," YouTube.
26. Paul J. D'Anieri, Introduction, in Kostenko, *Ukraine's Nuclear Disarmament*, 9–22; Scott D. Sagan, "Why Do States Build Nuclear Weapons?: Three Models in Search of a Bomb," *International Security* 21, no. 3 (Winter 1996–1997): 54–86, here 80–82.
27. "Central and Eastern European Nuclear-Weapon-Free Zone," Federation of American Scientists.

Chapter 26: The Return of Fear

1. "Comprehensive Nuclear-Test-Ban Treaty," Nuclear Threat Initiative.
2. Michael Krepon, "Looking Back: The 1998 Indian and Pakistani Nuclear Tests," Arms Control Association.
3. Official Press Statements, Ministry of External Affairs of India, New Delhi, May 11, 1998; www.wisconsinproject.org/india-pakistan-nuclear-weapon-update-1998/.
4. Molly MacCalman, "A.Q. Khan Nuclear Smuggling Network," *Journal of Strategic Security* 9, no. 1 (Spring 2016); David Armstrong and Joseph John Trento, *America and the Islamic Bomb: The Deadly Compromise* (Hanover, NH, 2007). "Pakistan Nuclear Weapons. A Brief History of Pakistan's Nuclear Program," Strategic Security Project; *Pakistan and the Bomb: Public Opinion and Nuclear Options*, ed. Samina Ahmed and David Cortright (Notre Dame, IN, 1998), 3–46, here 10–12.
5. "Pakistan Nuclear Weapons"; "India-Pakistan: Nuclear Weapon Update—1998," November 1, 1998, Wisconsin Project on Nuclear Arms Control.
6. "India and Pakistan on the Brink: The 1998 Nuclear Tests," Association for Diplomatic Study and Training; Fact Sheet: India and Pakistan Sanctions. Released by the Bureau of Economic and Agricultural Affairs, June 18, 1998, US Department of State; Nuclear Sanctions: Section 102(b) of the Arms Export Control Act and Its Application to India and Pakistan, December 9, 1999–October 5, 2000, Congressional Research Service Report.
7. Deok Ryong Yoon and Bradley O. Babson, "Understanding North Korea's Economic Crisis," *Asian Economic Papers* 1, no. 3 (July 2002): 69–89.
8. Lee Jae-Bong, "U.S. Deployment of Nuclear Weapons in 1950s South Korea & North Korea's Nuclear Development: Toward Denuclearization of the Korean Peninsula," *Asia-Pacific Journal* 7, issue 8, no. 3 (February 17, 2009).
9. William Burr, "The United States and South Korea's Nuclear Weapons Program, 1974–1976," Nuclear Proliferation International History Project/Cold War International History Project; Peter Hayes, "Park Chung Hee, the CIA & the Bomb," *Global Asia*.
10. William Burr, "Kissinger State Department Insisted That South Koreans Break Contract with French for Reprocessing Plant," National Security Archive; David Fischer, *History of the International Atomic Energy Agency* (Vienna, 1997), 261–62; Eunjung Lim, "South Korea's Nuclear Dilemmas," *Journal for Peace and Nuclear Disarmament* 2:1 (2019): 297–318.
11. Hayes, "Park Chung Hee, the CIA & the Bomb."
12. *North Korea: A Country Study*, Federal Research Division, Library of Congress, ed. Robert L. Worden, 5th ed. (Washington, DC, 2008), 203–6.
13. Jae-Bong, "U.S. Deployment of Nuclear Weapons in 1950s South Korea"; "Yongbyon 5MWe Reactor, North Korea Facilities," Nuclear Threat Initiative.
14. Kwang Ho Chun, *North Korea's Nuclear Question: Sense of Vulnerability, Defensive Motivation, and Peaceful Solution* (Carlisle, PA: Strategic Studies Institute, U.S. Army War College, 2010).
15. Jae-Bong, "U.S. Deployment of Nuclear Weapons in 1950s South Korea"; Andrew S. Natsios, *The Great North Korean Famine* (Washington, DC, 2001), 37–88; Jeffrey Lewis, "Revisiting the Agreed Framework," May 15, 2015, *38 North*, The Henry L. Stimson Center.

16. "DPRK Successfully Conducts Underground Nuclear Test," October 9, 2006; Richard L. Garwin and Frank N. von Hippel, "A Technical Analysis: Deconstructing North Korea's October 9 Nuclear Test," Arms Control Association.
17. Andrew Mack, "Potential, not proliferation: Northeast Asia has several nuclear-capable countries, but only China has built weapons," *Bulletin of the Atomic Scientists*, July 1, 1997.
18. Vitaly Fedchenko, "North Korea's Nuclear Test Explosion, 2009," *SIPRI Fact Sheet*, December 2009; "N. Korea Conducts Powerful Nuclear Test, Reportedly Fires Short-Range Missiles," *Fox News*, May 25, 2009.
19. "North Korea," Nuclear Threat Initiative; Zachary Cohen and Barbara Starr, "Trump condemns North Korean long-range missile launch," CNN, July 28, 2017; Michael D. Shear and David E. Sanger, "Trump Returns North Korea to List of State Sponsors of Terrorism," *New York Times*, November 20, 2017.
20. Matt Stevens, "Trump and Kim Jong-un, and the Names They've Called Each Other. President Trump and Mr. Kim, who have agreed to meet to discuss North Korea's nuclear program, have a history of colorful exchanges," *New York Times*, March 9, 2018; The White House, "Joint Statement of President Donald J. Trump of the United States of America and Chairman Kim Jong Un of the Democratic People's Republic of Korea at the Singapore Summit," June 12, 2018; Leo Byrne, "North Korean Missile Test did not Threaten U.S. or Allies: Pompeo," *NK News*, May 5, 2019; "North Korean Nuclear Negotiations, 1985–2024," Council on Foreign Affairs.

Chapter 27: Preemptive War

1. "September 11 Attacks," History.Com, September 11, 2020; Seth G. Jones, *In the Graveyard of Empires: America's War in Afghanistan* (New York, 2010), 86–150.
2. "Text of President Bush's 2002 State of the Union Address," January 29, 2002, *Washington Post*.
3. "U.N. Opposes U.S. Plan for Antimissile Defense," *Reuters*, December 2, 1999; Richard Dean Burns and Joseph M. Siracusa, *A Global History of the Nuclear Arms Race: Weapons, Strategy and Politics* (Santa Barbara, CA, 2013), vol. 2, 453–58, 484–91.
4. Wade Boese, "U.S. Withdraws from ABM Treaty; Global Response Muted," *Arms Control Today*, July/August 2002, Arms Control Association.
5. Iraq Survey Group Final Report, Weapons of mass destruction, GlobalSecurity.com; Netanel Avneri, "Iraq's oil war in the USA during the October 1973 War," *Middle Eastern Studies* 52, no. 5 (2016): 754–71.
6. Pierre Razoux and Nicholas Elliott, *The Iran–Iraq War* (Cambridge, MA, 2015).
7. Razoux and Elliott, *The Iran–Iraq War*, 126–27, 165–69; "Tuwaitha. Al Tuwaitha Nuclear Center al-Aseel / al-Diyalla Facility," Weapons of Mass Destruction, Federation of American Scientists; "Osiraq / Tammuz I," Weapons of Mass Destruction, Federation of American Scientists; Alexandra Evans, "A Lesson from the 1981 Raid on Osirak," Sources and Methods, Blog on the History and Public Policy Program, July 10, 2017.
8. Ronald Reagan, Diary Entry, 06/07/1981, The Ronald Reagan Presidential Foundation and Institute; Aluf Benn, "Where First Strikes Are Far from the Last Resort," *Washington Post*, November 10, 2002.

9. "Osiraq / Tammuz I," Weapons of Mass Destruction, Federation of American Scientists.
10. "Osiraq / Tammuz I," Weapons of Mass Destruction, Federation of American Scientists; "Iraqi Nuclear Weapons," Weapons of Mass Destruction, Federation of American Scientists.
11. Barton Gellman, "Iraq Inspections, Embargo in Danger at U.N. Council," *Washington Post*, December 22, 1998, A 25; "Iraq: A Chronology of UN Inspections," Arms Control Today, Arms Control Association.
12. Iraq Liberation Act of 1998 (Enrolled as Agreed to or Passed by Both House and Senate).
13. "Disarming Saddam—A Chronology of Iraq and UN Weapons Inspections From 2002–2003," Arms Control Association; "Text of Bush Speech on Iraq," *CBS News*, March 17, 2003.
14. Julian Borger, "There were no weapons of mass destruction in Iraq," *The Guardian*, October 7, 2004; Keith Krause, "Bodies count: the politics and practices of war and violent death data," *Human Remains and Violence* 3, no. 1 (2017): 90–115.
15. JoAnne Allen, "FBI says Saddam's weapons bluff aimed at Iran," *Reuters*, July 2, 2009; "Saddam Hussein executed in Iraq," *BBC News*, December 30, 2006.
16. "President Bush Admits Iraq Had No WMDs and 'Nothing' to Do With 9/11," *Independent Global News*, August 22, 2006.
17. "Iraq War in Figures," *BBC News*, December 14, 2011; "Iran: Syria part of 'axis of resistance,'" CNN, August 7, 2012.
18. Farhang Jahanpour, "Chronology of Iran's Nuclear Programme," Oxford Research Group.
19. Jahanpour, "Chronology of Iran's Nuclear Programme."
20. Jon Boyle, "Iran seen to need 3–8 yrs to produce bomb," Reuters, October 22, 2007; Sue Pleming, "Rice swipes at IAEA, urges bold action on Iran," Reuters, September 18, 2007; "Israel minister: Sack ElBaradei," *BBC News*, November 8, 2007.
21. Ken Dilanian, "U.S. intelligence chief sees limited benefit in an attack on Iran," *Los Angeles Times*, February 16, 2012; Kim Zetter, *Countdown to Zero Day: Stuxnet and the Launch of the World's First Digital Weapon* (New York, 2014).
22. Shahram Chubin, *Iran's Nuclear Ambitions* (Washington, DC, 2006); Dalia Dassa Kaye, Alireza Nader, and Parisa Roshan, *Israel and Iran: A Dangerous Rivalry*, National Defense Research Institute.
23. Kali Robinson, "What Is the Iran Nuclear Deal?" Council on Foreign Relations; "Iran nuclear deal: Key details," *BBC News*, June 11, 2019.
24. Isabel Kreshner, "Iran Deal Denounced by Netanyahu as 'Historic Mistake,'" *New York Times*, July 14, 2015.
25. Robinson, "What Is the Iran Nuclear Deal?"
26. Nahal Toosi, "U.S. tries to break Iran nuclear deadlock with new proposal for Tehran," *Politico*, March 29, 2021.
27. "Opposition party head appears to urge nuclear action to stop Iran's atomic program," *The Times of Israel*, July 4, 2024; Barak Ravid, "Scoop: Israel destroyed active nuclear weapons research facility in Iran, officials say," Axios, November 15, 2024, www.axios.com/2024/11/15/iran-israel-destroyed-active-nuclear-weapons-research-facility.

Epilogue

1. "Address by the President of the Russian Federation," February 24, 2022, President of Russia.
2. Serhii Plokhy, *The Russo-Ukrainian War: The Return of History* (New York, 2024), 224–26; Dan Sabbagh, "CIA boss says West should not be intimidated by Russia's nuclear threats," *The Guardian*, September 7, 2024, www.theguardian.com/world/article/2024/sep/07/cia-west-russia-nuclear-threats-putin; Brendan Cole, "China Told Putin Not to Use Nuclear Weapons, Blinken Says," *Newsweek*, January 4, 2025, www.newsweek.com/russia-nuclear-china-blinken-2009670.
3. John Erath, "Does Putin's New 'Oreshnik' Missile Transform Rules of Nuclear Warfare?," *Center for Arms Control and Non-Proliferation*, December 13, 2024, armscontrolcenter.org/does-putins-new-oreshnik-missile-transform-rules-of-nuclear-warfare.
4. Isabel van Brugen, "Mystery as Russia Abruptly Flips Nuclear Drill Scenario," *Newsweek*, June 13, 2024; Shannon Bugos, "Putin Orders Russian Nuclear Weapons on Higher Alert," March 2022, Arms Control Association; "Briefing: Russian state TV renews nuclear threats against West," *BBC Monitoring*, April 29, 2024; van Brugen, "Putin Issues Nuclear Warning to West," *Newsweek*, June 21, 2024; "Belarus' Defence Minister said nuclear drills showed high readiness of Belarusian military," *SB News*, June 27, 2024; "Lukashenka Reveals Plans for Russian Oreshnik Missile Deployment in Belarus," *Radio Free Europe/Radio Liberty*, December 10, 2024, www.rferl.org/a/lukashenka-russia-oreshnik-missile-deployment-belarus-putin-ukraine/33234767.html.
5. Maxim Starchak, "Russia's Nuclear Modernization Drive Is Only a Success on Paper," *Carnegie Politika*, January 21, 2024.
6. Paul Bracken, *The Second Nuclear Age: Strategy, Danger and the New Power Politics* (New York, 2012), 1–7; "The Anti-Ballistic Missile (ABM) Treaty at a Glance," Arms Control Association.
7. New START Treaty, U.S. Department of State; Ernesto Zedillo Ponce de León, "New Era of Nuclear Rearmament. Nuclear arms control is fast unraveling, and the United States leads the march—the Non-Proliferation Treaty could be at risk," *YaleGlobal Online*, February 21, 2019.
8. Rose Gottemoeller, "Russia Is Updating Their Nuclear Weapons: What Does That Mean for the Rest of Us?," Carnegie Endowment for International Peace, January 29, 2020; "Russia's Nuclear Weapons: Doctrine, Forces, and Modernization," US Congressional Research Service, Updated July 20, 2020, 20–26; Matthew Kroenig, Mark Massa, and Christian Trotti, *Russia's Exotic Nuclear Weapons and Implications for the United States and NATO*, Atlantic Council, March 2020.
9. "Nuclear Threats and the Role of Allies: Remarks by Acting Assistant Secretary of Defense for Space Policy Dr. Vipin Narang at CSIS," August 1, 2024, US Department of Defense.
10. Joe Gould, "China plans to double nuclear arsenal, Pentagon says," *DefenceNews*, September 1, 2020; Pranay Vaddi and Ankit Panda, "When it comes to China's nuclear weapons, numbers aren't everything," *DefenseNews*, March 13, 2021; Michael R. Gordon, "Possible Chinese Nuclear Testing Stirs U.S. Concern. Beijing might secretly be conducting small nuclear tests at its Lop Nur site, report

says," *Wall Street Journal*, April 15, 2021; "Which Countries have Nuclear Weapons?" International Campaign to Abolish Nuclear Weapons.
11. Willian Booth, "Boris Johnson's Vision for post-Brexit 'Global Britain' Includes More Nuclear Weapons," *Washington Post*, March 16, 2021; Alexander Yermakov, "Is France's Nuclear Shield Big Enough to Cover All of Europe?," Russian International Affairs Council, April 8, 2020; "France and Russia head forward with Nuclear Modernisation," *New Delhi Times*, March 8, 2021.
12. Vipin Narang, *Seeking the Bomb: Strategies of Nuclear Proliferation* (Princeton and Oxford, 2022), 1–6.
13. John Pickrell, "Introduction: The Nuclear Age," *New Scientist*, September 4, 2006.
14. Serhii Plokhy, *Chernobyl Roulette: War in the Nuclear Disaster Zone* (New York, 2024).
15. Plokhy, *Chernobyl Roulette*, 108–22.
16. Isabel van Brugen, "Putin's Nuclear Doctrine Questioned by Ally on TV," *Newsweek*, June 26, 2024.

INDEX

A-1 reactor ("Annushka") at Kyshtym, 140
"Able Archer" military exercise, 297
Acheson, Dean, 112–14, 173
Acheson-Lilienthal plan, 112–14, 164
Acton, Lord, 26
Advertising Council (US), 161
Advisory Committee on Uranium, 42
Aeroflot, 293–94
Afghanistan, 268, 282, 285, 299, 328, 330, 340
Africa
  Angolan civil war, 313–14
  the Border War, 313, 314
  decolonization after WWII, 84, 200
  Nuclear Weapons–Free Zone of, 325
  uranium from the Belgian Congo, 17–18, 21–25, 44, 113, 146–47
  See also specific countries and conflicts
African National Congress (ANC), 313, 315
Ahmadinejad, Mahmoud, 352
Aiken, Frank, 240
Alberta Project, 62
Aleksandrov, Semen, 115
Algeria, 201, 204, 247
Al-Qaeda, 340–41, 348
Alsop, Joseph, 187
Alsos Mission, 63
American nuclear monopoly, 109, 113–14, 120, 138–39, 144–47, 163–64, 178, 182, 238. *See also* Manhattan Project

American Physical Society, 30
Anderson, Sir John (Viscount Waverley), 66–69, 108
Andropov, Yurii, 286, 291–94, 298, 305, 306
Anglo-American nuclear research cooperation, 40–42, 48–51, 65–71, 75–78. *See also* Manhattan Project
Angola, 267, 313–14
anthrax, 341
anti-ballistic missile (ABM) systems, 260, 261–64, 285–86, 289–90, 299–300, 341–42, 360. *See also* Strategic Defense Initiative (SDI)
Anti-Ballistic Missile (ABM) Treaty of 1972, 265, 306, 341–42, 360
anticolonialism, 215, 219–20, 276, 281, 312
anticommunism, 155, 299
apartheid, 311, 312–15
Arab world, 250, 343. *See also specific countries and conflicts*
Argentina, 248, 310, 363
Arms Control and Disarmament Agency (US), 261
Arms Export Control Act (US), 330
Arzamas-16 complex at Sarov, 141–42
Aston, Francis, 26–27
atomic bombs
  clean bombs/neutron bombs, 189–90
  detonation devices, 8, 81, 128

explosive lenses, 63, 80, 83, 138, 174
filters, 129–30
gun-type fission model, 62–64
"hedgers, hiders, and sprinters," 362–63
on Hiroshima and Nagasaki, 2, 88–95, 100–107, 117, 128, 169, 173, 184, 265
hybrid "fission-fusion" bombs, 171, 175
implosion-type atomic bombs, 63, 138, 153
"strategic ambiguity" strategy, 253, 255, 312, 348
strategic bombers, 2, 91, 185, 208, 225, 227, 258–60, 286, 304, 308, 318
vulnerability to sabotage, 80–81
See also ballistic missiles; hydrogen bombs; nuclear physics
atomic cannons, 331
atomic club, 1–5, 102–4
  as an Anglo-Saxon club, 102
  new atomic states, 326–27
  renewed nuclear arms race, 358, 360–61
  solidarity among members, 354–55, 365
  See also information sharing
Atomic Research Center, India's Los Alamos, 275
atomic secrets/spies. *See* espionage

Atoms for Peace program, 183, 195, 207, 252, 270–73
atoms for peace used for war, 358, 364
Eisenhower's speech, 162–65, 182–83
in India, 273, 276–77
in Iraq, 349–50
in South Africa, 311, 314
Attlee, Clement, 107–10, 117–27, 130, 151–53, 188
Australia, 153, 362
"axis of evil," 341–42

B61-12 and -13 thermonuclear gravity bombs, 361
Ba'ath Party (Iraq), 342
Baby Teeth Survey, 232–33
Background to Modern Science (Aston), 26–27
Baker, James, 316–18
balance of terror, 6, 260, 298, 327–28
ballistic missiles, 3, 185–86. See also anti-ballistic missile (ABM) systems; Intercontinental Ballistic Missiles (ICBMs); Intermediate-Range Ballistic Missiles (IRBMs); Multiple Independently Targeted Reentry Vehicles (MIRVs); Submarine-Launched Ballistic Missiles (SLBMs)
"ban the bomb" movement, 232
Bangladesh, 281–82
Baotou Nuclear Fuel Component Plant, 215, 236–37
Baruch Plan, 113–16, 122, 126–27, 164, 276
Baruch, Bernard, 113–14
Bay of Pigs invasion, 219–20, 237
Beijing Nuclear Weapons Research Institute, 215
Belarus, 38, 308–9
denuclearization of, 311, 315–18, 323, 325, 336
in the Russo-Ukrainian war, 357, 359, 363
See also Lisbon Protocol
Belgium, 12, 17–18, 21–25, 44, 75, 113, 146–47, 333
Ben-Gurion, David, 243–44, 247–54, 313
Beria, Lavrentii, 97, 100, 101, 131–36, 142, 176, 181–82, 184
Berlin Wall, 219, 305

Bevin, Ernest, 108, 110, 122, 125, 127, 154
Bhabha, Homi Jehangir, 274–76, 278
Bhagavad Gita, 1
Bharatiya Janata Party (India), 328
Bhutto, Zulfikar Ali, 281–82
Biden, Joe, 339, 358, 360
Bikini Atoll, 115, 153, 168–69, 170, 177
biological warfare, 157, 341, 343–44, 346, 347
Black Sea Fleet, 320, 322, 324
Blinken, Antony, 358
blockade (of Cuba), 220, 222–23, 284
bluffing, 115, 186, 228, 329
Boeing aircraft, 88, 266, 293
Bohr, Niels, 19–21, 28, 30, 39, 46, 71, 76–78
bomb design. See atomic bombs; ballistic missiles; hydrogen bombs
Border War, 313, 314
Bormann, Martin, 72
Brazil, 310, 363
Brennan, Donald, 257
Brezhnev, Leonid, 225, 261–68, 285–86, 291, 341
Bridges, Sir Edward, 126–27
brinkmanship, nuclear, 6, 208, 228, 358, 364–65
British Commonwealth, 313
Brzezinski, Zbigniew, 267
Budapest Memorandum of 1994, 323–24, 336, 362
Bulganin, Nikolai, 35, 184
Bulgaria, 103, 104
Bulletin of the Atomic Scientists, 179, 181
Bundy, McGeorge, 236–37
Bunn, Matthew, 3–4
Burgess, Guy, 150
Burns, Bill, 358
Bush, George H. W., 307–8, 310, 316–17, 329
Bush, George W., 340–42, 345–48, 360
Bush, Vannevar, 41–43, 48–51, 52–55, 65–70, 82–83, 110, 134
Bushehr nuclear power station, Iran, 349–50
Byrnes, James, 84–85, 90, 103, 108, 110–12, 125

Cairncross, John, 136
Callaghan, James, 284–85
Cambridge Five, 97, 136–37, 150–51
Canada, 47–48, 66, 68, 108,

110, 120, 147–50, 206, 202, 280, 333
Chalk River National Research Experimental Reactor (NRX), 122, 276
deal with India, 276–77, 280
deal with South Korea, 333
Halifax, Nova Scotia, harbor explosion, 24, 27, 46, 80, 83
response to first Soviet nuclear test, 145
CANDU 6 heavy water reactors, 333
Carnegie Institution for Science, Washington, DC, 41–42
Carter, Jimmy, 266–69, 281–83, 285, 334
Casablanca Conference of 1943, 68
Casey, William, 297
Castle Bravo test of 1954, 2, 168–69, 177–79, 181, 183, 188, 202, 218, 275, 326
Castro, Fidel, 219–21, 226–28, 231
Catholics, 232, 287
Cavendish Laboratory at Cambridge University, 14–15, 77, 274
centrifuge separation in uranium enrichment, 53, 56, 98, 141, 346, 351–53
Chadwick, James, 13, 14–15, 18, 35–36, 44, 47, 66, 70, 80
chain reaction, 15, 20–25, 30–33, 38, 42–45, 55–58, 63, 81, 97, 99, 197
Chalk River National Research Experimental Reactor (NRX), Ottawa, 122, 276
Cheliabinsk-40 (Kyshtym industrial complex), 140–41, 192
chemical weapons, 10–11, 18, 115, 118, 157, 341, 344
Cheney, Dick, 345
Chernenko, Konstantin, 298
Chernobyl (Chornobyl) nuclear power plant, Ukraine, 301–3, 318–19, 322, 324, 363–64
Chiang Ching-kuo, 236
Chiang Kai-shek, 85–86
Chicago Pile-1, 55, 56, 58, 139

INDEX   411

Chicago, University of, 55–56, 58, 81–82, 104, 139
China, 2–4, 85–86, 103, 206–17, 237–40, 361, 362
  Baotou Nuclear Fuel Component Plant, 215, 236–37
  Beijing Nuclear Weapons Research Institute, China's Los Alamos, 215
  after the Cold War, 323, 330
  India and, 273–79, 328
  Jiuquan Atomic Energy Complex, 215
  during the Korean War, 155, 156–57, 166, 206–8
  Lop Nur nuclear test site, 215, 239, 361
  Nixon's opening to, 261, 332
  North Korea and, 332, 334–36, 338
  not afraid of nuclear weapons, 185–86, 211–12, 288
  Sino-Indian war of 1962, 277
  Sino-Soviet relations, 185, 208–9, 213–15, 218–28, 236, 237–39
  South Africa and, 313
  on test ban treaties, 215, 235–38, 326
  UN Security Council seat, 112, 346, 350
  US policy after 1949, 207–8, 216–17, 345
"China Syndrome," 212
Chornovil, Viacheslav, 319
Christian anti-communist crusades, 287–88
Churchill, Winston, 6
  attempt to ban testing via the UN, 188
  on the bomb, 78, 79, 108–9
  as postwar opposition leader, 108–9, 122
  reaction to hydrogen bomb, 179
  return to power, 151–54, 188, 210, 311
  Roosevelt and, 65–69, 75–77, 137, 151–52
  as wartime PM, 65–70, 75–78, 84–86, 311
CIA
  assessments of other nuclear programs, 253, 333–34
  assessments of Soviet nuclear capabilities, 143, 187, 192, 267, 292, 301

Bay of Pigs invasion, 219–20, 237
Iraq Survey Group, 348
operations in China, 236–37, 243
civil defense, 159–60, 178–80
clean bombs/neutron bombs, 189–90
Clinton Engineer Works. See Oak Ridge, Tennessee
Clinton Pile, 58–59, 63
Clinton, Bill, 321–22, 329, 336, 347
"clock of doom," 181
Cockcroft, John, 122, 128–30
"Cockcroft's folly," 130
Cold War, 3–4, 308–9, 360
  détente, 181, 259, 266, 267–68, 284
  nonaligned movement, 275
  nuclear umbrellas of the, 200, 202, 209, 214, 241
  superpower hegemony, 217
  See also nuclear weapons; specific Cold War hotspots
colonialism, 276, 313–14. See also anticolonialism; decolonization
Colorado detonations for natural gas, 271–72
Combined Policy Committee (US-UK), 69–70, 137
Commissariat à l'énergie atomique (France), 198, 201
Committee for a Sane Nuclear Policy (SANE), 189, 235, 286
Commoner, Barry, 189, 232
Commonwealth, 313
communism
  anticommunism in America, 155, 165–67, 287–88, 299
  in Cuba, 220
  in Eastern Europe, 111
  the end of the Cold War and, 330–32
  in France, 198
  leadership of the communist world, 191, 236
  among nuclear scientists, 36–37, 60–62, 136–39, 149–50, 198, 202
  Russian imperial nationalism and its brand of, 133–34
Communist Party of the Soviet Union

Politburo, 212, 263, 295, 300–305
  after Stalin's death, 181–84, 213, 215
Communist Party of the United States, 138
Comprehensive Nuclear Test Ban Treaty (1996), 326, 328, 330
Compton, Arthur H., 48–50, 54–56, 81–82
Conant, James B., 48, 49, 50, 52, 53, 56, 63, 68, 70, 82, 172
Congo (under Belgian rule), 17–18, 21–25, 44, 113, 146–47
conventional explosives, 30, 63, 80, 175
conventional warfare, 5, 158–59, 206, 212, 244, 265, 364
Council of foreign ministers meetings, 103, 108, 110
counterintelligence, 61–62, 78, 148–49
Cousins, Norman, 106–7, 189–90, 232–35
Crimea/Crimean Peninsula, 34, 307, 320–24, 357, 358, 362
Cripps, Stafford, 124
critical mass/criticality, 31–32, 36–37, 45, 47, 49–50, 62–63, 329
cruise missiles, 285–87, 292, 302, 308, 361
Cuba
  Bay of Pigs invasion, 219–20, 237
  the Cuban "mini-crisis," 268
  Khrushchev's "betrayal" of, 227–28
Cuban missile crisis, 4–5, 186, 218–28, 230–32, 235–38, 241, 258, 365
Curie, Irène, 11
Curie, Marie, 11
Cutler, Robert, 178
cyclotrons, 29, 32–33, 35, 43, 54, 57, 60, 62–63, 86–87, 100
Czechoslovakia, 24, 206–7, 325, 332

Daily Express (UK), 250
Dalton, Hugh, 124
Dardanelles, 84
Darwin, Charles Galton, 52, 65–66
Dautry, Raoul, 197
Day After, The (film), 296–96, 303
Dayan, Moshe, 254

de Gaulle, Charles, 192, 197–99, 201–5, 216–17, 244, 247–48
de Klerk, Frederik Willem, 315
decolonization, 84, 200, 199, 220, 247, 274, 281–82, 327
Defense Readiness Condition (DEFCON) levels, 225, 254–55
Defense Technical Accord of 1957, 213
Democratic Party (US), 144, 155, 188–89, 218, 234, 259
Democratic Republic of Germany. *See* East Germany
Denmark, 39, 44, 46, 71
denuclearization, 310–11, 323, 336, 339. *See also* nuclear disarmament
d'Estaing, Valéry Giscard, 285
de-Stalinization, 213, 215
détente, 181, 259, 266, 267–68, 284
deterrence. *See* nuclear deterrence
detonation devices, 8, 81, 128
deuterium (hydrogen-2), 171, 174, 175–77
Deutsche Physik, 29, 32
"Development of Substitute Materials" program. *See* Manhattan Project
Diebner, Kurt, 31
Dimona complex (Negev Nuclear Research Center), 246–53, 276
diplomacy
  back-channel, 261–62
  human rights in, 267, 316
  international recognition, 111, 315–16, 319, 321, 324
  isolation, 35, 332, 338
  two-track nuclear diplomacy, 182, 249, 269
  See also international relations
"dirty bombs," 79, 364
dissidents and defectors, 61–62, 120, 174, 263
Dnipropetrovsk missile factory, 186, 320, 359
Dobrynin, Anatolii, 224, 266, 292, 293
Dodd, Thomas E., 233–34
Dolan, Anthony R., 288
Dolezhal, Nikolai, 140
"doomsday machine," 257
Dostrovsky, Israel, 252
Dr. Strangelove (film), 256–57

"dual key" arrangement, 196, 240
dual-purpose reactors, 302, 335–36, 343
Duck and Cover (cartoon), 160
Dulles, Allen, 249
Dulles, John Foster, 158, 165–66, 202–3, 210
Dunkirk, 32, 34
DuPont Company, 59

East Germany, 206–7, 219, 234
"East River" project, 159–60
Eastern Europe, 25–26, 84–85, 103–4, 111, 207
  after the Cold War, 307, 325, 335
  Warsaw Pact, 206, 230, 285, 287, 296
Egypt
  as part of the United Arab Republic, 250, 251
  Suez crisis, 199–201, 244–46
  Yom Kippur War, 254, 272, 343
Einstein, Albert, 16–18, 22–27, 31–32, 42, 50, 79, 107, 179–80
Eisenhower, Dwight D., 155–67, 180, 185–94, 195, 208, 218, 221
  American nuclear energy under, 271
  Atoms for Peace speech, 162–65, 182–83
  civil defense under, 159–60, 178–80
  First and Second Taiwan Straits crises, 210, 214–15, 236–37
  at the Geneva conference of 1955, 185
  Gaither Report, 187
  on the Israeli nuclear program, 249–50
  Khrushchev and, 188–94
  Korean War, 156–58
  "massive retaliation," 165–67
  military spending and nuclear weapons under, 158–62, 187, 195–96
  NATO allies and, 195–96, 202–4
  his "New Look," 158–59
  nuclear weapons in South Korea, 331–32
  Science Advisory Committee, 187
  See also Atoms for Peace program

ElBaradei, Mohamed Mustafa, 351
electricity. *See* nuclear energy
electromagnetic separation in uranium enrichment, 53, 56–59, 141
Elizabeth of Bavaria, 17–18, 22
embargo, 272, 314, 343
Emergency Committee of Nuclear Scientists, 107
Eniwetok Atoll, 176. *See* "Ivy Mike"
Enola Gay (bomber), 88
Enterprise (battleship), 279
"escalation management," 358–59
Eshkol, Levi, 253
espionage
  Alsos Mission, 63
  Cambridge Five, 97, 136–37, 150–51
  the case of Klaus Fuchs, 71, 137–39, 148–51, 171
  the case of the Rosenbergs, 138, 149, 155
  double agents, 297
  the "nth country problem," 280
  Venona Project, 149
Euratom (European Atomic Energy Community), 199, 247
Euromissile crisis, 305–6
Europe. *See* NATO (North Atlantic Treaty Organization); Warsaw Pact; *specific countries and arrangement*
Evangelical Convention in Orlando, 283, 287–88
"evil empire" speech, 287–88, 306
ExCom (Executive Committee) of the National Security Council, 222
experimental physics, 29, 60, 78
explosive lenses, 63, 80, 83, 138, 174

F-1 uranium-graphite reactor. *See* Chicago Pile-1
F-16 Fighting Falcon, 344–45, 399
F-35 Stealth Strike Fighter, 361
fallout. *See* radioactive fallout
Fast Flux Test Facility at Hanford, 270, 272
"Fat Man," 64, 91, 115, 128
fear, 5–6, 106, 326–39, 354, 365
  the "balance of terror," 6, 260, 298, 327–28

INDEX   413

as diplomatic touchpoint, 185, 191, 219, 243–44, 260
early prophecies of nuclear war, 7–10
in Eisenhower's "New Look," 159–60
as motivation, 211, 215, 217
nuclear disasters, 2, 272, 364
See also terrorism
Federal Civil Defense Administration, 159
Federal Republic of Germany. *See* West Germany
Federation of Atomic Scientists, 189
Fermi, Enrico, 20–21, 23, 55–56, 58–59, 82, 83, 99, 170, 244
Fermi, Laura, 21
Finland, 103, 259, 267
First Taiwan Strait Crisis, 210
fission-fusion bombs, 171, 175
Force de frappe, 204–5
Ford, Gerald, 266, 267–68, 280
Fordo uranium enrichment plant, Iran, 352
Foreign Affairs (journal), 161, 321
Forrestal, James, 85
France, 2–3, 10–13, 195–99, 201–5, 217, 238, 242, 258, 361, 362
  Communist Party of, 198
  de Gaulle's return, 201–5
  end of empire, 199, 205, 247
  fall of Paris, 32, 41, 197
  Force de frappe, 204–5
  Israel's nuclear program and, 244–45, 247–48
  Paris talks of May 1960, 192, 196–99
  Suez crisis, 199–201, 244–46
  UN Security Council seat, 112
  during WWII, 26, 32–34, 68, 75, 85, 103, 112
Franck, James, 81
Frankfurter, Felix, 76–77
"freeze" movement, 260, 283, 285–87, 290
Frisch, Otto R., 19–20, 37, 44–47, 79
Fromm, Friedrich, 73
Fuchs, Klaus, 71, 137–39, 148–51, 171
Fukushima nuclear power plant, 363–64

Fulton, Robert, 26
fusion bombs. *See* hydrogen bombs

Gaddis, John Lewis, 259
Gaillard, Félix, 198, 199, 201
Gaither Report, 187
Gaither, Horace Rowan, 187
Gamow, George, 61–62
Gandhi, Indira, 277–79
Gandhi, Mahatma, 274
gaseous diffusion uranium enrichment, 53, 56–58, 67, 87, 123–26, 141, 147, 215, 353
Geiger counters, 20
Geiger, Hans, 13
Gen-163 Committee, 126–27
Gen-75 Committee ("Atomic Bomb Committee"), 117–18, 122, 126
Geneva conference of 1955, 184–85, 191, 192
Geneva Conference on the Peaceful Uses of Nuclear Energy, 275
Geneva summit of 1985, 299–300, 303
Germany, 2, 10–15, 16–39, 44–49, 52–53, 63–64, 71–75, 78, 96–102
  Ministry of Education, 30–31, 72, 73
  reunification of, 335
  See also East Germany; West Germany
glasnost ("openness"), 303
GLEEP (Graphite Low Energy Experimental Pile at Harwell), 129
Godzilla (film), 179
Goering, Hermann, 72, 74
Gold, Harry, 138, 149
Golos, Jacob, 138
Gorbachev, Mikhail, 298, 299–309, 314, 316–17
Gordievsky, Oleg, 297
Gouzenko, Igor, 120, 149
graphite reactors, 31, 42, 55, 58, 74, 99–100, 122, 129, 139, 199, 302
Grechko, Andrei, 261, 267
Greene, Graham, 300
Greenglass, David, 138, 149, 174
Grignard, Victor, 11
Gromyko, Andrei, 114–15, 242, 262, 295
Groves, Leslie R., 54–63, 78–83, 120
Gryphon ground-launched cruise missiles, 285
Guadeloupe Conference of 1979, 285
Guillemet, Pierre, 199

Gulag system of forced labor, 140, 232
Gulf War of 1990–1991, 346–47
gun-type fission model, 62–64

Haber, Fritz, 10, 12
Hague Convention on Land Warfare (1907), 10
Hahn, Otto, 18–20, 29, 31, 44–45, 72–73, 75
Halifax, Nova Scotia, harbor explosion, 24, 27, 46, 80, 83
Hall, Theodore, 138, 139
Hamas attack on Israel (October 2023), 354
Hanford (Clinton Engineer Works), Washington State, 58–59, 64, 122, 140, 270–72
Hankey, Maurice, 136
Harriman, Averill, 94–96, 234–36
Harrison, Katherine ("Kitty") Puening, 60
Harteck, Paul, 30, 170
Harwell Laboratory (UK), 129, 148–50
heavy water reactors, 31, 38, 99, 122, 198, 333, 353
heavy water supply and trade, 38, 52, 72, 197, 246, 276
Heisenberg, Werner, 15, 28–32, 35–39, 45, 47, 71–73, 75
Helsinki Accords (1975), 259, 267
Herter, Christian, 249–50
Hezbollah, 349
hibakusha (survivors), 169
Himmler, Heinrich, 29, 72
Hinton, Christopher, 128–29
Hirohito of Japan, 91–92
Hiroshima, bombing of, 2, 88–95, 100–107, 117, 169, 184, 265
Hitler, Adolf, 16–17, 71, 73–76, 93
HMS *Plym*, 153–54
Holocaust, 243, 246–47
Honest John atomic rockets, 331
Hopkins, Harry, 68–69
horizontal proliferation, 2, 195, 229, 240, 341
Hornig, Don, 80–81
Houtermans, Friedrich Georg, 36
Hudson Institute, 257
human rights, 267, 316
Humphrey, Hubert H., 234
Hungary, 103–4, 111, 213, 325

"Hurricane," 153
Hussein, Saddam, 342–48
Hyde Park memorandum of 1944, 75, 77, 119
hydrogen bombs, 2–3, 170–77, 189–90
   Castle Bravo test of 1954, 2, 168–69, 177–79, 181, 183, 188, 202, 218, 275, 326
   Chinese development of, 211–12, 239–41
   "Ivy Mike," 176–77, 181
   North Korean development of, 339
   political fallout from, 211–12, 275
   radioactive fallout, 43–44, 169, 177, 179–80, 189–90, 202, 235, 256, 272, 302
   Soviet "Layer Cake" design, 175–79, 182
   Soviet development of, 239
   See also plutonium

Imperial Japanese Army, 32–33, 86
implosion-type atomic bombs, 63, 138, 153
India, 2–4, 68, 216, 241, 250, 339, 362
   Atomic Research Center, India's Los Alamos, 275
   nuclear program, 272–80, 327–30
   Pakistan and, 276, 279, 281–82, 327–28
   Pokhran ("Smiling Buddha") tests, 279, 282, 328–29
   Sino-Indian war of 1962, 277
information sharing
   Anglo-American nuclear research cooperation, 40–42, 48–51, 65–71, 75–78
   leaked information, 66, 67, 77, 120, 145, 187, 239, 267, 278
   McMahon Act of 1946, 120–21, 128, 195–96, 202, 204
   See also espionage; intelligence; peaceful uses of atomic energy
inspections, 109, 112, 113–15, 192–93, 242, 265, 306
   in Cuba, 227–28
   in Iran, 350–51
   in Iraq, 340–41, 344–49, 350, 354
   in Israel, 248, 251–53

   in North Korea, 335–36
   in test ban talks, 189, 230–34
   underground nuclear facilities, 246, 248, 248, 332
   See also espionage; intelligence
Institute for Physical and Chemical Research (RIKEN), 32
intelligence
   counterintelligence, 61–62, 78, 148–49
   leaked information, 66, 67, 77, 120, 145, 187, 239, 267, 278
   overflights, 187–88, 192–94, 226, 265
   satellite surveillance, 265, 312
   U-2 spy planes, 187–88, 192–94, 226, 265
   See also espionage
Intercontinental Ballistic Missiles (ICBMs), 185–87, 220, 258–62, 264, 268, 284–85, 308, 323, 339, 361
Intermediate-Range Ballistic Missiles (IRBMs), 186, 196, 221, 268, 284–87, 306, 326, 359–61
Intermediate-Range Nuclear Forces (INF) Treaty, 286–87, 306, 360
International Atomic Energy Agency (IAEA), 164–65, 242, 250, 276, 281, 314, 335, 337–38, 346, 350–51
International Conference on the Peaceful Uses of Nuclear Energy, Geneva, 275
international control of nuclear weapons, 82, 85, 102–4, 109
   black market, 300, 337
   nuclear-free zones, 281–82, 325
   See also inspections; nuclear disarmament; United Nations
international relations
   adventurism in, 6, 225, 295
   the anarchic world of, 9–10
   balance of power, 216
   "blinking" in, 224
   embargo in, 272, 314, 343
   nuclear brinkmanship, 6, 208, 228, 358, 364–65
   nuclear multipolarity, 365

   preemptive war/strike, 4, 9–10, 25, 158, 257, 344–45, 347, 352, 354–55
   prestige in, 5, 125, 191, 199, 200, 245, 274
   realism and realpolitik, 9–10, 298
   recklessness in, 4, 290, 345
   See also diplomacy; sanctions
Interpretation of Radium (Soddy), 8
Ioffe, Abram, 35–36, 98, 131, 97
Iran, 3–4, 111, 166, 348–54
   in the "axis of evil," 341
   Iranian Revolution of 1979, 344, 350
   Obama's "Iran deal," 353
   Stuxnet virus, 351–53
   war with Iraq, 343–44
Iraq Liberation Act of 1998, 347
Iraq, 4, 341–49, 354
   Gulf War, 346–47
   Iraq War, 348–49, 354
   Osirak reactor, 343–46, 352
   war with Iran, 343–44
   in the "war on terror," 348–49, 354
   weapons of mass destruction in, 340–41, 344–49, 350, 354
Iraq Survey Group (CIA), 348
Israel, 2, 199–200, 201, 243–48, 253–55, 313–14, 350–55, 361
   Dimona complex (Negev Nuclear Research Center), 246–53, 276
   F-16 fighter jet deal with the US, 344–45, 399
   failure to deter, 253, 255
   France and, 244–45, 247–48
   Hamas attack (2023), 354
   Ministry of Defense, 245
   Operation Opera (1981), 344–45
   preemptive strikes and, 344–46
   Six-Day War, 253–54
   Suez crisis, 199–201, 244–46
   Yom Kippur War, 254, 272, 343
Italy, 11–12, 84, 196, 221, 282, 331, 360
Ivanenko, Dmitrii (Dmytro), 15, 35–36
"Ivy Mike," 176–77, 181

INDEX 415

Jackson, C. D., 162
Jackson, Henry, 266
Japan, 2, 32–33, 75–76,
   81–82, 84, 105, 168,
   241, 282
  since the Cold War, 335,
     338, 362
  attack on Pearl Harbor,
     50, 52, 60, 74
  Fukushima nuclear power
     plant, 363–64
  Hiroshima and Nagasaki,
     2, 88–95, 100–107, 117,
     128, 169, 173, 184, 265
  Imperial Japanese Army,
     32–33, 86
  Potsdam Declaration
     and the surrender of,
     86–92, 117
Jewish diaspora, 16–17, 19,
   22–23, 29, 32, 249–50
Jiuquan Atomic Energy
   Complex, 215
John Birch Society, 293
John XXIII (pope), 232
Johnson, Lyndon, 237, 242,
   253, 258–59
Joint Committee on Atomic
   Energy, 170, 189
Joint Comprehensive Plan of
   Action (2015), 352
Joint Congressional Committee on Atomic
   Energy (JCAE), 121
Joliot-Curie, Frédéric, 30, 36,
   44, 197–98
Juche (self-sufficiency) ideology, 334
Jüdische Physik, 29, 32
Jupiter medium-range ballistic missiles, 196, 331

Kahn, Herman, 257
Kaiser Wilhelm Institute for
   Physical Chemistry,
   Berlin, 10, 18–19, 24, 31
Kalckar, Herman M., 232
Kapitsa, Petr, 77, 131–34
Karzai, Hamid, 340
Kashmir, 327–28
Katzir, Ephraim, 255
Kazakh steppe, 142, 177
Kazakhstan, 239, 309, 311,
   317–18, 321–22
Kennan, George, 94–95,
   133–34
Kennedy, Edward, 290
Kennedy, John F., 3, 187, 193–
   94, 218–28, 229–39
  China, 215, 216, 235–39,
     345
  Israel, 250–53
  Khrushchev and, 218–28,
     229–36, 261
  as a senator, 187, 216

Kennedy, Robert, 222, 224,
   233
KGB, 286, 292, 297
Kharkiv memo, 36–37
Khlopin, Vitalii, 35, 37
Khomeini, Ayatollah,
   343–44, 350
Khrushchev, Nikita, 183–94,
   196, 204
  de-Stalinization, 213, 215
  Kennedy and, 218–28,
     229–36, 261
  Mao and, 185, 213–15,
     218–28, 236, 237–39
  ouster of, 261
  "peaceful coexistence,"
     182–84, 213
  Sputnik, 185–87, 202,
     213, 220
  U-2 incident, 192–94
  visit to the US, 190–91
Khrushchev, Sergei, 186
Khuzestan, 343–44
Kim Il-sung, 332, 334–37
Kim Jong-il, 336–37, 338
Kim Jong-un, 339
Kishinev pogrom of 1903, 23
Kissinger, Henry, 257,
   261–66, 267, 333
Kistiakowsky, George B., 49,
   63, 80, 83, 138, 174
Koniukhov, Stanislav, 320
Korea. *See* North Korea;
   South Korea
Korean Air Lines (KAL)
   flight 007, 292–95
Korean War, 152, 155–57,
   208–10, 331, 334, 336
Kosygin, Aleksei, 261
Kozlov, Frol, 225
Kozyrev, Andrei, 317–18
Kravchuk, Leonid, 316–17,
   319–23
Kuchma, Leonid, 322–23
Kurchatov, Igor, 36, 94,
   98–102, 131, 133–36,
   139, 142, 175–76, 178,
   183
Kuriles, 84–85
Kuwait, 345–46, 347
Kyshtym industrial complex,
   140–41, 192

Labour Party (UK), 108,
   124, 130, 145, 151–52,
   154, 188
Lange, Fritz, 36–37
Lanzhou Gaseous Diffusion
   Plant, 215
Latin America, 220
Laurence, William, 34–35,
   105
Lawrence Livermore
   National Laboratory,
   176, 289–90

Lawrence, Ernest O., 43,
   48–49, 54–60, 82,
   172, 173
Layer Cake design, 175–79,
   182
League of Nations, 10
Lehman Brothers, 23
Leipunsky, Aleksandr, 98
Lenard, Phillip, 12
Lend-Lease, 40, 124
light-water reactors, 336,
   343, 344
Lilienthal, David, 112
"limited deterrence," 312
Lindemann, Frederick, 77,
   152
liquid-drop model of the
   nucleus, 19–20
Lisbon Protocol, 317–18,
   321–22
lithium deuteride/deuterium, 175–77
lithium-6, 176
"Little Boy," 64, 88, 91
Lockheed, 187, 192
"Long Telegram," 133–34
Lop Nur nuclear test site,
   215, 239, 361
Los Alamos, New Mexico,
   59–64
  British scientists at,
     70–71, 78–79, 128–29
  building the hydrogen
     bomb, 170–74
  Soviet spies at, 137–39,
     148–51, 171, 178
Los Angeles Times, 144
Lucky Dragon, 168–69, 177,
   179
Luftwaffe, 47
Lukashenka, Aliaksandr, 359

MacArthur, Douglas, 207–8
Mackenzie King, William
   Lyon, 108–9, 119, 145
Maclean, Donald, 137, 150
Macmillan, Harold, 184,
   192, 234
Maiak (Lighthouse). *See*
   Kyshtym industrial
   complex
Malenkov, Georgii, 182–84
Malinovsky, Rodion, 223,
   228
Mandela, Nelson, 313, 315
Manhattan Project, 52–64
  the Alsos Mission, 63
  atomic bombs on Hiroshima and Nagasaki,
     2, 88–95, 100–107, 117,
     128, 169, 173, 184, 265
  Chalk River National
     Research Experimental
     Reactor (NRX), 122,
     276

Manhattan Project (*continued*)
  Chicago Pile-1, 55, 56, 58, 139
  counterintelligence, 61–62
  enter Oppenheimer, 59–62
  "Fat Man," 64, 91, 115, 128
  the first reactors, 53–56
  going thermonuclear, 170–77
  Hanford, Washington State, 58–59, 64, 122, 140, 270–72
  "Little Boy," 64, 88, 91
  Military Policy Committee, 68
  Oak Ridge, Tennessee, 56–59, 63–64, 67, 83, 88, 129, 141, 302
  the plutonium project, 56–59
  technical problems in the, 62–64
  "Trinity" test, Alamogordo, 80–83, 84, 88, 91, 137, 142, 265
  "Tube Alloys" project, 77, 137
  University of Chicago Metallurgical Lab, 55, 81–82, 104
  See also Los Alamos, New Mexico
Mao Zedong, 185, 206–17
  Khrushchev and, 185, 213–15, 218–28, 236, 237–39
  North Korea and, 332
  not afraid of nuclear weapons, 185–86, 211–12, 288
  Sino-Soviet split and, 220–21, 231, 236, 239–40
Marcoule nuclear site, France, 199
Markey, Edward, 283
Marsden, Ernest, 13
Maslov, Viktor, 36–37, 96
"massive retaliation," 165–67
MAUD Committee, 46–47, 50, 65, 80, 97, 122
McCone, John, 248–49
McDonald, Larry, 293
McMahon Act of 1946, 120–21, 128, 195–96, 202, 204
McMahon, Brian, 121, 172
McNamara, Robert, 222, 257, 258
Mearsheimer, John, 321, 323
Medvedev, Dmitry, 358, 360
Meir, Golda, 254

Meitner, Lise, 18–20, 44
meltdowns, 2, 272, 364
Merkulov, Vsevolod, 131
MI5 (British counterintelligence), 148–49
Mikoyan, Anastas, 222–23, 227
militarization, 182, 308, 334, 334
Ministry of Aircraft Production (UK), 47
Ministry of Defense (Israel), 245
Ministry of Defense (UK), 146, 201
Ministry of Education (Germany), 30–31, 72, 73
Ministry of Supply (UK), 123, 129, 151
Ministry of the Interior (USSR), 101, 141
Minuteman III missiles, 220, 361
"missile gap," 187–88, 220, 286
"Modern Man Is Obsolete" (Cousins), 106, 189
Mollet, Guy, 199, 201
Molotov, Viacheslav, 90, 94–98, 103–4, 108, 110, 115, 133, 144, 183
Molotov-Ribbentrop Pact, 25–26, 36, 38
Montebello Islands, near Australia, 153
moratorium on testing, 39, 188, 192, 229–30, 275, 295, 322
Moscow Conference of 1944, 110–11
Moseley, Henry, 13
Multilateral Force (MLF), 241
Multiple Independently Targeted Reentry Vehicles (MIRVs), 266, 268, 284, 310, 320
"mustard gas," 11
mutual assured destruction (MAD), 256–60, 264–66, 269, 270, 289–90, 295, 298

Nagasaki, bombing of, 2, 91, 94, 100–107, 117, 128, 169, 173, 184, 265
Namibia, 313, 314
Narang, Vipin, 361, 362
Nassau Point, Long Island, 16–18, 22, 23
Nasser, Gamal Abdel, 200, 245, 343
Natanz nuclear facility, Iran, 351–53
National Association of Evangelicals, 283, 287

National Defense Research Committee, 41–42, 48, 63
National Security Council (NSC), 157, 158, 161, 161, 222, 249, 331
national sovereignty, 180, 307, 318, 323, 362
nationalism, 86, 120, 133, 210, 263, 328, 342
nationalization of industries, 200, 219–20, 245
NATO, 202–4, 283–87
  "Able Archer" military exercise, 297
  deployment of US nuclear weapons in, 162, 196, 240–41, 283–87, 361–62
  "dual key" arrangement, 196, 240
  expansion, 325
  F-35 Stealth Strike Fighter, 361
  France's withdrawal from, 217
  Multilateral Force (MLF), 241
  Warsaw Pact and, 230, 285
natural gas, 271–72, 284
Nature (journal), 15, 44
Nazarbayev, Nursultan, 317
Negev Nuclear Research Center (Dimona complex), 246–53, 276
Nehru, Jawaharlal, 188, 213–14, 274–77, 328
neptunium-239, 58
Netanyahu, Benjamin, 353
Neutrality Act of 1939, 40
neutron-proton model of the atom, 35–36
Nevada Proving Grounds, 179
"New Look," 158–59
New Mexico, 1, 61, 80
New START Treaty of 2010, 360
New York Times, 14, 29, 34, 105, 166, 235, 250, 253, 322
Nichols, Dennis D., 54
Nishina, Yoshio, 32–33, 87
Nixon, Richard
  back-channel diplomacy of, 261–62
  election of 1960, 187, 193, 216
  impeachment of, 266, 347
  opening to China, 261, 332
  Plowshares program, 271–72, 312, 314

INDEX   417

as president, 257, 259–66
  SALT I talks, 259–66
as vice president, 216, 249
nonaggression pacts, 230–31, 234–35, 264
nonaligned movement, 275
Non-Proliferation Act (US), 281
Non-Proliferation Treaty of 1968, 4, 242, 259, 273, 278–82
  later signatories to the, 315, 317, 319, 321–25, 333, 335, 338, 349, 350
  non-nuclear states signing the, 322, 333
  See also Lisbon Protocol
nonproliferation
  failure of, 240–42
  "freeze" movement, 260, 283, 285–87, 290
  non-nuclear states, 5, 240–42, 273–74, 280, 317–19, 321–24, 328–30, 335, 354, 358–59, 362, 364
  See also Lisbon Protocol; nuclear disarmament
North Korea, 3–4, 207–8, 326–27, 330–39, 351–52, 360, 362–63
Norway, 38, 52, 72, 197, 246
Norway, Nevil S. ("Nevil Shute"), 188
NSC 162/2, 158
nuclear age, 1–6, 25, 93, 213, 233, 311, 354–55, 357–58, 364–66
nuclear arms race, 1–6, 358, 360–61
  "missile gap," 187–88, 220, 286
  parity, 186, 258, 284, 287, 291, 305, 308
  See also international control of nuclear weapons; nuclear disarmament; testing
nuclear deterrence, 4–5, 27, 46
  American deterrence, 27, 158–59, 162–63, 165, 173, 331
  blackmail and, 6, 69–70, 219, 228, 254, 291, 359–60, 365
  brinkmanship and, 6, 208, 228, 358, 364–65
  British deterrence, 108–9, 118, 121, 147–48, 152
  China's lack of fear, 185–86, 211–12, 288
  Chinese deterrence, 215
  Israel's failure to deter, 253, 255

"nuclear sword of Damocles," 218–19
  in regional contexts, 277–79, 312, 320–24, 337, 348
  security umbrellas, 200, 202, 209, 214, 241
  Soviet deterrence, 221, 239, 305–6, 308–9
  threat of a "nuclear holocaust," 2, 243, 246–47, 252
  See also mutual assured destruction (MAD)
nuclear disarmament
  of Belarus, 311, 315–18, 323, 325, 336
  of South Africa, 311–16
  of Ukraine, 308–9, 311, 315–25
  See also Anti-Ballistic Missile (ABM) Treaty; Budapest Memorandum of 1994; inspections; Lisbon Protocol; New START Treaty of 2010; Non-Proliferation Treaty of 1968; Strategic Arms Limitation Talks (SALT) I and II; Strategic Arms Reduction Treaty (START) I and II
nuclear energy, 109–14, 119–20, 161–64, 270–76.
  See also nuclear reactor design
Nuclear Folly (Plokhy), 5
nuclear physics, 14–20
  critical mass/criticality, 31–32, 36–37, 45, 47, 49–50, 62–63, 329
  cyclotron development, 29, 32–33, 35, 43, 54, 57, 60, 62–63, 86–87, 100
  meltdowns, 2, 272, 364
  neutron-proton model of the atom, 35–36
  patents, 15, 21, 36
  quantum mechanics and, 19, 28
  science of nuclear chain reaction, 15, 20–25, 30–33, 38, 42–45, 55–58, 63, 81, 97, 99, 197
  theoretical vs. experimental physics, 29, 60, 73, 78, 96
  Washington Conference on Theoretical Physics of 1939, 20
  weaponization of nuclear research, 30–31
  See also uranium enrichment

nuclear reactor design
  channel–type reactors, 302
  dual-purpose reactors, 302, 335–36, 343
  graphite reactors, 31, 42, 55, 58, 74, 99–100, 122, 129, 139, 199, 302
  heavy water reactors, 31, 38, 99, 122, 198, 333, 353
  heavy water supply and trade, 38, 52, 72, 197, 246, 276
  light-water reactors, 336, 343, 344
  meltdowns, 272, 364
  militarization, 182, 308, 334, 334
  "Wigner growth," 129–30
Nuclear Suppliers Group, 280, 282
nuclear weapons
  clean bombs/neutron bombs, 189–90
  cruise missiles, 285–87, 292, 302, 308, 361
  "dirty bombs," 79, 364
  multiplicity of roles played by, 324
  as "offensive weapons," 153, 222, 227, 290, 291, 303–4
  strategic bombers, 2, 91, 185, 208, 225, 227, 258–60, 286, 304, 308, 318
  strategies for states to acquire, 362–63
  tactical, 196, 208, 223, 225, 227, 319–21, 358–59
  targeting, 37, 45–46, 75, 87–92, 157, 208, 225, 258–59
  See also atomic bombs; hydrogen bombs
nuclear-free zones, 281–82, 325

Oak Ridge, Tennessee, 56–59, 63–64, 67, 83, 88, 129, 141, 302
Obama, Barack, 353, 360
Office of Scientific Research and Development (US), 41, 48
oil crisis of the 1970s, 272, 343
Okinawa, US base on, 208
Oliphant, Mark, 44–46, 48, 50, 66, 170
On the Beach (novel), 188
"open skies" policy, 188, 193
Operation Alert, 180

Operation Candor, 159, 161, 163
Operation Desert Storm, 345–46
Operation Opera (1981), 344–45
Operation RYaN, 292, 294, 297
Operation Tooth Club, 233
Oppenheimer, J. Robert, 1, 59–64, 76, 78, 80–83, 85, 112, 160–61, 171–73, 175
Oreshnik missiles in Belarus, 359
Orsagen, Lars, 30
Osirak reactor, Iraq, 343–46, 352
outer space
  anti-ballistic missile (ABM) systems in, 288–90, 291, 294, 299, 305
  satellite surveillance, 265, 312
  since the end of the Cold War, 360–61
  Soviet space program, 185–87, 202, 213, 220
  testing in, 191, 229, 231, 233, 234
  weaponization of, 299
overflights, 187–88, 192–94, 226, 265
Ozersk, Russia. *See* Kyshtym industrial complex

Pahlavi, Mohammad Reza, 343–44, 349–50
Pakistan, 3, 4, 216, 281–82, 326–30, 339, 362
  Afghanistan and, 329
  American military presence in, 192, 226
  attack on US embassy in Islamabad, 282
  India and, 276, 279, 281–82, 327–28
  nuclear black market, 300, 337
Paris talks of 1960, 192–93, 196–97, 215, 229, 263
Park Chung-hee, 332–34
Partial Test Ban Treaty of 1963, 191, 216–17, 233, 235, 237, 279, 330
patents, 15, 21, 36
Pavlov, Vladimir, 85, 93
"peaceful coexistence," 182–84, 213
peaceful uses of atomic energy
  for infrastructure and industry, 270–72

nuclear energy, 109–14, 119–20, 161–64, 270–76
"peaceful nuclear explosives," 277, 279–81, 312
Plowshares program, 271–72, 312, 314
See also Atoms for Peace program; information sharing
Pearl Harbor, 50, 52, 60, 74
Peierls, Rudolf, 45–47, 79
Penney, W. G., 128–29
Peres, Shimon, 245–47, 248, 249
perestroika ("restructuring"), 305, 307
Perestroika: New Thinking for Our Country and the World (Gorbachev), 305
Pershing II ballistic missiles, 268, 283–87, 292, 305
Petrov, Stanislav, 294–95
Philby, Kim, 150
physics. *See* nuclear physics
Pipes, Richard, 267
Pivdenmash missile factory, Ukraine, 323
Pliev, Issa, 223, 226
Plowshares program, 271–72, 312, 314
plutonium, 32, 43, 55, 58–59, 62–64, 69–72, 91, 99
  dual-purpose reactors, 302, 335–36, 343
  isotopes of, 43, 58, 63, 99
  producing, 62–63, 71, 122–24, 147
  See also hydrogen bombs
Podgorny, Nikolai, 261, 263
"poison gas," 10–11, 18, 115, 118, 157
Pokhran ("Smiling Buddha") tests, 279, 282, 328–29
Poland, 25–26, 30–31, 33, 116, 126, 325
Polaris missiles, 241
Pontecorvo, Bruno, 149–50, 151
pool-type reactors, 249
Portal, Charles, 123–24, 126–27, 128
Potsdam Conference of 1945, 83–86, 93, 103
Potsdam Declaration, 86–92, 117
Powers, Gary, 192–93
Pravda, 291, 295
preemptive war/strike, 4, 9–10, 25, 158, 257, 344–45, 347, 352, 354–55
proliferation, 2, 4, 9, 195, 229, 240, 253–56, 341
  See also information sharing; peaceful uses of atomic energy

Putin, Vladimir, 357–59

Quebec Agreement of 1943, 69–70, 119, 121, 148, 151–52
Quemoy (Jinmen) and Matsu (Matzu) Islands, 210

R-12 and R-14 missiles, 186, 221, 225–26, 284, 286
Radford, Arthur W., 158
radioactivity, 8, 11–12, 18, 58
  Baby Teeth Survey, 232–33
  clean bombs/neutron bombs, 189–90
  filters, 129–30
  radiation poisoning, 89, 177–78
  radioactive fallout, 43–44, 169, 177, 179–80, 189–90, 202, 235, 256, 272, 302
  radon, 11
  thorium, 8, 13, 113
  See also plutonium; uranium
Radium Institute, 35, 37, 98
Rapoport, Yakov, 140
Ray, Maud, 46
RBMK channel–type reactors, 302
Reagan, Ronald, 296–309, 345, 360
  "evil empire" speech, 287–88, 306
  Gorbachev and, 299–309
  "Star Wars" speech, 288–90, 291, 298–99, 306
  Strategic Defense Initiative (SDI), 288–92, 295, 297, 299–300, 304–6, 308, 342
realism and realpolitik, 9–10, 298
recklessness, 4, 290, 345
Republican Party (US), 144, 187, 218, 257, 336
Reston, James, 165–66
Revolutionary Guards (Iran), 352
Reykjavik summit of 1986, 303–5
Rhee, Syngman, 331
Ribbentrop, Joachim von, 25, 36, 38
Roberts, Owen J., 107
Roentgen, Wilhelm, 12, 35
Rogers, William P., 264
"rogue" states, 342
Romania, 103, 111, 207
Roosevelt, Franklin D.
  as assistant secretary of the US Navy, 24, 41
  campaigning in 1933, 23

Churchill and, 65–69, 75–77, 137, 151–52
during WWII, 40–41, 50, 53, 65–66, 134–35
Interim Committee, 81–82, 85, 88
Operation Candor, 161
and the postwar world, 85
Szilard-Einstein letter, 16–24, 26–27, 31–32, 79
See also Manhattan Project
Rosenberg, Julius and Ethel, 138, 149, 155
Rotblat, Joseph, 77–78
Rothman, David, 16
Royal Air Force, 126
Royal Navy, 153–54
RSD-10 Pioneer ballistic missile, 284
Rukh independence movement, 319
Rumelin, Gustav, 13
Rusk, Dean, 224
Russell, Bertrand, 180
Russell-Einstein manifesto, 179–80
Russia. *See* Russian Federation; Soviet Union
Russian Federation, 307
denuclearization of post-Soviet republics, 308–9, 311, 315–25
"deterrence forces," 358
in international nonproliferation, 346, 350, 352, 353
North Korea and, 328, 338–39
withdrawal from the START II, 342
Russo-Ukrainian war, 3–5, 357–61
annexation of Crimea, 320–24, 357, 358, 362
Belarus in the, 357, 359, 363
Black Sea Fleet, 320, 322, 324
Chernobyl redux, 363–64
Rutherford, Ernest, 8, 12–15, 18, 20, 170
RYKEN (Institute for Physical and Chemical Research), 32

Sachs, Alexander, 23–24, 26–27
Safari-1 reactor, South Africa, 311–12
Sakhalin, 94
Sakharov, Andrei, 175–77, 178, 179, 182

SALT I and II. *See* Strategic Arms Limitation Talks (SALT)
sanctions
on India, 330, 339
on Iran, 350–53
on Iraq, 346
on North Korea, 336–39
on Pakistan, 327–30, 339
on South Africa, 313–15
by the UN, 114, 200, 313–15
by the US, 4, 282, 313–15, 327–30, 336–39, 346, 350–53
on the USSR, 282
Sarov, the Soviet Los Alamos, 141, 176
satellite surveillance, 265, 312
satellites. *See* outer space
Saturday Review of Literature, 106, 189
Schmidt, Heinrich, 13
Schmidt, Helmut, 284–85
scientific community, 10–14, 16–18, 22–33
communism among, 36–37, 60–62, 136–39, 149–50, 198, 202
feeling of responsibility to society, 81
role scientists should play, 132
Russell-Einstein manifesto, 179–80
Szilard-Einstein letter, 16–24, 26–27, 31–32, 79
See also information sharing; nuclear physics; prominent nuclear scientists
Scribner, Fred, 249
Seaborg, Glenn T., 43, 55, 231–32
Second Taiwan Strait Crisis, 214–15
Semipalatinsk test site, Kazakh steppe, 142, 177
Sentinel ICBM program, 361
September 11, 2001, attacks, 330, 340, 348
Serber, Robert, 62
Sèvres Conference of 1956, 245
"Shakti-98" tests, 329
Sharpeville massacre of 1961, 313
Shastri, Lal Bahadur, 277
Shatt al-Arab, Iraq, 343–44
Shcherbytsky, Volodymyr, 263
Shelest, Petro, 263
Shevardnadze, Eduard, 300

Shiota, Kikue, 89
Shpinel, Vladimir, 36–37
Shultz, George, 295
Shushkevich, Stanislav, 317
Shute, Nevil, 188
Sino-Indian war of 1962, 277
Sino-Soviet relations, 185, 208–9, 213–15, 218–28, 236, 237–39
Six-Day War of 1967, 253–54
Slipy, Yosyf, 232
"Smiling Buddha" tests, India, 279, 282, 328–29
Smith, Gerard D., 261, 264
Smuts, Jan C., 311
Soddy, Frederick, 8, 13, 14, 47
solid-fuel missile development, 220, 339
Somervell, Brehon S., 54
Sordin, Moshe, 244
Soustelle, Jacques, 247
South Africa, 311–16, 325, 362
South Korea, 3, 155–57, 207, 292–95, 331–39
Soviet nuclear program, 33–38, 93–104, 131–33, 139–45
Academy of Sciences, 34–35, 96, 98, 135
Arzamas-16 complex at Sarov, 141–42
British reaction to Soviet nuclear program, 136–37, 145–48
CIA assessments of, 143, 187, 192, 267, 292, 301
going thermonuclear, 174–77
after Hiroshima and Nagasaki, 100–102
Kharkiv memo, 36–37
Kyshtym industrial complex, 140–41, 192
Layer Cake design, 175–79, 182
Novouralsk, the Soviet Oak Ridge, 141
nuclear deterrence, 221, 239, 305–6, 308–9
reaction to the Castle Bravo test, 183
under the security police, 131–33
security umbrella, 209, 214
Semipalatinsk test site, 142
spies at Los Alamos, 137–39, 148–51, 171, 178
Truman's announcement of Soviet atomic bomb test, 143–45

420   INDEX

Soviet Union
  anti-alcohol campaign, 301
  collapse of the, 296–309, 310–11, 316–18
  coups in the, 182, 228, 307, 316
  Defense Commissariat, 36–37
  dissidents and defectors from the, 61–62, 120, 174, 263
  economy, 291–92, 300–302
  Gulag system of forced labor, 140, 232
  invasion of Afghanistan, 268, 282, 285, 299, 328
  its postwar spheres of influence, 84–85, 96, 325
  KGB, 286, 292, 297
  military intelligence (GRU), 150, 292
  Ministry of the Interior, 101, 141
  Red Army, 71, 75, 90–91, 97
  Sino-Soviet relations, 185–86, 191, 207, 208–16, 239–41
  space program, 185–87, 202, 213, 220
  Special Committee, 101, 131–33
  after Stalin's death, 181–83, 213, 215
  State Defense Council, 97–98, 101
  UN Security Council seat, 112
  the White Sea canal, 140
  during WWII, 33–34, 94
Spain, 282
Spanish Civil War, 60
Speer, Albert, 72–74, 93
spheres of influence, 84–85, 96, 325
spies. *See* espionage
Sputnik (satellite), 185–87, 202, 213, 220
SS *Mont-Blanc*, 24, 27, 46, 80, 83
Stalin, Joseph, 25, 31, 35–36, 84–88, 93–98, 100–103, 131–42
  death of, 181–83, 213, 215
  at the end of WWII, 84–88, 93–96, 103
  Great Terror, 36
  Mao and, 208–9
  Soviet economy under, 33, 259
Star Wars (film), 288

"Star Wars" speech (Reagan), 288–90, 291, 298–99, 306. *See also* Strategic Defense Initiative (SDI)
START I and II. *See* Strategic Arms Reduction Treaty (START)
State Defense Council (USSR), 97–98, 101
Stevenson, Adlai, 155, 188–89
Stimson, Henry L., 50, 67–69, 87, 88, 90, 102–4
Stockholm Peace Appeal (1950), 209
Strassmann, Fritz, 18–19, 44
Strategic Air Command (SAC), 225
"strategic ambiguity," 253, 255, 312, 348
Strategic Arms Limitation Talks (SALT) I and II, 258–68, 282–85, 291
Strategic Arms Reduction Treaty (START) I and II, 307–9, 310, 316–19, 321, 342, 360
strategic bombers, 2, 91, 185, 208, 225, 227, 258–60, 286, 304, 308, 318
Strategic Defense Initiative (SDI), 288–92, 295, 297, 299–300, 304–6, 308, 342
"strategic parity," 284, 305
Strauss, Lewis, 169, 171–73, 190
strontium-90, 232–33
Stuxnet virus, 351–53
Submarine-Launched Ballistic Missiles (SLBMs), 259–62, 266
Sudoplatov, Pavel, 62
Suez crisis of 1956, 199–201, 244–46
Sukhoi S-15 supersonic interceptor, 293
Sundstrom, Carl-Johan (Cay), 211
Suzuki, Kantaro, 87
Suzuki, Tatsusaburo, 33
Sweden, 11–12, 71, 240–41, 349, 363
Switzerland, 75, 363
Syria, 250, 251, 254, 343, 344, 349
Szilard, Leo, 15, 17–27, 30–32, 42–43, 50, 55–56, 61–62, 81–83, 129, 197

tactical nuclear weapons, 196, 208, 223, 225, 227, 319–21, 358–59

Taiwan, 210, 214–15, 236–37
Talbott, Strobe, 308, 321–22
Taliban, 340
Tammuz reactor (Osirak reactor), Iraq, 343–46, 352
TASS (State Information Agency), 144
Tata Institute for Fundamental Research, 274
Teller, Edward, 21, 23, 42, 82, 129, 170–78, 181, 190, 289–90
Terletsky, Yakov, 95
terrorism, 4, 341, 347–49, 351, 353
  September 11, 2001, attacks, 330, 340, 348
  "war on terror," 340–42, 348
testing
  Comprehensive Nuclear Test Ban Treaty (1996), 326, 328, 330
  moratorium on, 39, 188, 192, 229–30, 275, 295, 322
  in outer space, 191, 229, 231, 233, 234
  Partial Test Ban Treaty of 1963, 191, 216–17, 233, 235, 237, 279, 330
  underground, 191, 230–34, 265, 271, 278, 326, 328, 330, 337
  underwater, 231, 233, 234
Thatcher, Margaret, 297
theoretical physics, 29, 60, 73, 96
thermal diffusion uranium enrichment, 56–59, 87, 98, 141
thermonuclear bombs. *See* hydrogen bombs
Thompson, Llewellyn, 242
Thomson, George Paget, 33, 44, 46
Thor intermediate-range ballistic missiles, 196
thorium, 8, 13, 113
Three Mile Island, 2, 272
Tibbets, Paul, Jr., 88
Time (magazine), 308
Tizard, Henry, 146
"To the Civilized World" (an appeal), 12
Tombaugh, Clyde W., 43
Tongxian Uranium Mining and Hydrometallurgy Institute, 214–15
Trident nuclear-armed submarines, 361
"Trinity," Alamogordo, 80–83, 84, 88, 91, 137, 142, 265
tritium, 171, 174, 175

Truman, Harry S., 81, 83–94, 101–3, 108–15, 120–21, 152–53, 164–66, 170–74
 announcement of Soviet atomic bomb test, 143–45
 civil defense under, 159
 the Korean War, 156–59, 173, 207–8
 Potsdam Declaration, 86–92, 117
Trump, Donald, 339, 353–54, 360
TU-114 turboprop, 191
"Tube Alloys" project, 77, 137
Turkey, 196, 221, 226–27, 228, 231, 331

U-2 incident, 192–94
U-2 spy planes, 187–88, 192–94, 226, 265
Ukraine, 3–5, 38, 263, 301, 307–9, 310–11, 315–18
 Catholicism in, 232, 233
 Chernobyl nuclear power plant, 301–3, 318–19, 322, 324, 363–64
 denuclearization of, 308–9, 311, 315–25
 Dnipropetrovsk missile factory, 186
 independent government of 1918, 49
 nuclear science in, 34–37, 49, 96–98
 Zaporizhia nuclear power plant, 363–64
 See also Lisbon Protocol; Russo-Ukrainian war
Ulam, Stanislav, 174, 176–78
underground testing, 191, 230–34, 265, 271, 278, 326, 328, 330, 337
underwater testing, 231, 233, 234
United Arab Republic, 250, 251
United Kingdom, 43–48, 117–30
 Anglo-American nuclear research cooperation, 40–42, 48–51, 65–71, 75–78
 Brexit, 361
 the case of Klaus Fuchs, 71, 137–39, 148–51, 171
 Churchill's return, 151–54
 Combined Policy Committee, 69–70, 137
 Gen-163 Committee, 126–27
 Gen-75 Committee ("Atomic Bomb Committee"), 117–18, 122, 126

"Global Britain" program, 361
"Hurricane," 153
independence of Scotland, 361
international control, 107–12
MAUD Committee, 46–47, 50, 65, 80, 97, 122
Ministry of Aircraft Production, 47
Ministry of Defense, 146, 201
Ministry of Supply, 123, 129, 151
nuclear deterrence, 108–9, 118, 121, 147–48, 152
reaction to Soviet nuclear program, 136–37, 145–48
Suez crisis, 199–201, 244–46
Thor intermediate-range ballistic missiles, 196
"Tube Alloys" project, 77, 137
UN Security Council seat, 112
view of the American nuclear monopoly, 109, 127
Windscale reactors, 129–30, 151
United Nations (UN), 105–16, 122, 162, 329, 332, 346
 Acheson-Lilienthal plan, 112–14, 164
 Atomic Development Authority (proposed), 112, 114
 Atomic Energy Commission, 110–16, 126
 Baruch Plan, 113–16, 122, 126–27, 164, 276
 Disarmament Commission, 188
 General Assembly, 107–8, 110–12, 133, 162–63, 216, 240, 326
 International Atomic Energy Agency (IAEA), 164–65, 242, 250, 276, 281, 314, 335, 337–38, 346, 350–51
 Korean War, 152, 155–57, 208–10, 331, 334, 336
 Security Council, 110–13, 114, 338, 347, 349–52
United States, 2–4, 41–43, 52–64, 98, 109

America's nuclear monopoly, 109, 113–14, 120, 138–39, 144–47, 163–64, 178, 182, 238
Anglo-American nuclear research cooperation, 40–42, 48–51, 65–71, 75–78
civil defense, 159–60, 178–80
on Communist China, 207–8, 216–17, 261, 332, 345
counterintelligence, 61–62, 78
invasion of Afghanistan, 330, 340
nuclear deterrence, 27, 158–59, 162–63, 165, 173, 331
reaction to Soviet nuclear program, 143–45
UN Security Council seat, 112
See also Manhattan Project; US Congress; *specific federal agencies and departments*
United States Forces Korea, 331
University of California, Berkeley, 43, 48, 54–55, 60
University of Chicago, 55–56, 58, 81–82, 104, 139
Uranium Club, 30–31
uranium enrichment, 29, 46, 53, 56–58
 by centrifugal separation, 53, 56, 98, 141, 346, 351–53
 in China, 215
 by electromagnetic separation, 53, 56–59, 141
 by gaseous diffusion, 53, 56–58, 67, 87, 123–26, 141, 147, 215, 353
 in Iran, 350–54
 in Iraq, 345–46
 in Japan, 87
 in North Korea, 335
 in South Africa, 311–12, 314–15
 by thermal diffusion, 56–59, 87, 98, 141
 in the UK, 66–67, 69, 123–24, 129
 in the USSR, 132–33, 141
uranium hexafluoride, 57, 141
uranium oxide (yellowcake), 55, 248

uranium
    from the Belgian Congo, 17–18, 21–25, 44, 113, 146–47
    isotopes of, 31–32, 34, 44–47, 50, 53, 56–59, 64, 67, 72, 88, 99, 151, 175, 353
    naturally occurring, 44, 45, 56, 199
    sources of ore, 33, 56–57, 68, 214
    See also plutonium
Uranverein. See German nuclear program
Uri, Harold, 57, 67, 141
US Air Force, 80, 156, 225
US Army, 53–54, 58, 61–62
US Atomic Energy Commission (AEC), 169, 171–74, 179, 190, 231–32, 251–52
US Congress, 40–41, 109, 120–21, 144, 146, 202–4, 230–35, 266–68, 281, 347
    congressional sanctions on South Africa, 315
    Joint Committee on Atomic Energy, 170, 172, 189
    Joint Congressional Committee on Atomic Energy (JCAE), 121
    McMahon Act of 1946, 120–21, 128, 195–96, 202, 204
    in the Senate, 121, 187, 216, 233–34, 266–67, 282, 290, 361
US Department of State, 144, 182, 193, 252
US Department of War, 56
US Navy, 40, 115, 224, 262
USS *Augusta* (cruiser), 89–90
USSR. See Soviet Union
Ustinov, Dmitrii, 291, 295

Vajpayee, Atal Bihari, 328
Vannikov, Boris, 100–101, 132
Venona Project, 149
Vernadsky, George, 34–35
Vernadsky, Vladimir (Volodymyr), 34–35, 98
vertical proliferation, 2, 195, 240, 341

Vienna summit of 1979, 268
Vienna talks of 1961, 219, 261
Vietnam War, 259, 260–64, 284
Vietnam, 330
Vladivostok summit of 1974, 266–68
von Neumann, John, 257
von Weizsäcker, Carl-Friedrich, 32, 43
Votintsev, Yurii, 294

Wallace, Henry A., 41, 50
War in the Air (Wells), 7
"war on terror," 340–42, 348
warfare
    biological, 157, 341, 343–44, 346, 347
    chemical, 10–11, 18, 115, 118, 157, 341, 344
    cyber, 351–52
    over nuclear sites, 45, 363–64
    weapons of mass destruction, 10–11, 340–41, 344–49, 350, 354
    See also nuclear weapons; specific conflicts
Warsaw Pact, 206, 230, 285, 287, 296
Washington Conference on Theoretical Physics of 1939, 20
Washington Declaration of 1945, 109–10, 119
Washington summit of 1941, 66, 68–69
Watson, Edwin ("Pa"), 26, 27
weaponization of nuclear research, 30–31
weapons of mass destruction, 10–11, 340–41, 344–49, 350, 354
Wells, H. G., 7–9, 10, 14, 25, 42, 76, 79, 82, 107
West Germany, 196, 199, 247, 284–85, 305, 362
Westinghouse, 281
"When the Atom Bomb Fell" (song), 105–6
Wien, Wilhelm, 12
"Wigner growth," 129–30
Wigner, Eugene, 17–18, 21–22, 25, 42, 56, 62

Wilson, Robert R., 78
Wilson, Woodrow, 41
Windscale reactors, 129–30, 151
World Set Free, The (Wells), 7–9, 10, 14, 76, 79
World War I, 10–14, 115–16
World War II
    atomic bombs on Hiroshima and Nagasaki, 2, 88–95, 100–107, 117, 128, 169, 173, 184, 265
    Casablanca Conference, 68
    Dunkirk, 32, 34
    on the eastern front, 38–39
    fall of Paris, 32, 41, 197
    Japan's attack on Pearl Harbor, 50, 52, 60, 74
    Potsdam Conference, 83–86, 93, 103
    Second Front in Sicily, 68
    Yalta Conference, 84–85, 92, 103
    See also Manhattan Project
World War III, 144

X-10 Graphite Reactor at Oak Ridge, 58, 129
X-ray lasers, 290
X-ray revolution, 11–12

Yakovlev, Aleksandr, 300
Yalta Conference of 1945, 84–85, 92, 103
Yangel, Mikhail, 186
Yasuda, Takeo, 32–33
"yellowcake," 248
Yeltsin, Boris, 307, 308, 310, 316–17, 322, 342
Yom Kippur War of 1973, 254, 272, 343
Ypres, Battle of, 10
Yugoslavia, 281

Zambia, 313
Zaporizhia nuclear power plant, Ukraine, 363–64
"zero-zero option," 286–87, 306
Zhou Enlai, 210–12, 215–16, 237
Zhukov, Georgii, 94
Zorin, Valerian, 204